MW00845018

INTEGRATED PHOTONICS

INTEGRATED PHOTONICS

CLIFFORD R. POLLOCK
School of Electrical and Computer Engineering
Cornell University

MICHAL LIPSON
School of Electrical and Computer Engineering
Cornell University

Kluwer Academic Publishers
Boston/Dordrecht/London

Distributors for North, Central and South America:
Kluwer Academic Publishers
101 Philip Drive
Assinippi Park
Norwell, Massachusetts 02061 USA
Telephone (781) 871-6600
Fax (781) 871-6528
E-Mail <kluwer@wkap.com>

Distributors for all other countries:
Kluwer Academic Publishers Group
Post Office Box 322
3300 AH Dordrecht, THE NETHERLANDS
Telephone 31 78 6576 000
Fax 31 78 6576 474
E-Mail <orderdept@wkap.nl>

Electronic Services <http://www.wkap.nl>

Library of Congress Cataloging-in-Publication

CIP info or:

Title: Integrated Photonics
Author (s): Clifford R. Pollock and Michal Lipson
ISBN: 1-4020-7635-5

Printed on acid-free paper.

Printed in the United States of America

Contents

Preface

This book is directed at the issues of integrated photonics. Four major topics are covered: 1) fundamental principles of electromagnetic theory; 2) waveguides; 3) simulation of waveguide modes, and 4) photonic structures. The emphasis is slightly heavier into optical waveguides and numerical simulation techniques because advances in optical communication will be based on nanostructured waveguide structures coupled with new materials and structures. This text is targeted for students and technical people who want to gain a working knowledge of photonics devices. The text is designed for the senior/1st year graduate student, and requires a basic familiarity with electromagnetic waves, and the ability to solve differential equations with boundary conditions.

The first part of the text explores the basis for optical propagation and establishes the use of the MKS system, discussing the wave equation and the properties of materials such as attenuation and dispersion. The next section explores the operation of optical waveguides. We start with planar slab waveguides, then systematically advance to more complicated structures, such as graded index waveguides, circular waveguides, and rectangular waveguides. The details of coupling light between and within waveguide modes is clearly described, and applied to optoelectronic devices such as modulators and switches. The final section of the text discusses the examination of photonic bandgap crystals and optical devices such as ring resonators. These topics are very active areas of research today, and are likely to increase in significance as they mature.

From the beginning this text introduces numerical techniques for studying non-analytic structures. Most chapters have numerical problems designed for solution using a computational program such as Matlab or Mathematica. An entire chapter is devoted to one of the numeric simulation techniques being used in optoelectronic design (the Beam Propagation Method), and provides opportunity for students to explore some novel optical structures without too much effort. Small pieces of code are supplied where appropriate to get the reader started on the numeric work.

We wish to express our sincere gratitude to our graduate students who have helped by developing code and in critiquing text. We are also grateful for their enthusiasm for the subject, and for providing the experimental data and simulations presented in the final section of the text (the design of the cover of this text by Vilson Almeida is especially appreciated). Special thanks go to Professor Lionel Kimerling for stimulating discussions and ideas concerning high index-contrast waveguides. We also wish to thank Melissa for her help and assistance in putting the text and figures together.

Clifford Pollock
Michal Lipson
Ithaca, New York

Chapter 1

INTRODUCTION AND OVERVIEW

1. A Brief History of Telecommunications

In 1837, the era of electrical communication began with the demonstration of the telegraph by *Samuel Morse.* The telegraph passed information at a data rate, in modern terminology, of a few bits per second, but the speed of propagation was essentially infinite compared to the message itself. The transmission medium was wire cable. The telegraph was followed by *Alexander Graham Bell's* invention of the telephone. The first telephone exchange was operated in New Haven, Connecticut, USA, in 1878. At approximately 4 kHz bandwidth, the telephone represented a major increase in the effective bandwidth of a moderate distance communication system.

Arguably, the greatest technological achievement of the 19th century was *James C. Maxwell's* elucidation [1] of "Maxwell's Equations," in 1878. These equations mathematically describe the propagation of electromagnetic waves. Maxwell's equations, and applications derived from them, are the foundation of electrical machines and electronic devices which form what we now characterize as "high technology." Based on the predictions of Maxwell, *Heinrich Hertz* [2] demonstrated long radio waves in 1888, and in 1895, *Guglielmo Marconi* demonstrated radio (communication without wires!) based on electromagnetic waves.

Since these pioneering efforts, scientists and engineers have made steady progress toward better and faster communication technologies. The trend has been toward higher frequency carrier waves with proportionally increased modulation bandwidths for carrying information. Early radio, which carried voice signals with 15 kHz bandwidth, operated in the 0.5-2 MHz range. Television, which requires about 6 MHz bandwidth, raised carrier frequencies to 100 MHz. During the 1940's, radar research pushed frequencies to the gigahertz domain (microwaves). Low power microwave technology is now widely used in the 2.4 and 5 GHz region for cellular phone and wireless links between laptop computers, and high power applications include terrestrial and satellite communication links operating at 18 GHz.

The push toward higher frequencies took a giant leap forward with the invention of the laser in 1960.[3] The first laser operated at a wavelength of 694 nm, which corresponds to a carrier frequency of approximately 5×10^{14} Hz. People noted that if even 1% of this bandwidth could be realized in a communication system, it would represent a signal channel with 5 THz bandwidth (a terahertz is 10^{12} Hz). Such a system could carry approximately 10^6 analog video channels at 6 MHz per channel, or $\approx 10^9$ telephone calls at 5 KHz per call. However, progress toward using this tremendous bandwidth was limited by two factors:

1. Electronic components did not operate at such frequencies or speeds. Since most information today is ultimately converted to electronic form, the speed of the electronics determines the realizable bandwidth of any communication link.

2. There was no dependable transmission media for light.

Truly incredible progress has been made in using optical carrier waves for communication over the four decades since the first demonstration of the laser. The electronic speed bottleneck is still a challenge to the direct use of the full bandwidth of the optical carrier, but creative optical methods have been developed which circumvent some of these limits. The drive for increased bandwidth has led to faster electronic components which have switching times approaching several picoseconds. One can now install transmitters which operate at 40 GigaBits/second (giga = 10^9), and 120GB/s transmitters are under development. It is always dangerous to define limits, but many feel that electronic devices are reaching their practical speed limits. Creative research into optically-based information systems where for example, information is carried by many wavelengths simultaneously or by ultrashort optical pulses called solitons, is providing cost-effective access to the full bandwidth of the optical carrier. Without doubt, the biggest research task in the next decade will be the development of optical switches and devices and in better communication architectures.

The lack of suitable transmission media has been creatively addressed over the past 40 years. The field of optical waveguides is well established and is now a thriving industry. The first part of this text will develop the theory and application of the optical waveguide, both for long distance communication (optical fibers) and for integrated optic applications.

To address these problems, a new discipline has emerged called *Photonics*. Using light to convey information requires special technologies. Information must be put on the light beam using extremely high speed modulation. Once

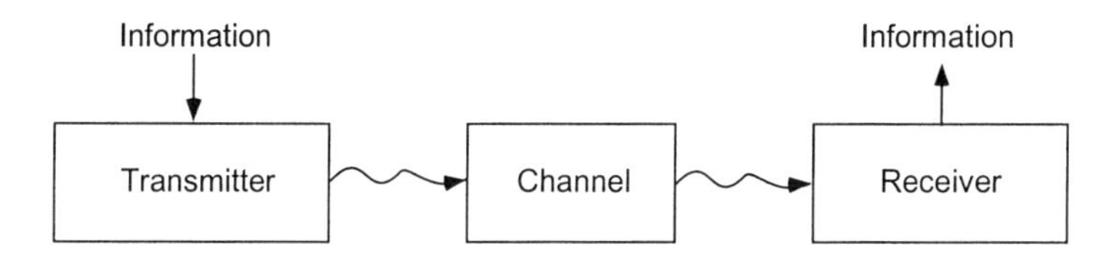

Figure 1.1. A communication channel consists of a transmitter, transmission channel, and a receiver. The transmitter converts information into an energy form appropriate for the transmission channel. The channel carries the energy, but also distorts the signal and adds noise. Following detection, the receiver regenerates the information in (hopefully) a nearly identical form to the original signal.

modulated, the light must be carried over sizable distances, and directed or switched to the desired receiver. When it arrives at the receiver, the information must be extracted from the light. Consider the simplified block diagram of a communication system shown in Fig. 1.1.

1. The *Transmitter* couples information onto a transmission channel in the form of a suitable signal.

2. The *Channel* is a medium bridging the distance between the transmitter and receiver. For electromagnetic signals, the channel might be a wire, a waveguide, or free space. As the signal travels through the medium it is progressively attenuated and distorted.

3. The *Receiver* extracts a weakened signal from the channel, and amplifies it. A semblance of the original information (audio, video, etc.) is generated from the modulated signal .

2. Development of the Optical Waveguide

Researchers rapidly discovered that free space propagation of laser beams was not suitable for reliable communication links. Problems of bad weather (precipitation), flying objects that interrupted the beam, and the need for "line-of-sight" links complicated reliable transmissions. Laser beams are also distorted and randomly aberrated by propagation through turbulent air (this effect is called scintillation). There are a few specialized applications for free space communication, such as communication between satellites in orbit, but most terrestrial applications require a protected transmission channel.

A solution to these problems is to propagate light through a waveguide which both protects the beam from interruptions, and counters diffraction (the tendency of a wave to spread as it propagates). One scheme that is easy to analyze

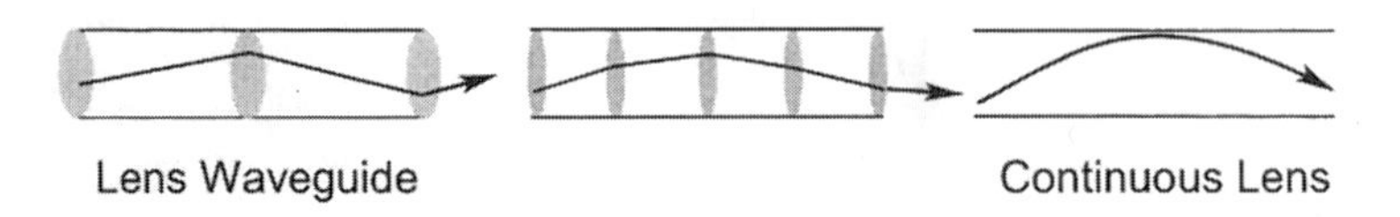

Figure 1.2. The optical waveguide is a natural extension of a lens waveguide

and visualize is the periodic lens waveguide shown in Fig.1.2. The lenses periodically refocus the light, countering diffraction, and also allow light to travel around gentle bends. It is fairly straightforward to select a focal length and lens spacing that will provide a stable beam path [1]. Manufacturing difficulties and surface reflection losses rule out such a structure as a practical solution for waveguiding. However an extension of this idea is the continuous lens. Consider making each lens continually weaker in focussing power, while increasing the number of lenses so that the light stays confined. In the limit of weaker yet more frequent lenses, one gets a continuous lens.

In the continuous lens,the ray of light is constantly bent back toward the center of the waveguide. The continuous lens consists of a glass fiber with a higher index of refraction at the core than at the outer perimeter. This waveguide structure solves the problem of surface reflections and also allows the waveguide to be bent. We will develop an understanding of these guiding structures in the next chapters.

What about transmission losses? Is it possible to send light through kilometers of dense material without excess attenuation? Glass is an obvious candidate for making an optical waveguide; it is commonly available, easy to draw into fibers, and looks transparent. However, our common experience with glass usually involves looking through plates no thicker than a few millimeters. What happens if light passes though a kilometer of glass? Early measurements indicated that the attenuation of near infrared light in glass was about 1000 dB per kilometer. To fully appreciate this number, recall that a dB is defined as a logarithmic ratio:

$$\text{Power (dB)} = 10 \log_{10} \frac{P_{out}}{P_{in}} \tag{1.1}$$

A 1000 dB loss represents $P_{out} = 10^{-100} P_{in}$! Considering that there are approximately 10^{19} photons per watt of visible light, this remarkable number implies that it would be necessary to launch approximately 10^{80} W of light to get 1 photon through one kilometer length of glass. To put this in perspective, the sun radiates only (!) 3.8×10^{26} W.

In view of this incredible attenuation, it is impressive that far sighted researchers pursued optical fiber waveguides. In 1966, *K. Charles Kao* of Standard Telecommunications Laboratories, in Harlow, England, suggested that the high loss was due to impurities, and not an intrinsic property of glass. *Kao* went on to propose using fibers as a transmission medium, describing many of the fundamental modes that such a fiber would carry in his classic paper [5]. In 1970, *Kapron, Keck*, and *Maurer*[3] of Corning Glass Works confirmed this prediction by making an ultrapure glass using chemical vapor deposition techniques that displayed only 20 dB/km attenuation. Today, ultrapure glass now displays intrinsic attenuation less than 0.2 dB/km near a wavelength of 1.5 μm.

Once the attenuation problem was mastered, optical fiber communication become a major factor in communication systems, for both long distance and local area networks. Motivations for using optical communication include:

1. Optical communication links have a wider bandwidth than copper or microwave links, so more information can be carried on a given link. The effective bandwidth of current optical fibers is approximately 30 THz. (30 THz!!)

2. Attenuation in glass fibers is much less than experienced in copper or microwave systems. Fewer repeaters are required, and longer distances can be spanned more cost effectively.

3. Optical systems are smaller and lighter, giving them an advantage in crowded ducts or aircraft.

4. Optical waveguides are difficult (but not impossible) to tap or monitor, so data security is higher.

5. Optical waveguides are immune from electromagnetic interference (EMI), ground loops, induced cross talk, etc.

6. Finally, and perhaps most important, semiconductor technology has developed a family of lasers, detectors, and other integrated optical devices that are compatible with optical fibers in power, wavelength, and size.

3. Types of Optical Communication Systems

Two types of optical communication system have developed. Historically, long distance telecommunication (Telecom) was the first optical communication application. Telecom primarily involves point-to-point links, such as a long distance telephone link between two cities, which carry vast numbers of multiplexed signals. Optical fiber has become the standard for telecommunication links for the reasons listed above, not the least of which is cost. All of the advanced technology that appeals to engineers and scientists would never be installed were it not the most cost-effective solution to most telecommunications problems. The major cost component in telecom is installing the optical fiber.

The second type of optical communication system is data communication (Datacom). Datacom is used to link information devices such as computers, memory banks, data bases, and workstations together in a local area network that may span thousands of connections and hundreds of kilometers. Datacom applications are different from telecom, in that the cost pressure is no longer on the carrier medium (the optical fiber), but on the associated hardware such as transmitters, connectors, switches, filters, and receivers. The need for cost-effective devices for datacom applications is driving much of today's research in optoelectronics. Requests for "Fiber to the desk" are not unusual in new local area network installations, but for data rates below 1 GigaBit/second, fiber is not cost-competitive with copper right now. "Fiber to the home" faces the same cost disadvantage: unless the bandwidth requirements of the typical home increase dramatically, or the cost of neighborhood-level optical networks decreases, it will be many years before optical fiber replaces residential copper coaxial cables and telephone wires. Certainly optical fiber has the bandwidth to support such a network, but the necessary support hardware is still not developed to the point of cost-effective implementation.

4. Opportunities in Optoelectronics

To transfer information from one point to another, whether between two workstations or two cities, communications systems will require switches, connectors, amplifiers, filters, etc. While such devices are well developed for copper-based and microwave communication links, the optical analogs of these devices are still expensive and difficult to manufacture in an integrated fashion.

Unlike electronic communication systems that have a limited bandwidth, optical *systems* can almost be treated as having infinite bandwidth. Most present optical communication systems use simple digital modulation schemes ("on" and "off"), much like Marconi's early radio. Researchers are now exploring frequency and phase modulation schemes called "coherent detection". These offer improvements in signal-to-noise ratios. These "new" optical techniques are really tried and true modulation schemes that are presently used in modern radio communication systems. This new technology has also opened the door to new phenomena, such as *optical solitons*, which exploit nonlinear properties of optical fibers to make potentially better communication systems. A system designer today will ask, "Should the huge bandwidth available in optical links be exploited through time division multiplexing, using extremely short temporal pulses? Or should Wavelength Division Multiplexing be used, where each signal is transmitted at its own wavelength?" Most people today would reply that wavelength division multiplexing is the answer. However disruptive technology could easily appear that is better than both time domain and wavelength domain multiplexing. Optical networks are still in a nascent stage, and so there is great opportunity for new ideas.

Noting that this book is primarily devoted to describing the hardware aspects of optoelectronics, readers interested in the systems aspect of optical communications should also refer to the excellent book by Paul Green [7] on fiber optic networks (datacom) and the books by Stuart Personik [8] on telecommunications. These books serve as excellent companions to this one in terms of complementary material and concept development.

References

[1] J. C. Maxwell, *A Treatise on Electricity and Magnetism*, 2 vol. Unabridged 3rd ed., Dover Publications, New York (1954).

[2] Heinrich Rudolph Hertz, *Electric Waves*, 2d ed. Macmillan and Co., London, New York (1900).

[3] T. Maimann, "Stimulated optical radiation in ruby", *Nature*, vol. 187, pp.493-494 (1960).

[4] J. Verdeyen, *Laser Electronics, 2nd Ed.*, Chap. 2, Prentice-Hall, Inc., Englewood Cliffs, NJ, (1989).

[5] K. C. Kao and G. A. Hockman, *Proc. IEE*, vol. 133, pp. 1151-1158 (1966).

[6] F. P. Kapron, D. B. Keck, and R. D. Maurer, *Appl. Phys. Lett.*, vol. 17, pp. 423-425 (1970).

[7] Paul E. Green, Jr., *Fiber Optic Networks*, Prentice Hall, Englewood Cliffs, New Jersey (1993).

[8] Stuart Personik, *Optical Fiber Transmission Systems*, Plenum Press, New York (1981), and *Fiber Optics Technology and Applications*, Plenum Press, New York (1985).

Chapter 2

FUNDAMENTAL TOOLS OF OPTOELECTRONICS: MAXWELL'S EQUATIONS

1. Introduction: The Tools of the Trade

There are many tools available to analyze or design an optical device. Imagine trying to characterize the optical behavior of a simple magnifying lens. We might first project an image of an object onto a screen. Using a ray picture, we could describe the magnification, focal length, principal planes, and so forth (Fig. 1a) of the lens. This characterization is called *geometric optics* [1]. If we were very perceptive, we might notice that different colors form images at slightly different distances from the lens (Fig. 1b). To adequately describe this effect, we would have to understand and explore the material and dispersion properties of the lens [2]. This is called *physical optics*.

If we could shrink the diameter of the lens to dimensions on the order of the wavelength of light, we would notice that the image begins to blur. Fringes

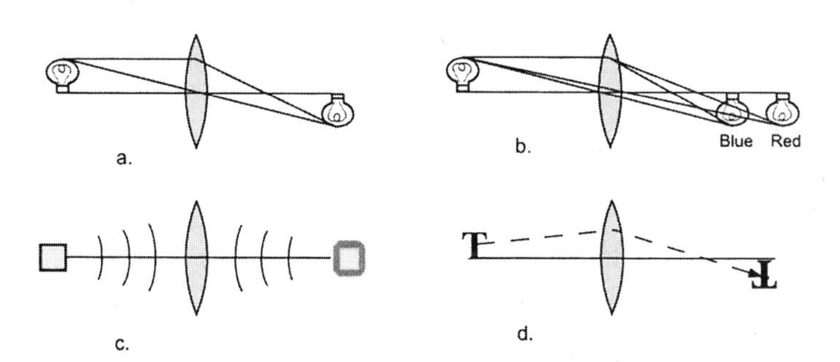

Figure 2.1. Four (of many) ways to describe a lens. a) An image is formed through ray tracing. b) The image position may vary with wavelength, due to dispersion in the lens. c) Diffraction can blur the image of a small object. d) An image being formed one photon at a time.

begin to appear that are not apparent in the object. These *diffraction phenomena* are most notable at small dimensions, where the geometrical optics assumption that light travels in straight lines begins to fail. The proper description of these effects requires wave theory based on Maxwell's equations, and is called *wave optics*[4].

If we repeat our measurements in very dim illumination, we will reach a point where the corpuscular nature of light is evident. For sufficiently low intensity, one *photon* at a time passes through the lens. The individual photons arrive with certain statistical patterns (such as a Poisson distribution)[2]. Different sources of light can have different statistical properties, and these properties will affect the quality of optical information that can be extracted from the input signal. In this photon realm, *quantum optics* is needed to describe the system [5].

Each of these optical methods is appropriate within a certain domain. When dealing with the modes of a waveguide, we use physical optics. When dealing with optical detection of signals, we use quantum optical concepts. In general, we resort to the technique that leads to the most direct solution of a given problem. In this chapter, we will introduce *Maxwell's Equations*, which are the optical tools needed to describe the propagation of light in optical waveguides. Using Maxwell's equations, we will derive and solve the wave equation in an isotropic media. Following solution of the wave equation, we will explore refraction and reflection. Total internal reflection is addressed in the following chapter.

2. Maxwell's Equations

Maxwell's equations are, arguably, the most significant scientific development of the 19th century. It is impressive to realize that the same equations can be applied from 0 Hz (DC) to frequencies exceeding 10^{18} Hz (in fact, an upper frequency limit on the validity of the equations has never been shown). The four equations can be presented in differential or integral form. They are listed below in both forms.

$$\nabla \times \mathbf{E} = \frac{-\partial \mathbf{B}}{\partial t} \qquad \oint \mathbf{E} \cdot dl = -\frac{\partial}{\partial t} \int \mathbf{B} \cdot d\mathbf{S} \tag{2.1}$$

$$\nabla \times \mathbf{H} = \mathbf{J} + \frac{\partial \mathbf{D}}{\partial t} \qquad \oint \mathbf{H} \cdot dl = \int \mathbf{J} \cdot d\mathbf{S} + \frac{\partial}{\partial t} \int_{area} \mathbf{D} \cdot d\mathbf{S} \tag{2.2}$$

$$\nabla \cdot \mathbf{B} = 0 \qquad \int \mathbf{B} \cdot d\mathbf{S} = 0 \tag{2.3}$$

$$\nabla \cdot \mathbf{D} = \rho \qquad \int \mathbf{D} \cdot d\mathbf{S} = Q_{enclosed} \tag{2.4}$$

where **S** is the unit normal to a surface, and the surface integrals extend only over the area enclosed by the path of the line integral. In this text we will use *MKS* units, with the exception that most physical distances will be related in centimeters where it is convenient and obvious. The units that describe an optical field in *MKS* units are listed in Table 2.1 below.

Note that **E**and **H** are *amplitudes* which describe the strength of a field at a given point in space. **D** and **B** are *fluxes*. Unfortunately, due to the different systems often used for electromagnetic quantities, the distinction between amplitude and flux is sometimes lost. This is especially true for the magnetic field. Amplitudes and fluxes are both vectorial in nature, which means that direction and magnitude are important.

These quantities are continuous functions of space and time, with continuous derivatives. All real solutions will be bounded (no infinities exist in physical situations) and will be single-valued at all points. At surfaces the distribution of charge or current can be changed, so boundary conditions are used to connect solutions in adjacent regions.

The integral form of the equations are listed more for reference than for potential application in this text. They are, however, useful to establish boundary conditions. The integral forms of the curl equations are readily derived from the differential forms by application of Stokes Theorem. This theorem relates the curl of a vector function, **A**, into a line integral of the function

$$\int_{area} (\nabla \times \mathbf{A}) \cdot d\mathbf{S} = \oint_{loop} \mathbf{A} \cdot d\mathbf{l} \tag{2.5}$$

where $d\mathbf{S}$ and $d\mathbf{l}$ are unit vectors oriented normal to the surface, or tangential to the loop, respectively. For the divergence equations, Gauss' divergence theorem

$$\int_{closed\ surface} \mathbf{D} \cdot d\mathbf{S} = \int_{volume\ enclosed} \nabla \cdot \mathbf{D} dv \tag{2.6}$$

Table 2.1. Electromagnetic Units in the MKS Format

Quantity	Description	Units	
E	Electric field amplitude	Volts/meter	(V/m)
H	Magnetic field amplitude	Amps/meter	(A/m)
D	Electric flux density	Coulombs/meter^2	(C/m^2)
B	Magnetic flux density	Webers/meter^2	(Wb/m^2)
J	Current density	Amps/meter^2	(A/m^2)
ρ	Charge density	Coulombs/meter^3	(C/m^3)
Q	Charge	Coulombs	(C)

relates the two forms. The integral forms are used only when fields have a high degree of symmetry. This is not the typical case in optical waveguides, so the differential forms of Maxwell's equations are usually required.

3. Constitutive Relations

The flux densities, **D** and **B**, are related to the field amplitudes **E** and **H** by the *constitutive relations*. The nature of the medium defines the functional form of the relationship. For linear, isotropic media, the relations are simply given by

$$\mathbf{B} = \mu \mathbf{H} \qquad\qquad \mathbf{D} = \epsilon \mathbf{E} \tag{2.7}$$

where ϵ is the electric permittivity of the medium, with units of *Farads/meter*, and μ is the magnetic permeability of the medium, with units of *Henrys/meter*. A linear medium is one where the permittivity, ϵ, and permeability, μ, are independent of field strengths. The assumption of linearity is valid only for low intensities, where **E** is, for example, much less than the Coulomb fields that bind electrons to the central nucleus. Since these binding fields are on the order of 10^{10} V/cm, nonlinear effects are only observed using high intensity light. Nonlinear effects can be exploited for various applications, and we will deal with them in later chapters. Vacuum is a linear medium since there are no binding fields to be distorted by intense fields. (However, under extreme intensities, e.g. $> 10^{23}$W/cm^2, it is possible to spontaneously create an electron-positron pair from vacuum, so even vacuum cannot be considered strictly linear. Such field strengths are well beyond the realm of interest for present electrooptical devices). The vacuum values of ϵ and μ are symbolically denoted as ϵ_0 and μ_0, respectively, and have values

$$\epsilon_0 = 8.854 \times 10^{-12} \text{ Farads/m} \tag{2.8}$$

$$\mu_0 = 4\pi \times 10^{-7} \text{ Henrys/m} \tag{2.9}$$

In non-vacuum media, the general expressions for permittivity and permeability are not necessarily scalar quantities. Since the field quantities are vectorial, the constitutive relationships must be described by a tensor. The electric flux density, **D**, is properly described by

$$D_i = \epsilon_{ij} E_j = \sum_{j=1}^{3} \epsilon_{ij} E_j \tag{2.10}$$

Einstein notation of repeated indices is used, and ϵ_{ij} is the permittivity tensor. A similar tensor expression exists that relates the magnetic flux, **B**, to **H** [6]. The components of ϵ_{ij} depend on the properties of the material. Crystals with low degrees of symmetry generally have tensorial permittivity. Highly symmetric crystal structures, such as NaCl or silicon, and amorphous material such as glass, are isotropic in permittivity, unless the symmetry of their structure is perturbed through a strain.

Example 2.1 Permittivity tensor of two crystalline materials

Two optical crystals are shown below, GaAs and Calcite. GaAs has the diamond structure, and has an isotropic permittivity. The three crystalline axes are chosen to lie along the three $\langle 100 \rangle$ axes of the crystal. The permittivity tensor near $\lambda = 1\mu$m is given by

$$\epsilon = \epsilon_0 \begin{pmatrix} 11.56 & 0 & 0 \\ 0 & 11.56 & 0 \\ 0 & 0 & 11.56 \end{pmatrix}$$

If the GaAs crystal were rotated 90°, it would look and act exactly the same as before the rotation.

Calcite has a less symmetric structure, as shown in the Fig. 2.2. If calcite is rotated 90° about x-axis, the crystal will look quite different to a beam of light travelling through it. For linearly polarized light propagating in the z-direction in Calcite, the electric field can be x or y polarized, or some mixture of both. The index of refraction for these waves is identical, and is called the *ordinary* index of refraction. For polarized light along the z-axis, the index of refraction is called the *extraordinary* index. The permittivity tensor for calcite is

$$\epsilon = \epsilon_0 \begin{pmatrix} 2.75 & 0 & 0 \\ 0 & 2.75 & 0 \\ 0 & 0 & 2.21 \end{pmatrix}$$

Calcite is called a negative uniaxial crystal, because the extraordinary index is less than the ordinary index. For a good review of crystal optics, see Chapter 4 of Yariv and Yeh. [7]

In an isotropic medium, the permittivity is independent of orientation, and is described accurately by the scalar relation $D = \epsilon E$. But beware! "Isotropic"

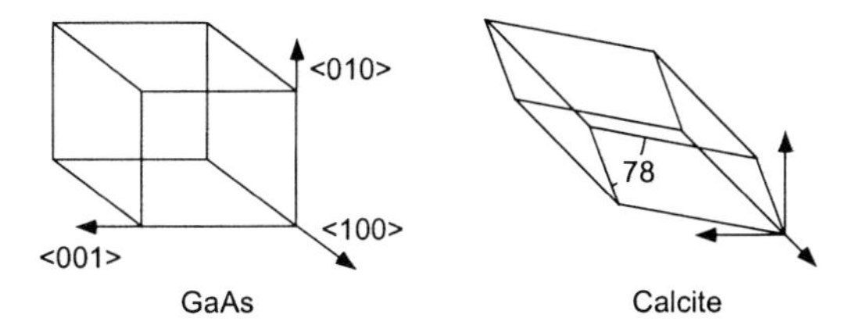

Figure 2.2. GaAs and Calcite have different regular structures. GaAs is a cube which looks identical along each axis. Calcite has a different length crystal axis along each direction, and the axes meet at non-right angles.

does not necessarily mean homogeneous. The permittivity can be a function of position, $\epsilon(r)$. In an inhomogeneous medium, the electric field will encounter a different permittivity, ϵ, depending upon spatial location in the material. A graded index waveguide, discussed in Chapter 7, is a good example of an inhomogeneous medium.

For most optical dielectric materials, μ is effectively μ_0. We can ignore magnetic effects except when dealing with special magnetic optical materials, such as Yttrium Iron Garnet (YIG), used as an optical isolator between waveguides and sources. Unless otherwise stated, it is safe to assume that the permeability, μ, is that of free space, μ_0. We will discuss the frequency dependence of μ and ϵ in Chapter 7.

4. The Wave Equation

The electromagnetic wave equation comes directly from Maxwell's equations. Derivation is straightforward if we assume conditions that are reasonable for optical wave propagation. These conditions are that we are operating in a *source free* ($\rho = 0$, $J = 0$), *linear* (ϵ and μ are independent of E and H), and *isotropic* medium. Eqs. 2.1 - 2.4 become

$$\nabla \times \mathbf{E} = -\partial \mathbf{B}/\partial t \tag{2.11}$$

$$\nabla \times \mathbf{H} = \partial \mathbf{D}/\partial t \tag{2.12}$$

$$\nabla \cdot \mathbf{D} = 0 \tag{2.13}$$

$$\nabla \cdot \mathbf{B} = 0 \tag{2.14}$$

These simple looking equations completely describe the electromagnetic field in time and position. Are the assumptions reasonable? Sure, at high frequencies (e.g. $\nu > 10^{13} Hz$) free charge and current are generally not the source of electromagnetic energy. The typical sources of optical energy are electric or magnetic dipoles formed by atoms and molecules undergoing transitions. Maxwell's equations account for these sources through the bulk permeability and permittivity constants.

Eqs.2.11- 2.14 are strongly coupled first-order differential equations. To decouple the two curl equations we follow the usual technique of creating a single second order differential equation by first taking the curl of both sides of Eq. 2.11.

$$\nabla \times (\nabla \times \mathbf{E}) = \nabla \times \frac{-\partial \mathbf{B}}{\partial t} = \nabla \times \frac{-\partial \mu \mathbf{H}}{\partial t} \tag{2.15}$$

Assuming that $\mu(r,t)$ is independent of time and position, Eq. 2.15 becomes

$$\nabla \times \nabla \times \mathbf{E} = -\mu\left(\nabla \times \frac{\partial \mathbf{H}}{\partial t}\right) \tag{2.16}$$

Since the functions are continuous, the order of the curl and time derivative operators can be reversed:

$$\nabla \times \nabla \times \mathbf{E} = -\mu \frac{\partial}{\partial t}(\nabla \times \mathbf{H}) \tag{2.17}$$

Substituting $\nabla \times \mathbf{H} = \partial \mathbf{D}/\partial t$ into Eq.2.17 and assuming ϵ is time invariant

$$\begin{aligned} \nabla \times \nabla \times \mathbf{E} &= -\mu\partial/\partial t \frac{\partial \mathbf{D}}{\partial t} \\ &= -\mu\epsilon\partial^2 \mathbf{E}/\partial t^2 \end{aligned} \tag{2.18}$$

Now we have a second order differential equation with only one variable, **E**. The $(\nabla \times \nabla \times)$ operator is usually simplified using a vector identity

$$\nabla \times \nabla \times \mathbf{E} = \nabla(\nabla \cdot \mathbf{E}) - \nabla^2 \mathbf{E} \tag{2.19}$$

The ∇^2 operator should not be confused with the *scalar Laplacian* operator. The ∇^2 operator in Eq. 2.19 is the *vector Laplacian* operator that acts on a vector, in this case **E**. For a rectangular coordinate system, the vector Laplacian can be written in terms of the scalar Laplacian as

$$\nabla^2 \mathbf{E} = \nabla^2 E_x \hat{x} + \nabla^2 E_y \hat{y} + \nabla^2 E_z \hat{z} \tag{2.20}$$

where $\hat{x}$, $\hat{y}$, and $\hat{z}$ represent unit vectors along the three axes. The ∇^2's on the right hand side of Eq. 2.20 are scalar, given by

$$\nabla^2 = \frac{\partial^2}{\partial x^2} + \frac{\partial^2}{\partial y^2} + \frac{\partial^2}{\partial z^2} \tag{2.21}$$

in cartesian coordinates. Solution of the vector wave equation requires that we first break the equation into the orthogonal vector components, which is sometimes extremely difficult, and then combine the individual vector field solutions together.

What about the term, $\nabla \cdot \mathbf{E}$? It is not necessarily equal to zero, as is often assumed. We know only that $\nabla \cdot \mathbf{D} = 0$. Simple calculus leads to an expression for $\nabla \cdot \mathbf{E}$:

$$\begin{aligned} \nabla \cdot \mathbf{D} &= 0 \\ &= \nabla \cdot \epsilon \mathbf{E} \\ &= \nabla \epsilon \cdot \mathbf{E} + \epsilon \nabla \cdot \mathbf{E} \end{aligned} \tag{2.22}$$

Solve for $\nabla \cdot \mathbf{E}$

$$\nabla \cdot \mathbf{E} = -\mathbf{E} \cdot \frac{\nabla \epsilon}{\epsilon} \tag{2.23}$$

Plugging this value into the linear wave equation for electromagnetic waves yields

$$\nabla^2 \mathbf{E} - \mu\epsilon \frac{\partial^2 \mathbf{E}}{\partial t^2} = -\nabla\Big(\mathbf{E} \cdot \frac{\nabla\epsilon}{\epsilon}\Big) \tag{2.24}$$

The right hand side deserves special consideration. It is non-zero when there is a gradient in the permittivity of the medium. Index gradients are quite common in guided wave optics, since most guided wave structures use a graded permittivity. So how do we deal with this extra term? Well, we ignore it! Fortunately in most structures, the term is negligibly small. (Problem 2.2. explores the limits of $\nabla\epsilon/\epsilon$, showing that it is almost always negligible). Neglecting this term, the wave equation reduces to its *homogeneous* form

$$\nabla^2 \mathbf{E} - \mu\epsilon \frac{\partial^2 \mathbf{E}}{\partial t^2} = 0 \tag{2.25}$$

Had we started with Eq.2.12 instead of Eq.2.11 we could have derived a similar wave equation in terms of the magnetic field amplitude (see Prob. 2.9.),

$$\nabla^2 \mathbf{H} - \mu\epsilon \frac{\partial^2 \mathbf{H}}{\partial t^2} = 0 \tag{2.26}$$

5. Solutions to the Wave Equation

Consider the units of each term in either Eq. 2.25 or Eq. 2.26. The ∇^2 term has units of $1/(\text{distance})^2$. The second order time derivative clearly has units of $1/(\text{sec})^2$. In order to make physical sense, the units of $\mu\epsilon$ must be $(\text{sec/m})^2$. We will show, in a later section, that $\sqrt{1/\epsilon\mu}$ is the phase velocity of light in a medium. Notice that the speed of propagation is determined by the material parameters. In free space, $\sqrt{1/\mu_0\epsilon_0} = 2.998 \times 10^8 m/sec$, or c, the speed of light in vacuum.(The speed of light is now defined (not measured) to be exactly 299,792,458 m/sec. The meter is thus defined in terms of the speed of light, being the distance light travels in 1/299,792,458 second, where the second is now the primary standard.) We will discuss the speed of propagation more thoroughly in the next section.

Eqs 2.25 and 2.26 are vector equations. The equations can be simplified by rewriting them in terms of the components of the field. In rectangular coordinates, the vector Laplacian breaks into three uncoupled components. The scalar component equations become

$$\nabla^2 E_i - \mu\epsilon \frac{\partial^2 E_i}{\partial t^2} = 0 \tag{2.27}$$

Here the subscript indicates the ith component, where i stands for x, y, or z, and ∇^2 is the scalar Laplacian given in Eq. 2.21. Since the symbol for the

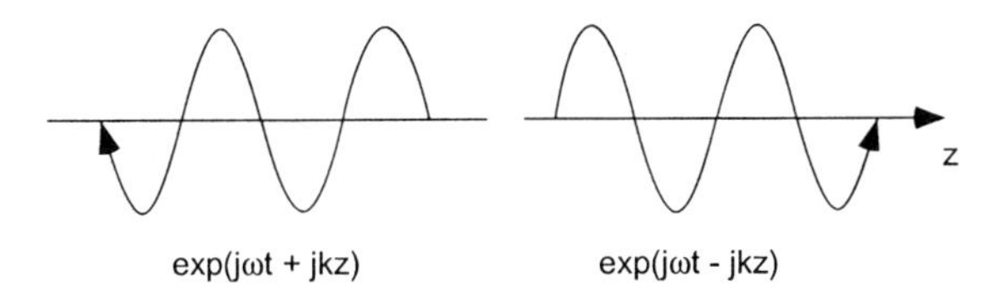

Figure 2.3. The general solution to the wave equation in a linear homogeneous medium leads to plane waves. Depending on the relative sign, the wave will travel left or right.

vector and scalar Laplacian look the same, we rely on context to distinguish the operators.

The choice of coordinate system is critical to solving the wave equation. For example, choosing rectangular coordinates to describe a wave in a cylinder leads to inseparable coupling upon reflection at the cylindrical surface. We seek a coordinate system with no coupling between the orthogonal components, and in such a case the individual equations can be written as scalar wave equations. The scalar wave equation is written as

$$\nabla^2 E_i - \mu\epsilon \frac{\partial^2 E_i}{\partial t^2} = 0 \tag{2.28}$$

where E_i stands for one of the orthogonal amplitude components.

To find a valid solution to the wave equation, we use the separation of variables technique to get

$$\begin{aligned} E_i(\mathbf{r}, t) &= E_i(\mathbf{r})E_i(t) \\ &= E_0 \exp(j\mathbf{k}\cdot\mathbf{r})\exp(j\omega t) + c.c. \end{aligned} \tag{2.29}$$

The term E_0 is the electric field amplitude: the separation constant, $\mathbf{k}$, is called the *wavevector* (in units of rads/meter); and ω is the *angular frequency* of the wave (in units of rads/sec). We will use the wavevector as the primary variable in most waveguide calculations. The magnitude of the wavevector is defined in terms of the angular frequency and the phase velocity:

$$|\mathbf{k}| = \omega\sqrt{\mu\epsilon} = k \tag{2.30}$$

The wavevector $\mathbf{k}$ points in the direction of travel for the plane wave. The magnitude of $|\mathbf{k}|$ describes how much phase accumulates as a plane wave travels a unit distance. Think of k as a *spatial frequency*.

Through proper choice of sign for each term, one can describe a wave that travels in the forward or backward direction along the axis of propagation. Fig. 2.3 shows the two cases.

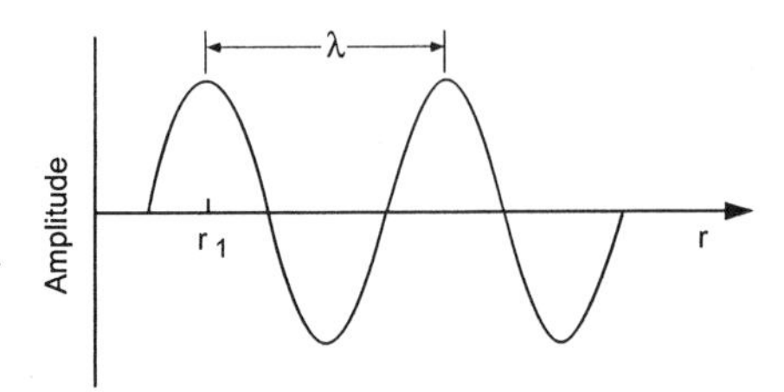

Figure 2.4. Basic description of the wavelength. The wave accumulates 2π of phase after travelling one wavelength.

In optics, it is common to describe optical fields by their *wavelength*. The waveform in Fig. 2.4 shows the real part of the spatial component of the plane wave, $E(r) = E_0 e^{jkr}$. The distance between two adjacent peaks in amplitude is called a *wavelength*, λ. The amplitude of the wave at the first peak, e^{jkr_1}, is the same as the amplitude at the peak located one wavelength away, $e^{jk(r_1+\lambda)}$. We can find a relation between k and λ:

$$e^{jkr_1} = e^{jk(r_1+\lambda)} \tag{2.31}$$

$$= e^{jkr_1} e^{jk\lambda} \tag{2.32}$$

This equality holds only if $e^{jk\lambda} = 1$, which requires that $|k\lambda| = 2\pi$. Solving for k

$$|k| = 2\pi/\lambda \tag{2.33}$$

6. Transverse Electromagnetic Waves and the Poynting Vector

Assume that a plane wave is propagating along the $\hat{z}$-direction and that the electric field is polarized along the $\hat{x}$-axis, $\mathbf{E}(r,t) = \hat{x}E_0 \cos(\omega t - kz)$. In complex notation, this would be described as

$$\begin{aligned} \mathbf{E}(r,t) &= \hat{x}E_0 \frac{1}{2}(e^{-j(kz-\omega t)} + e^{+j(kz-\omega t)}) \\ &= \hat{x}\frac{E_0}{2} e^{-j(kz-\omega t)} + c.c. \end{aligned} \tag{2.34}$$

We use complex notation because derivative and integral operations do not change the functional form. We must be careful to take the *real part* of expressions like Eq. 2.34 when we want to describe the physical wave.

The magnitude of the magnetic amplitude can be derived from the electric amplitude using Maxwell's equations. Plug the electric amplitude (Eq. 2.34) into Eq. 2.1 and use Eq. 2.9, and Eq. 2.29 to show

$$\mathbf{H}(r,t) = \hat{y}\frac{k}{\mu\omega}\frac{E_0}{2} e^{-jkz} e^{j\omega t} + c.c.$$

$$= \hat{y}\frac{\omega\sqrt{\mu\epsilon}}{\mu\omega}\frac{E_0}{2}e^{-jkz}e^{j\omega t} + c.c$$

$$= \hat{y}\frac{1}{\eta}\frac{E_0}{2}e^{-jkz}e^{j\omega t} + c.c. \tag{2.35}$$

where η is called the characteristic impedance of the medium,

$$\eta = \sqrt{\frac{\mu}{\epsilon}} \tag{2.36}$$

In vacuum, the characteristic impedance is $\eta_0 = 377\Omega$. The magnitude of the magnetic amplitude is directly proportional to the magnitude of the electric amplitude. Note that **E** is perpendicular to **H**.

A useful concept for characterizing electromagnetic waves is the measure of power flowing through a surface. This quantity is called the *Poynting vector*, defined as

$$\mathbf{S} = \mathbf{E} \times \mathbf{H} \tag{2.37}$$

S represents the instantaneous intensity (W/m^2) of the wave. The Poynting vector points in the direction of power flow, which is perpendicular to both the **E** and **H** fields. The time average intensity for a harmonic field (i.e. sinusoidal waveform) is often given using phasor notation

$$\langle \mathbf{S} \rangle = \frac{1}{2}Re[\mathbf{E} \times \mathbf{H}^*] \tag{2.38}$$

where $\mathbf{H}^*$ is the complex conjugate of **H**. The total electromagnetic power entering into a volume is determined by a surface integral of the Poynting vector over the entire area of the volume. In waveguides we are usually interested in the average energy flow in one direction, e.g. along the axis of the waveguide. In such cases, the dot product of the Poynting vector with the unit direction vector must be evaluated,

$$\langle S_z \rangle = \frac{1}{2}Re[\mathbf{E} \times \mathbf{H}^* \cdot \hat{z}] \tag{2.39}$$

This value of $\langle S_z \rangle$ is a function of position, so it is necessary to integrate the Poynting vector over the cross section of the guide.

7. Phase Velocity

Two velocities describe the propagation of electromagnetic waves: the *phase velocity*, and the *group velocity*. We will consider phase velocity first. Consider the sinusoidal electromagnetic wave plotted in Fig. 2.5 , travelling in the $\hat{z}$ direction. A point is attached to the top of one of the amplitude crests. How fast must this point move to stay on the crest of the wave? Since this crest

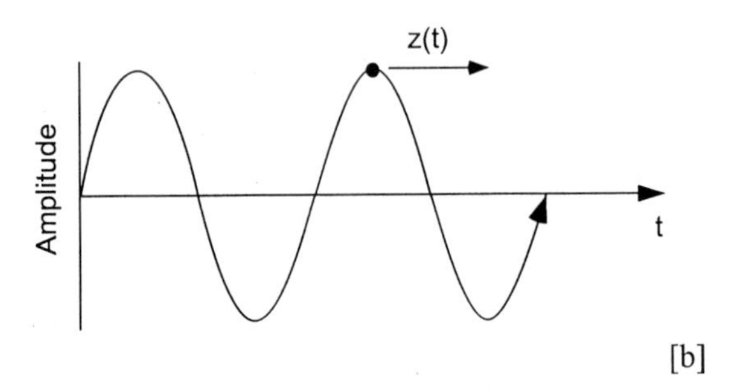

Figure 2.5. The phase velocity is determined by the speed necessary for a point to ride the crest of a wave.

represents a specific phase of the wave, the point must move at a speed such that

$$e^{-j(kz-\omega t)} = \text{constant} \tag{2.40}$$

which is satisfied if $kz - \omega t = $ constant. It is easy to see $z(t)$ must satisfy

$$z(t) = \frac{\omega t}{k} + \text{constant} \tag{2.41}$$

We can differentiate $z(t)$ with respect to time to find the phase velocity, $v(t)$

$$\frac{dz}{dt} = \frac{\omega}{k} = v_p \tag{2.42}$$

Also, recall from Eq. 2.30 that $\omega = k/\sqrt{\mu\epsilon}$, so

$$v_p = 1/\sqrt{\mu\epsilon} \tag{2.43}$$

This is the same velocity that we derived in Eq. 2.25, so the "speed of light" that comes from the wave equation is the phase velocity. If permittivity $\epsilon > \epsilon_0$, then v_p is less than c, the speed of light in a vacuum. Except for unusual circumstances, such as propagation in plasmas or x-rays in a certain frequency range, most materials have a permittivity, ϵ, that is greater in magnitude than ϵ_0. Do not be alarmed that the phase velocity can exceed c in certain situations. Such instances are results of collective action by an oscillating medium. We define the *index of refraction*, n, of a medium as the ratio of the phase velocity of light in a vacuum to the velocity in the medium,

$$n \equiv \frac{c}{v_p} \tag{2.44}$$

or using Eq. 2.43

$$n = \frac{\sqrt{\mu\epsilon}}{\sqrt{\mu_0\epsilon_0}}$$

$$= \sqrt{\frac{\epsilon}{\epsilon_0}} \quad \text{when } \mu = \mu_0. \tag{2.45}$$

The index of refraction is an important parameter in optical design and material characterization. We will explore its dependence on wavelength in later chapters. The ratio ϵ/ϵ_0 is called the *dielectric constant*. The index of refraction, n, is the square root of the dielectric constant.

We often write the wavevector, k, in terms of the *vacuum wavevector*, k_0, and the index of refraction. The vacuum wavevector is the magnitude of the wavevector in a vacuum, and is given by $k = 2\pi/\lambda$. Using the relation $k = \omega\sqrt{\mu_0\epsilon}$, we can rewrite this as

$$k = \omega\sqrt{\mu_0\epsilon} = \omega\sqrt{\mu_0\epsilon_0}\sqrt{\frac{\epsilon}{\epsilon_0}} = \omega\sqrt{\mu_0\epsilon_0}n = k_0 n \tag{2.46}$$

Once we know the vacuum wavevector, we can define the magnitude of the wavevector in all media based on the index of refraction.

To summarize, many parameters change inside a dielectric: the wavelength is reduced by $1/n$, k increases to $k_0 n$, and the phase velocity reduces to c/n. One parameter that stays constant is the angular frequency, ω, which is identical in all media. This follows from conservation of energy, where Planck's relation, $E = \hbar\omega$, describes the energy in the wave.

8. Group Velocity

Except in regions of high attenuation, energy in an electromagnetic wave travels at the *group velocity*, v_g. Information, which is carried by modulation on a light wave, also travels at the group velocity. The group velocity describes the speed of propagation of a pulse of light. A simple construction allows us to develop an expression for the group velocity through a superposition of two waves with different frequencies. With the frequencies assigned

$$\omega_1 = \omega + \Delta\omega \qquad \omega_2 = \omega - \Delta\omega \tag{2.47}$$

the two associated wavevectors will have values

$$k_1 = k + \Delta k \qquad k_2 = k - \Delta k \tag{2.48}$$

Assuming the waves have equal amplitudes, E_0, the superposition can be described as

$$\begin{aligned} &E_1 + E_2 \\ &= E_0(\cos[(\omega + \Delta\omega)t - (k + \Delta k)z] + \cos[(\omega - \Delta\omega)t - (k - \Delta k)z]) \end{aligned} \tag{2.49}$$

Using the trigonometric identity

$$2\cos x \cos y = \cos(x + y) + \cos(x - y) \tag{2.50}$$

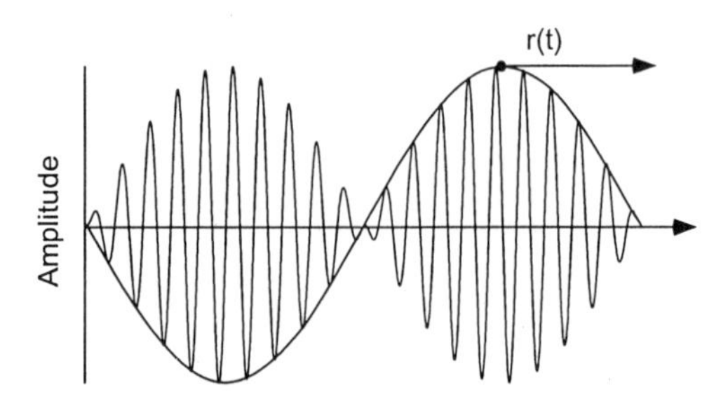

Figure 2.6. Two waves of similar frequency will form a beat pattern. The envelope of the beat travels at the group velocity.

the electric field superposition can be rewritten as

$$E_1 + E_2 = 2E_0 \cos(\omega t - kz) \cos(\Delta\omega t - \Delta kz) \tag{2.51}$$

This superposition of two waves at different frequencies leads to a temporal beat at frequency $\Delta\omega$ and a spatial beat with period Δk. Fig. 2.6 shows the superposition of the two waves. The envelope of the amplitude clearly depicts the beat frequency.

The group velocity is the speed at which a pulse, or in this case, the envelope, travels. The envelope is described by the $\cos(\Delta\omega t - \Delta kz)$ term of Eq. 2.51. We again attach a point to the crest of the envelope, and ask what speed, $v(t)$, is required to stay on the crest of the envelope. Following the arguments used to derive the phase velocity, we set the phase argument of the envelope, $\Delta\omega t - \Delta kz = constant$. Solving for $z(t)$,

$$z(t) = \frac{\Delta\omega t}{\Delta k} + constant \tag{2.52}$$

The group velocity is the derivative of this

$$v_g = \frac{dz}{dt} = \frac{\Delta\omega}{\Delta k} \quad \text{becomes} \quad v_g = \lim_{\Delta\omega \to 0} \frac{\Delta\omega}{\Delta k} = \frac{d\omega}{dk} \tag{2.53}$$

The group velocity, v_g, depends on the *first derivative* of the angular frequency with respect to the wavevector. In free space, where $\omega = kc$, the relation is simple and leads to $d\omega/dk = c$. In a vacuum, the phase and group velocities are identical. The relation is more complicated in other media. The constitutive constants, especially ϵ, usually depend on frequency. Recall that $\omega = kv_p = kc/n$. Then

$$\begin{aligned} v_g &= \frac{d\omega}{dk} = \frac{d}{dk}\left(\frac{kc}{n}\right) = \frac{c}{n} - \frac{kc}{n^2}\frac{dn}{dk} \\ &= \frac{c}{n} - \lambda\frac{dn}{d\lambda} \end{aligned} \tag{2.54}$$

The last relation can be confirmed with a simple calculation. The group velocity is nearly equal to the phase velocity, but is reduced or increased by a small term proportional to the change of index of refraction with wavelength. This change in index is called *dispersion*. In regions of *regular* dispersion, $dn/dk > 0$, the group velocity is less than the phase velocity, c/n. *Anomalous* dispersion occurs when $dn/dk < 0$. As we will see in subsequent chapters, both dispersions will play roles in the propagation of pulses in an optical fiber.

9. Boundary Conditions for Dielectric Interfaces: Reflection and Refraction

When two different media are adjacent to one another, the wave solutions in the two regions must be connected at the interface. The rules for connecting solutions are called *boundary conditions*. In general, if there is an index difference between two media, there will be a reflection. This is called a Fresnel reflection, after the French scientist, A. J. Fresnel (1788-1827).

Consider the interface shown in Fig. 2.7. The **k** vector of an electromagnetic wave propagates from one medium into another (accompanied by a partial reflection back into the originating media). The wave has frequency ω, and is incident on the interface from region 1 at an angle of incidence, θ_i. The two regions have indices of refraction n_1 and n_2, respectively. We want to determine the amplitudes of the transmitted and reflected waves, E_t and E_r, and their respective wavevectors, $\mathbf{k}_t$ and $\mathbf{k}_r$.

We must first solve the wave equation (2.25) in each region. This is straightforward and yields,

$$\mathbf{E}_q(r,t) = E_q e^{-j(\mathbf{k}_q \cdot \mathbf{r} - \omega t)} \tag{2.55}$$

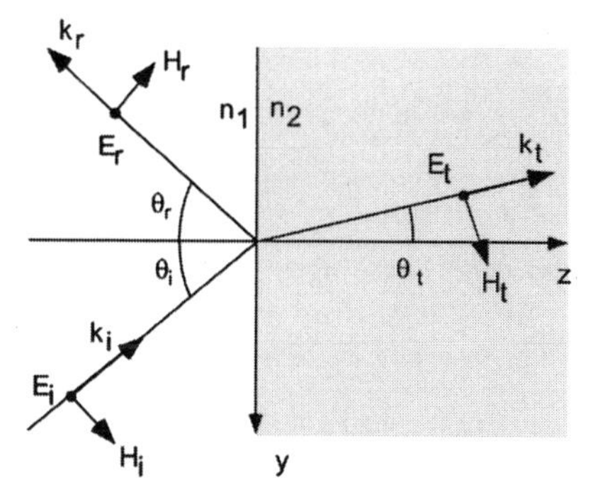

Figure 2.7. A ray incident on an interface at angle θ_i will reflect and refract into two different rays. The electric field in this figure is directed out of the page for all waves.

where E_q is the amplitude. The subscript, q, refers to the three different fields that will arise. The tough part of the problem is connecting these solutions at the interface. The boundary conditions that apply to this situation can be derived from the integral form of Maxwell's equations. In a medium where there are no sources, $(\rho, \mathbf{J} = 0)$, the boundary conditions are

$$\begin{aligned} \hat{s} \times (\mathbf{E}_2 - \mathbf{E}_1) &= 0 \quad \text{tangential } E \text{ is continuous} && (2.56) \\ \hat{s} \times (\mathbf{H}_2 - \mathbf{H}_1) &= 0 \quad \text{tangential } H \text{ is continuous} && (2.57) \\ \hat{s} \cdot (\mathbf{B}_2 - \mathbf{B}_1) &= 0 \quad \text{normal } B \text{ is continuous} && (2.58) \\ \hat{s} \cdot (\mathbf{D}_2 - \mathbf{D}_1) &= 0 \quad \text{normal } D \text{ is continuous} && (2.59) \end{aligned}$$

Where $\hat{s}$ refers to the unit normal to the interface.

There are two possible orientations for the electric field with respect to the interface. The field can be perpendicular or parallel to the *plane of incidence*. The plane of incidence contains both the **k** vector and $\hat{s}$. When the electric field is perpendicular to the plane of incidence, it is called a *Transverse Electric*, or TE wave. Fig. 2.7 shows the specific case of a TE wave incident on an interface at an angle θ_i.

Fig. 2.7 shows there are six field amplitudes ($E_i, E_t, E_r, H_i, H_t, H_r$), three wavevectors ($k_i, k_t, k_r$), and three angles ($\theta_i$, θ_t, and θ_r). Some of these, like E_i and θ_i, are initial conditions of the problem while the others are dependent variables. It is convenient to first relate the angle of incidence to the angle of reflection:

$$\theta_i = \theta_r \tag{2.60}$$

Justification is straightforward: we can apply Fermat's principle (Prob. 2.4.), or conservation of photon momentum (Prob. 2.5.)

The general description of the $\hat{x}$-polarized incident electric field is

$$\mathbf{E}_i = E_i \hat{x} e^{-j\hat{k}_i \cdot \hat{r}} e^{j\omega t} \tag{2.61}$$

The wavevector $\mathbf{k}_i$ is described in terms of its vector components

$$\hat{k}_i = (\hat{z} \cos\theta_i - \hat{y} \sin\theta_i) k_0 n_1 \tag{2.62}$$

where k_0 is the *vacuum wavevector* (ω/c). Position, $\mathbf{r}$, is also described in vector form,

$$\mathbf{r} = x\hat{x} + y\hat{y} + z\hat{z} \tag{2.63}$$

Substituting these terms into Eq. 2.61, the complete description of the incident field is

$$\begin{aligned} \mathbf{E}_i(x, y, z, t) &= \hat{x} E_i e^{-jk_0 n_1 (\hat{z}\cos\theta_i - \hat{y}\sin\theta_i)\cdot(x\hat{x}+y\hat{y}+z\hat{z})} e^{j\omega t} \\ &= \hat{x} E_i e^{-jk_o n_1 (z\cos\theta_i - y\sin\theta_i)} e^{j\omega t} \end{aligned} \tag{2.64}$$

The incident field is completely defined in terms of direction, frequency, and polarization. The frequency term, $e^{j\omega t}$, can be dropped from the explicit formulation because it is the same in all regions. The other electric fields in Fig. 2.7 are similarly described.

$$\begin{aligned} \mathbf{E}_t(x,y,z) &= \hat{x}E_t e^{-j\mathbf{k}_t\cdot\mathbf{r}} \\ &= \hat{x}E_t e^{-jk_0 n_2(z\cos\theta_t - y\sin\theta_t)} \end{aligned} \quad (2.65)$$

$$\begin{aligned} \mathbf{E}_r(x,y,z) &= \hat{x}E_r e^{-j\mathbf{k}_r\cdot\mathbf{r}} \\ &= \hat{x}E_r e^{-jk_0 n_1(-z\cos\theta_r - y\sin\theta_r)} \end{aligned} \quad (2.66)$$

We have assumed that the electric field will continue to point out of the page for each component. This may or may not be true: in some cases the phase of the field advances by 180°, and the direction would reverse (i.e. point into the page). If this happens, when we have completed our solution, one of the components will be multiplied by a negative sign. So do not be too concerned about choosing the proper orientations initially, as these problems will solve themselves.

We will need to describe the magnetic fields for the three waves. The appropriate k-vector for each magnetic field is the same as for the electric field. Note that for TE waves, the H fields have two vector components, a z-component and a y-component. The *magnitude* of the magnetic field is related to the electric field through the impedance, η, of the medium (Eq. 2.35)

$$|H| = |E|/\eta \quad (2.67)$$

where $\eta_i = \sqrt{\mu/\epsilon_i}$. Using trigonometry, the correct expressions for the H-field components are

$$\mathbf{H}_i = (E_i/\eta_1)\,(\hat{z}\sin\theta_i + \hat{y}\cos\theta_i)e^{-jn_1k_0(z\cos\theta_i - y\sin\theta_i)} \quad (2.68)$$

$$\mathbf{H}_t = (E_t/\eta_2)\,(\hat{z}\sin\theta_t + \hat{y}\cos\theta_t)e^{-jn_2k_0(z\cos\theta_t - y\sin\theta_t)} \quad (2.69)$$

$$\mathbf{H}_r = (E_r/\eta_1)\,(\hat{z}\sin\theta_r - \hat{y}\cos\theta_r)e^{-jn_1k_0(-z\cos\theta_r - y\sin\theta_r)} \quad (2.70)$$

With a complete description of the field in all regions, (Eqs. 2.64 -2.66 and Eqs. 2.68- 2.70), we can connect the solutions at the interface, yielding formulae for transmission and reflection. First, apply the condition that the tangential component of E must be continuous across the interface,

$$\hat{z}\times(\mathbf{E}_i + \mathbf{E}_r)|_{z=0} = \hat{z}\times\mathbf{E}_t\big|_{z=0} \quad (2.71)$$

The tangential E field at the interface is the E_x component. Expanding this at $z = 0$, and using the fact that $\hat{z}\times\hat{x} = \hat{y}$ and $\theta_i = \theta_r$ yields

$$\hat{y}E_i e^{(jk_0n_1y\sin\theta_i)} + \hat{y}E_r e^{(jk_0n_1y\sin\theta_i)} = \hat{y}E_t e^{(jk_0n_2y\sin\theta_t)} \quad (2.72)$$

Combining terms of equal phase

$$\hat{y}(E_i + E_r)e^{(jk_0 n_1 y \sin\theta_i)} = \hat{y}E_t e^{(jk_0 n_2 y \sin\theta_t)}. \tag{2.73}$$

For this equation to hold, it must be true for all values of y. At $y = 0$ the equation becomes simply

$$E_i + E_r = E_t \qquad \text{(continuity of magnitude)} \tag{2.74}$$

Substituting this into Eq. 2.73 and cancelling common terms yields

$$e^{(jk_0 n_1 y \sin\theta_i)} = e^{(jk_0 n_2 y \sin\theta_t)} \tag{2.75}$$

which can only be true if

$$k_0 n_1 y \sin\theta_i = k_0 n_2 y \sin\theta_t \tag{2.76}$$

Cancelling common terms on both sides we arrive at Snell's Law

$$n_1 \sin\theta_i = n_2 \sin\theta_t \tag{2.77}$$

From Snell's law, the direction of the transmitted wave can be found. This leaves only the amplitudes, E_t, E_r, H_t and H_r to be determined. To determine the amplitude of E_r in terms of E_i, we resort to the magnetic boundary conditions. The continuity of tangential H requires that

$$\hat{z} \times (\mathbf{H}_i + \mathbf{H}_r)|_{z=0} = \hat{z} \times \mathbf{H}_t|_{z=0} \tag{2.78}$$

In this case, $\mathbf{H}_i$ has both z and y components, so we must be careful to carry only the y component through the cross product. Using Eq. 2.68- 2.70, $\hat{z} \times \hat{y} = -\hat{x}$, and $\theta_r = \theta_i$,

$$\hat{z} \times \mathbf{H}_i = (-\hat{x}E_i \cos\theta_i/\eta_1)e^{-jk_0 n_1(-y\sin\theta_i)} \tag{2.79}$$
$$\hat{z} \times \mathbf{H}_t = (-\hat{x}E_t \cos\theta_t/\eta_2)e^{-jk_0 n_2(-y\sin\theta_t)} \tag{2.80}$$
$$\hat{z} \times \mathbf{H}_r = (+\hat{x}E_r \cos\theta_i/\eta_1)e^{-jk_0 n_1(-y\sin\theta_i)} \tag{2.81}$$

where E_i, E_r, and E_t represent magnitudes, not vectors. Adding the terms according to Eq. 2.78, and applying Snell's law (Eq. 2.77), we get

$$(E_i - E_r)\cos\theta_i/\eta_1 = E_t \cos\theta_t/\eta_2 \tag{2.82}$$

Since $E_t = E_i + E_r$, we can replace E_t in terms of the other variables

$$(E_i - E_r)\cos\theta_i/\eta_1 = (E_i + E_r)\cos\theta_t/\eta_2 \tag{2.83}$$

and solve for the ratio of E_r/E_i

$$E_r/E_i = \frac{(\eta_2 \cos\theta_i - \eta_1 \cos\theta_t)}{(\eta_2 \cos\theta_i + \eta_1 \cos\theta_t)} \tag{2.84}$$

Similarly, we could eliminate E_r from Eq. 2.83 and solve for the ratio E_t/E_i

$$E_t/E_i = \frac{2\eta_2 \cos\theta_i}{(\eta_2 \cos\theta_i + \eta_1 \cos\theta_t)} \tag{2.85}$$

It is more common to deal with the index of refraction, n_i, than with impedance, η_i, for a material (be careful to distinguish η from n). If $\mu = \mu_o$, then η can be rewritten as

$$\eta_i = \sqrt{\frac{\mu_o}{\epsilon_i}} = \sqrt{\frac{\mu_o \epsilon_0}{\epsilon_i \epsilon_0}} = \frac{\eta_0}{n_i} \tag{2.86}$$

Substituting this expression into Eqs. 2.84 and 2.85 generates the more familiar forms of the amplitude transmission and reflection formulae for a transverse electric field. In these formulae, the field is incident from the n_1 side, entering into the n_2 side.

$$E_r/E_i = \frac{n_1 \cos\theta_i - n_2 \cos\theta_t}{n_1 \cos\theta_i + n_2 \cos\theta_t} \tag{2.87}$$

$$E_t/E_i = \frac{2n_1 \cos\theta_i}{n_1 \cos\theta_i + n_2 \cos\theta_t} \tag{2.88}$$

The expressions for transmission and reflection of a wave which has the magnetic field, H, perpendicular to the plane of incidence (the so-called *Transverse Magnetic* or TM wave) are significantly different. Their derivation is left as an exercise to show

$$E_r/E_i = \frac{n_1 \cos\theta_t - n_2 \cos\theta_i}{n_2 \cos\theta_i + n_1 \cos\theta_t}. \tag{2.89}$$

$$E_t/E_i = \frac{2 \cos\theta_i}{(n_2/n_1) \cos\theta_i + \cos\theta_t} \tag{2.90}$$

One word of caution about the Fresnel formulae: they describe the *amplitude* of the transmitted and reflected field, and not the *power* of the fields. In some circumstances, the magnitude of the transmitted electric field is *larger* than that of the incident electric field. This dilemma is resolved when total power is accounted for in the solution. One must account for geometric change of area between the incident and transmitted beams and the impedance change. We can also develop expressions for the H components, but these can be found simply and directly through the impedance relationships.

Example 2.2 Normal reflection from a glass interface

The most common example of Fresnel reflection is that which occurs when light strikes a glass-air interface. Let's apply the reflection formula to this problem to illustrate the magnitude of the effect, and the phase shift which occurs.

A beam of light is incident normally on a glass-air interface as shown in Fig.2.8 What is the intensity of the reflected light if the glass has an index of refraction of $n = 1.5$?

Solution: Plugging numbers into Eq. 2.88, noting that $\cos\theta = 1$ in this case, we get

$$E_r/E_i = \frac{1-1.5}{1+1.5} = -0.2$$

The reflected amplitude is 20% of the incident amplitude. The negative sign indicates that the reflected wave is 180° out of phase with the incident wave (when light strikes a higher index, the phase of reflected wave will always be reversed). Now, what is the intensity? Using the Poynting vector and the fact that $|H| = |E|/\eta$, we find the incident intensity is

$$S_{inc} = \frac{1}{2}\frac{E_0^2}{\eta}$$

while the reflected intensity is only

$$S_{ref} = \frac{1}{2}\frac{(0.2E_0)^2}{\eta} = 0.04S_{inc}$$

Thus, only 4% of the incident power is reflected by the glass interface. This reflection can become a significant loss in certain applications. For example, a camera lens often will consist of three or more separate lenses, representing six glass-air interfaces. The total transmission for such a system would be $T = (0.96)^6 = 0.78$ if the lenses are not modified. This represents a significant

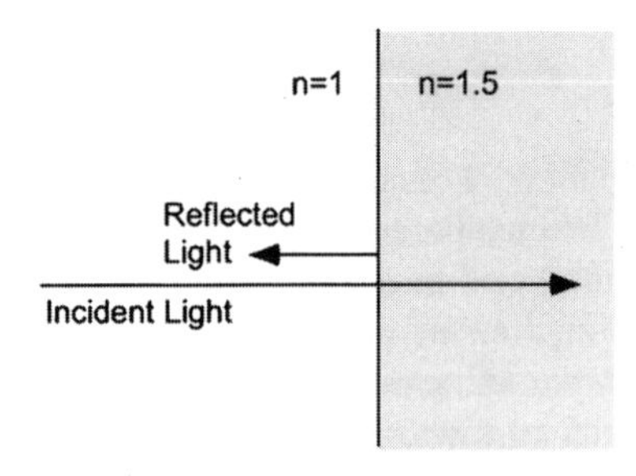

Figure 2.8. A beam of light strikes a glass interface normally, causing a small reflection.

loss of power in an application where light collection efficiency is critical. Not only would the reflections require larger apertures and longer exposure times, but they also could contribute ghost images on the film. These problems are overcome by putting an anti-reflection (AR) coating on each surface. The AR coating is basically a stack of $\lambda/4$ thick layers of dielectric material which interferometrically reduce the total reflection coefficient.

10. Total Internal Reflection

An important physical process in guided wave optics is Total Internal Reflection. We will look at total internal reflection from two perspectives: ray tracing, and the wave equation. Ray tracing is useful when the dimensions of the optical element are large compared to the wavelength of light. Ray tracing is useful for concepts such as the numerical aperture. The wave picture provides a complete description of the phase shifts and evanescent fields that accompany total internal reflection.

Ray Tracing

Ray tracing models light as rays travelling in straight lines between optical elements. The only action of an optical element is to redirect the ray. The angle of incidence of the ray, and the properties of the optical element establish the degree to which the ray is redirected.

The important operational rules for ray tracing are *Snell's Law*

$$n_1 \sin\theta_1 = n_2 \sin\theta_2 \tag{2.91}$$

and the *Law of Reflection*,

$$\theta_{incidence} = \theta_{reflected} \tag{2.92}$$

illustrated in Fig. 2.9. Using these two simple equations, a powerful calculus can be developed for designing and evaluating lenses and optical systems. Many excellent references [1], [8], [15] elaborate on the application of ray tracing to optical design. Numerical matrix techniques have been developed based on these simple laws which allow the engineer to design complex linear optical systems. The ray tracing analysis is usually of limited use for guided wave optical design, however, because the size of the waveguide is often comparable to the wavelength of guided light. Ray tracing's most common application is to describe graded index waveguides, and to define the numerical aperture.

Total Internal Reflection using ray tracing

Total Internal Reflection (TIR) is the phenomenon where light is completely reflected at a dielectric interface without the help of reflective coatings. TIR is often exploited to make efficient achromatic reflectors. For example, right-angle prisms are often used to redirect light from imaging systems such as

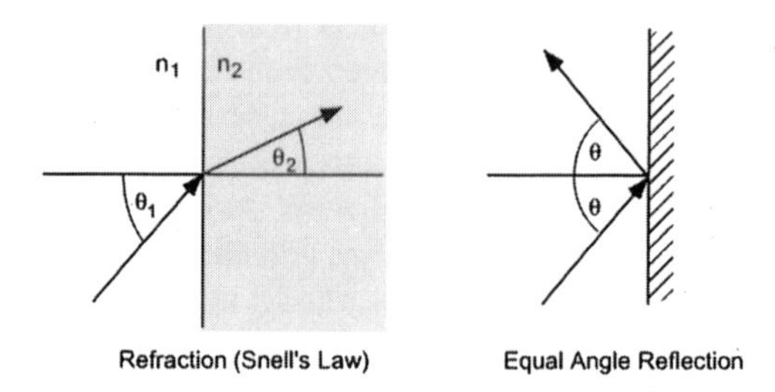

Figure 2.9. The two principle laws of ray tracing. The left figure shows Snell's Law. The right figure illustrates that the angle of incidence equals the angle of reflection.

binoculars, or to serve as rugged mirrors for high powered lasers. Here we want to consider optical waveguides. Fig. 2.10 illustrates the ray picture of a right-angle prism and of a waveguide. The key requirement for TIR is that the light must be incident on a dielectric interface from the high index side. Thus an optical waveguide must consist of a layer of high index dielectric surrounded by material with a lower index.

Total internal reflection occurs over a certain range of angles. Fig. 2.11 shows a wave incident at an angle, θ_1, on a dielectric interface from the *high index side*. The refracted ray in the low index medium, n_2, exits at angle θ_2. The exit angle is

$$\theta_2 = \sin^{-1}(\frac{n_1}{n_2}\sin\theta_1) \tag{2.93}$$

As the angle of incidence, θ_1, increases, the angle of refraction, θ_2, must also increase to satisfy the equality. But because $n_1/n_2 > 1$, the refraction angle, θ_2, will reach a value of $90°$ before θ_1 does. This occurs when

$$\sin\theta_1 = \frac{n_2}{n_1} \tag{2.94}$$

This value of θ_1 is known as the *critical angle.* For angles of incidence larger than the critical angle, θ_2 must be a complex number (see Prob. 2.11.) to

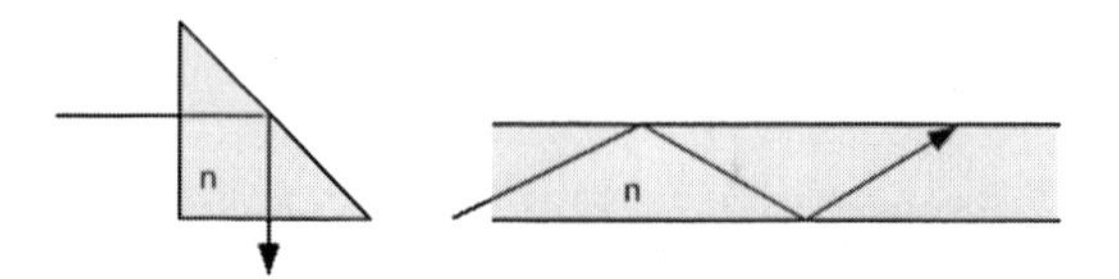

Figure 2.10. Total internal reflection can be implemented in many ways. The right-angle prism, and the optical waveguide both use total internal reflection to redirect or trap light, respectively. Note that the light is incident from the high index side of the interface in all cases of TIR.

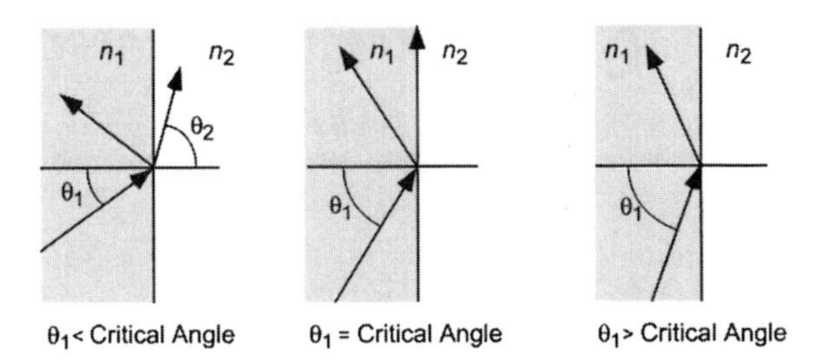

Figure 2.11. Three cases where the angle of incidence is below, at, and above the critical angle respectively.

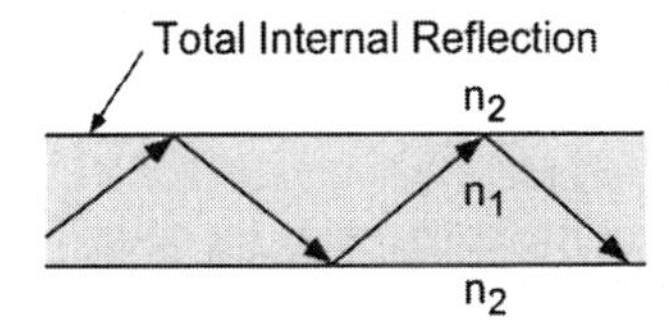

Figure 2.12. A waveguide can be formed when total internal reflection traps a wave between two surfaces.

satisfy Snell's Law. A complex angle in the expressions for transmission (for example, Eq. 2.88) leads directly to a complex amplitude in the low index region. Complex amplitudes simply mean that a phase shift occurs. While we will not prove it here (see problems 2.4. and 2.5.), as with all simple reflections, the angle of reflection is equal to the angle of incidence of the ray.

Total internal reflection is the key to optical waveguiding. Consider the dielectric structure shown in Fig. 2.12. A dielectric slab of index n_1 is surrounded by a lower index dielectric. A ray travelling within the high index material will be total-internal-reflected at the upper and lower interfaces of this structure *if* the angle of incidence at the interface exceeds the critical angle. This is a simplified picture, as the actual ray picture of a waveguide is more subtle in terms of allowed directions for the rays (to be fully developed in the next chapter). However, the essential idea behind the optical waveguide is that light is trapped in a high index media through total internal reflection.

11. Wave Description of Total Internal Reflection

We claim that the ray became totally reflected for angles beyond the critical angle, yet the only evidence we offered to support this claim is fact that the trigonometric identity is impossible to rationalize using real angles. We can put the description on a more physical basis by examining total internal reflection

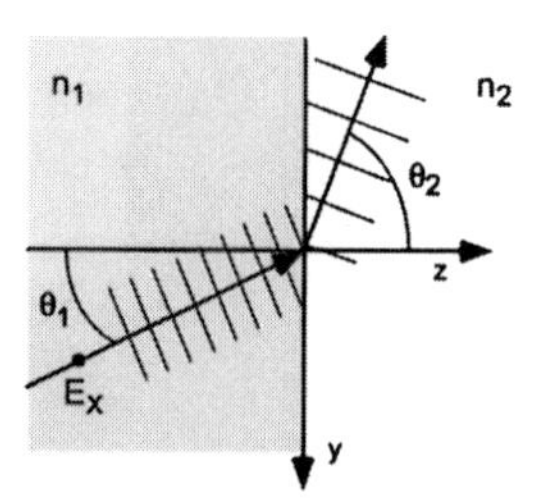

Figure 2.13. A plane wave incident on a dielectric interface at angle θ_1 will refract at an angle θ_2 in the second medium. The reflected ray is not shown for clarity.

using electromagnetic waves. The wave picture provides a physical explanation of the reflection, and yields information on the phase shift caused by reflection.

Consider a TE plane wave, polarized along the $\hat{x}$-axis with amplitude, E_0, incident on a dielectric interface, as shown in Fig. 2.13. The angle of incidence is less than θ_c. Since the time behavior is identical for both, only the spatial descriptions of the two waves are considered:

$$\begin{aligned} \mathbf{E}_1(y,z) &= \hat{x}E_0 e^{-jk_0 n_1(z\cos\theta_1 - y\sin\theta_1)} + c.c. \\ \mathbf{E}_2(y,z) &= \tau\hat{x}E_0 e^{-jk_0 n_2(z\cos\theta_2 - y\sin\theta_2)} + c.c. \end{aligned} \tag{2.95}$$

where τ is the amplitude transmission coefficient (from Eq. 2.85). The angles θ_1 and θ_2 are related by Snell's Law.

$$\begin{aligned} \sin\theta_2 &= \frac{n_1}{n_2}\sin\theta_1 \\ \cos\theta_2 &= \sqrt{1 - \frac{{n_1}^2}{{n_2}^2}\sin^2\theta_1} \end{aligned} \tag{2.96}$$

Substituting these values into Eq. 2.65, we get an expression for the transmitted amplitude, $\mathbf{E}_2$, that is a function of the incident angle, θ_1,

$$\mathbf{E}_2 = \tau\hat{x}E_0 \exp\left\{-jk_0 n_2\left(z\sqrt{1 - \frac{{n_1}^2}{{n_2}^2}\sin^2\theta_1} - y\frac{n_1}{n_2}\sin\theta_1\right)\right\} \tag{2.97}$$

Physically, we can understand refraction by considering what happens to the wavefronts at the interface. On the incident side, the wavefront strikes the interface and is partially reflected and partially transmitted. If $\theta_1 < \theta_c$, the wavefronts must be continuous across the interface. The node where these two wavefronts connect travels along the interface with a velocity, v_{node} as shown in Fig. 2.14. The velocity of this intersection, v_{node}, is simply

$$v_{node_1} = v_{p1}/\sin\theta_1 \tag{2.98}$$

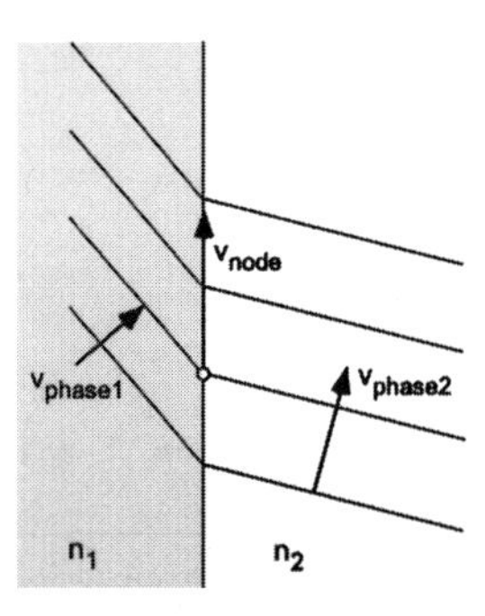

Figure 2.14. The plane waves on either side of the interface must connect as they cross the interface. These connecting nodes travel along the interface at a velocity that depends on the angle of incidence.

where v_{p_1} is the phase velocity, c/n_1 in the first medium. The transmitted wave, $\mathbf{E}_2$, must travel in such a direction that the velocity of the nodes of its phase front is identical to that of the incident field. Since the phase velocity in medium n_2 is different, the only way the node velocities can be matched is if the direction of the transmitted field refracts to angle θ_2 such that

$$v_{p_1}/\sin\theta_1 = v_{p_2}/\sin\theta_2 \tag{2.99}$$

This is simply a restatement of Snell's law.

As the angle of incidence θ_1 increases, the transmitted waves must make a larger angle θ_2 to maintain the proper velocity of the intersection at the interface. At $\theta_1 = \theta_{cr}$, $\cos\theta_2$ goes to zero, and the transmitted field contains only one component,

$$\mathbf{E}_2 = \tau\hat{x}E_0 e^{(jk_0 n_2 y)} + c.c. \quad \text{at} \quad \theta_1 = \theta_{cr} \tag{2.100}$$

This is the description of a plane wave travelling parallel to the interface in the $\hat{y}$ direction. This direction will yield a node velocity that is as slow as can be achieved in medium n_2. In the ray picture, we would say that the transmitted ray is parallel to the plane of incidence. Fig. 2.15 shows this condition. The plane waves on either side of the interface must connect as they cross the interface. These connecting nodes travel along the interface at a velocity that depends on the angle of incidence.

What happens as θ_1 increases *beyond* the critical angle? The radical in Eq. 2.96 which describes $\cos\theta_2$, becomes imaginary, so the transmitted electric amplitude is described as

$$\mathbf{E}_2 = \tau\hat{x}E_0 e^{-k_0 n_2\sqrt{(n_1^2/n_2^2)\sin^2\theta_1 - 1}\, z} e^{jk_0 n_1 \sin\theta_1 y} \tag{2.101}$$

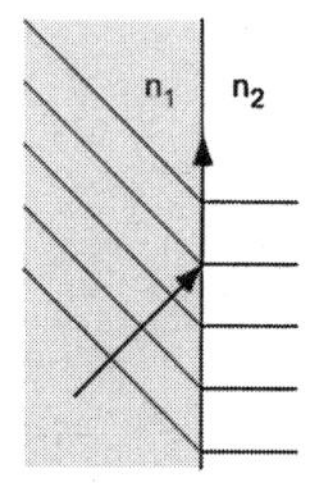

Figure 2.15. At the critical angle, the transmitted plane waves travel parallel to the interface.

where we choose the proper sign of the radical to ensure that the amplitude decays as distance from the interface increases. This cumbersome form is often written as

$$\mathbf{E}_2 = \tau \hat{x} E_0 e^{-\gamma z} e^{j\beta y} \tag{2.102}$$

where γ represents the *attenuation coefficient* (units: cm^{-1}),

$$\gamma = k_0 n_2 \sqrt{\frac{n_1^2}{n_2^2} \sin^2 \theta_1 - 1} \tag{2.103}$$

and β represents the *propagation coefficient* (units: rads/cm),

$$\beta = k_0 n_1 \sin \theta_1 \tag{2.104}$$

Inspection of Eq. 2.103 shows that the field amplitude decays exponentially away from the interface. This field is called the *evanescent* field. The evanescent field contains real values of **E** and **H**, but they are 90° out-of-phase with each other. The evanescent field contains *reactive power*, not real power. In reactive power, no work is done, but energy is stored. This evanescent field is very important to device applications. It is possible to tap some of the energy away using special structures. We will see many such devices in later chapters concerning switches, modulators, and couplers.

Returning to the physical picture, when θ_1 is increased beyond the critical angle, the node velocity in n_1 is slower than the minimum possible velocity of nodes in medium n_2. In medium n_2, the phase fronts advance beyond their generating counterparts in n_1. As the transmitted wave fronts travel ahead, they run up on wavefronts emitted from earlier nodes. At a certain distance, the fronts in n_2 will be 180° out of phase with the nodes of n_1 and destructive interference will occur. The larger the angle of incidence, θ_1, the slower the node velocity in n_1 will be. Destructive interference will occur sooner, leading to increased attenuation. We see from Eq. 2.103 that the attenuation coefficient, γ, increases as the angle of incidence is increased.

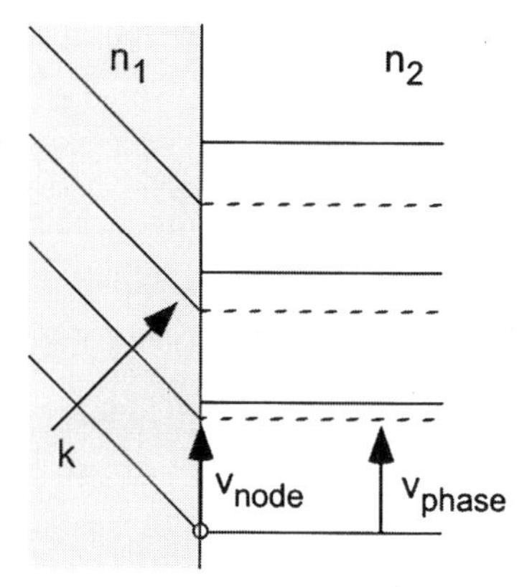

Figure 2.16. Beyond the critical angle, the plane waves on the low index side of the interface travel faster than the nodes due to the incident field. They get ahead of their source nodes, and then react back against them.

12. Phase Shift Upon Reflection

A more subtle, yet critically important effect that occurs in TIR is the phase shift of the light upon reflection. These phase shifts help determine which modes propagate in a waveguide. After reflection, the optical signal slightly lags in phase compared to the incident wave. One can view this phase shift as being due to the extra distance the light travels when going into and returning from the low index media during its evanescent phase (this is called the *Goos-Hänchen* shift, see Prob. 2.17. and Appendix A: The Goos-Hänchen Shift), or one can view the phase shift as occurring due to the mixing of two waves that are slightly out of phase (the reflected and evanescent wave).

How big is the phase shift? For a TE wave, the phase shift can be determined directly by writing the the amplitude reflection formula, Eq. 2.88, in polar form

$$\frac{E_r}{E_i} = \frac{(n_1 \cos\theta_1 - n_2 \cos\theta_2)}{(n_1 \cos\theta_1 + n_2 \cos\theta_2)} = |r|\, e^{j2\phi} \tag{2.105}$$

The reflection coefficient is described in terms of its magnitude, $|r|$, and phase shift, 2ϕ. Beyond the critical angle, $\cos\theta_2$ becomes pure imaginary ($\theta = \sqrt{1 - n_1^2/n_2^2 \sin^2\theta_1}$). Letting $\alpha = n_1 \cos\theta_1$, and $j\beta = n_2 \cos\theta_2$, Eq.2.105 can be rewritten as

$$\frac{E_r}{E_i} = \frac{\alpha - j\beta}{\alpha + j\beta} \tag{2.106}$$

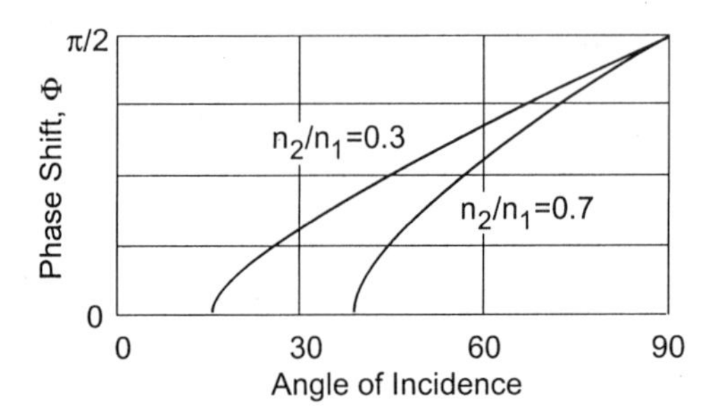

Figure 2.17. Plot of the phase shift, Φ, as a function of the angle of incidence, θ. Note that the phase shift below the critical angle is zero.

Substituting the value of $\cos\theta_2$ from Eq.2.96, the phase of this transfer function is

$$\begin{aligned} 2\Phi_{TE} &= \tan^{-1}\left(\frac{-\beta}{\alpha}\right) - \tan^{-1}\left(\frac{\beta}{\alpha}\right) \\ &= 2\tan^{-1}\left(\frac{-\beta}{\alpha}\right) \\ &= 2\tan^{-1}\frac{-\sqrt{n_1^2\sin^2\theta_1 - n_2^2}}{n_1\cos\theta_1} \end{aligned} \quad (2.107)$$

This equation is only valid for $\theta_1 > \theta_{cr}$. The magnitude of E_r/E_i is obviously unity. We leave it as an exercise to show that the correct formula for TM waves is given by

$$\Phi_{TM} = \tan^{-1}\left(\frac{-n_1^2}{n_2^2}\frac{\sqrt{n_1^2\sin^2\theta_1 - n_2^2}}{n_1\cos\theta_1}\right) \quad (2.108)$$

Figure 2.17 shows the dependence of Φ_{TE} as a function of the angle of incidence θ_1 for two ratios n_1/n_2. The ratios 0.3 and 0.7 correspond to the approximate values of a GaAs-air and glass-air interface, respectively. The phase shift for the TM case is similar.

For angles of incidence below the critical angle, there is no phase shift upon reflection (actually, the phase shift can be 0 or π, depending on the relative indices).

13. Summary

This chapter reviewed Maxwell's equations, using them to establish a set of units (MKS), and several important quantities and concepts. We derived the wave equation, and solved it in homogeneous media. From the solution, we developed expressions for phase and group velocity. The concept of the wavevector was introduced and related to the angular frequency of a wave. Using boundary conditions, we developed expressions for the reflection and refraction of electromagnetic waves from a dielectric interface.

We then explored total internal reflection. Snell's law was used to illustrate the ray picture of total internal reflection. While Snell's law, if used with complex angles, can give a total description of the evanescent fields associated with these reflections, the wave description based on Maxwell's equations provides a clearer picture. Using the wave picture, we used the Fresnel formulae for reflection and transmission at a dielectric interface to develop expressions for phase shift associated with TIR. This phase shift always accompanies TIR, and plays a unique role in establishing which rays will be allowed inside an optical waveguide.

The material parameters, μ and ϵ, play a critical role in determining the action of a wave at a dielectric interface. We alluded to the frequency dependence of these material parameters in the discussion of group velocity. This will be further developed in Chapter 8.

References

[1] F. Jenkins and H. White, *Fundamentals of Optics*, (New York) McGraw-Hill Co., (1957).

[2] E. Hecht, *Optics*, Addison-Wesley Publishing Company, Massachusetts, (1990).

[3] S. G. Lipson and H. Lipson, *Optical Physics*, Cambridge University Press, New York (1981).

[4] W. Louisell, *Radiation and Noise in Quantum Electronics*, McGraw-Hill Book Company, New York (1964).

[5] R. Loudon, *Quantum Theory of Light, 2nd Ed.*, Clarendon Press-Oxford, Great Britain (1983).

[6] See for example, W. J. Tabor, "Magneto-Optic Materials", *Laser Handbook Vol. 2*, p.1009, F. T. Arecchi and E. O Schulz-Dubois, eds., North-Holland Pub. Co., Great Britain, (1972).

[7] A. Yariv and P. Yeh, *Optical Waves in Crystals*, John Wiley and Sons, New York (1984).

[8] A. Nussbuam and R. Phillips, *Contemporary Optics for Scientists and Engineers*, Prentice-Hall, Inc., New Jersey (1976)

[9] Warren Smith, *Modern Optical Engineering, 2nd Ed.*, McGraw-Hill, Inc., New York (1990)

Practice Problems

1. Derive the Fresnel amplitude reflection and transmission coefficients for an electromagnetic wave that is polarized with the electric field in the plane of incidence (TM wave).

2. We simplified Eq. 2.24 by assuming that the term

$$-\nabla(\mathbf{E}\cdot\nabla\epsilon/\epsilon)$$

is negligible. Determine how small $\nabla\epsilon$ must be for this assumption to be reasonable. Starting from the exact wave equation (with the above term included), use separation of variables to solve for the one-dimensional wave (i.e. $E = Z(z)T(t)$). Solve for $T(t)$ in terms of separation constant k and $(\mu\epsilon)^{1/2}$. From the resulting equation for $Z(z)$, find a rough value for $\nabla\epsilon$ over a characteristic distance of one wavelength of the field. How small must $\frac{\Delta\epsilon}{\epsilon}$ be to make it negligible (say less than 1% in magnitude) compared to the other terms in the wave equation?

3. Show that for an harmonic wave, the average value $\langle\mathbf{S}\rangle = \frac{1}{2}\langle\mathbf{E}\times\mathbf{H}\rangle = \frac{\mathbf{k}}{|k|}\frac{k}{2\omega\mu}E_0^2$, for a wave with wavevector, k, and electric amplitude, E_0.

4. Fermat's Principle states that if a light ray travels between two points, it follows the path that takes the least time. Use Fermat's principle to 1) verify that the angle of incidence equals the angle of reflection for a simple plane mirror, and 2) derive Snell's law for a ray crossing a dielectric interface.

5. Use conservation of momentum and the fact that a photon has momentum given by

$$\mathbf{p} = \hbar\mathbf{k} = \hbar n\mathbf{k}_0$$

where $\mathbf{k}_0$ is the vacuum wavevector of the photon, to 1) show that the angle of incidence equals the angle of reflection for a simple plane mirror, and 2) derive Snell's law for a ray crossing a dielectric interface.

6. Derive the four boundary conditions, Eqs. 2.57-2.59, relating $\mathbf{E}, \mathbf{H}, \mathbf{D}$, and $\mathbf{B}$ across a dielectric boundary. Use the integral form of Maxwell's equations.

7. Consider the situation shown in the Fig. 2.18 below. A TE wave (polarized along $\hat{y}$) with a wavelength of 1 μm is incident from air onto the GaAs-air interface, at an angle of incidence of 45°. The index of refraction of GaAs equals 3.4 at 1 μm. Describe the electric fields in all regions surrounding the interface.

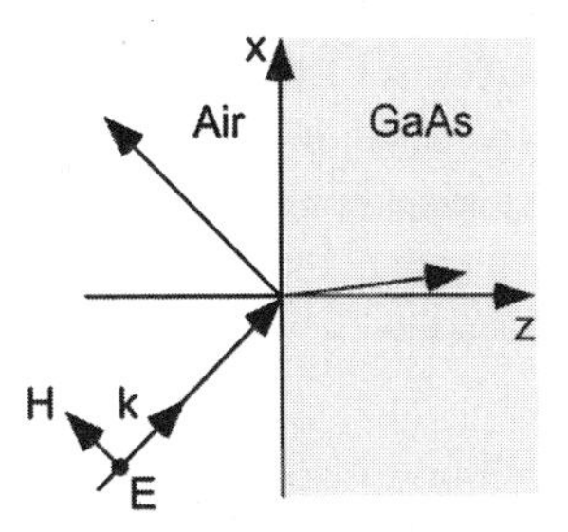

Figure 2.18. A wave incident at 45° on a dielectric interface from the low index side. The wave is TE polarized, and has a wavelength of 1 μm.

8. Using a computer and a program like *Mathematica*, plot the amplitude and power reflection and amplitude and power transmission coefficients as a function of angle for both TE and TM waves for a) a glass-air interface ($n_{glass} = 1.5$), and b) a GaAs-air interface ($n_{GaAs} = 3.4$). Assume the light is incident from the air side.

9. Beginning with E. 2.12, derive the homogeneous wave equation in terms of the magnetic field amplitude, H.

10. From the definition of the Poynting vector, Eq.2.37, show that Eq. 2.38 follows when the E field is a sinusoidal function of time.

11. It is easy to understand trigonometric identities as θ becomes complex: use the Euler identity, $\sin\theta = 1/2(e^{j\theta} + e^{-j\theta})$, and let θ become a complex number, $\alpha + j\beta$. Show that $\cos^2\theta + \sin^2\theta = 1$.

12. Confirm Eq. 2.108 for the phase shift that occurs for a TM wave upon total internal reflection.

13. Consider a TE wave with amplitude E_0 incident on an air-dielectric interface from the dielectric side. The dielectric has an index of refraction, n_1, of 1.6. The angle of incidence is 5° larger than the critical angle. $\lambda = 1\mu$m.

    (a) What is the critical angle for this interface?

    (b) Determine the electric field amplitude for all points on the air side of the interface.

    (c) What is the phase shift, 2Φ, for the reflected light?

14. The transmission coefficient τ defined by Eq. 2.88 becomes a complex number when the angle of incidence exceeds the critical angle. What does it mean physically when the transmission coefficient becomes complex?

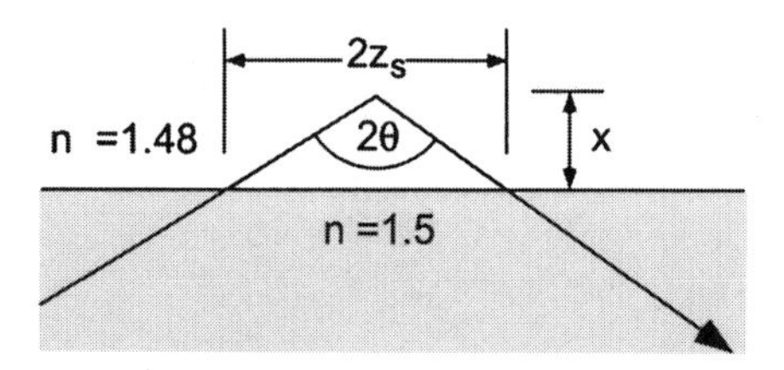

Figure 2.19. Schematic representation for Prob. 2.17.

15. Use a computer and programming language such as *Pascal* or *Mathematica* to generate and plot phase shift curves for the TM wave, similar to those in Fig. 2.17. Generate curves for dielectric-air interfaces, where the dielectrics have indices of 3.34 (GaAs) and 1.45 (glass).

16. On a hot day on the highway, the distant road sometimes appears to be a shining reflective pool of water. This phenomenon is really an example of total internal reflection. The air directly above the surface of the road has a lower index of refraction than that of the surrounding layers. Assume the index of refraction of air at 273° is n= 1.0003, and the index is directly proportional to the density of the air. If air follows the ideal gas law (PV=NRT), and the surface layer of air on the highway is 30° higher than the surrounding layer, what is the critical angle of incidence for Total Internal Reflection at this interface?

17. In the wave picture, we know that upon Total Internal Reflection at an interface, the guided wave undergoes a phase shift, 2Φ. However, in the ray picture, we can interpret a phase shift as a lateral displacement of the reflected wave. This is known as the *Goos-Hänchen effect.* The lateral displacement arises because wave energy actually penetrates beyond the interface into the lower index media before turning around. Consider Fig.2.19 showing a ray penetrating a surface. Assume that the incident ray is striking the interface at an angle exactly 1° larger than the critical angle. Referring to Appendix A, what is the depth of penetration for this interface? What is the lateral displacement, $2z_s$? How does the depth compare to the "depth" of the evanescent field from this structure?

18. A waveguide has a core index of 1.457 and a cladding index of 1.454.

    (a) What is the critical angle for this interface?

    (b) How far does the field extend into the cladding if excited by 1μm light at 88°?

19. Using the ray tracing picture of total internal reflection, and the Fresnel expressions for transmission and reflection of the electric and magnetic

fields, show that the electric and magnetic field are 90° out of phase in the low index region near the interface.

20. Confirm that the reflection coefficient for the electric field, E_r/E_i is unity for all angles greater than the critical angle for TE waves.

Chapter 3

THE PLANAR SLAB WAVEGUIDE

1. Introduction

In this chapter we establish the fundamental concepts of guided waves. It is perhaps the most important chapter in the book, as almost everything else will build on these concepts. This chapter includes many numerical examples and problems. The homework problems offer many opportunities to test your understanding of the concepts. Due to the transcendental nature of the eigenvalue equations, a computer with numerical analytical software is required to make the computations feasible. You are strongly advised to develop a set of standard programs that can quickly evaluate the basic waveguide parameters of a given structure, based on the material in this chapter. You will find these programs useful as we explore different types of waveguide, mode coupling, and device construction later in the text.

2. The Infinite Slab Waveguide

The simplest optical waveguide structure is the step-index planar waveguide. The slab waveguide, shown in Fig. 3.1, consists of a high-index dielectric layer surrounded on either side by lower index material. The slab is infinite in extent in the yz-plane, and finite in the x direction. The index of refraction of the guiding slab, n_f, must be larger than that of the cover material, n_c, or the substrate material, n_s, in order for total internal reflection to occur at the interfaces. If the cover and substrate materials have the same index of refraction, the waveguide is called "symmetric", otherwise the waveguide is called "asymmetric." The symmetric waveguide is a special case of the asymmetric waveguide.

We will always choose the direction of propagation to be along the z-axis. The slab waveguide is clearly an idealization of real waveguides, because real waveguides are not infinite in width. However, the one-dimensional analysis is

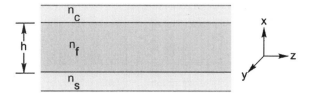

Figure 3.1. The planar slab waveguide consists of three materials, arranged such that the guiding index (n_f), is larger than the surrounding substrate (n_s) and cover (n_c) indices.

directly applicable to many real problems, and the techniques form the foundation for further understanding. We will begin by solving the wave equation using boundary conditions for the slab waveguide structure. This will lead naturally to the concept of *modes*. We will then develop formal mode concepts such as orthogonality, completeness, and modal expansion. We will see that a waveguide structure can support only a discrete number of guided modes. The mode picture is very powerful, and will be used extensively as we delve deeper into the subject of wave propagation in structures.

3. Electromagnetic Analysis of the Planar Waveguide

Consider the waveguide structure shown in Fig. 3.1. The three indices are chosen such that $n_f > n_s > n_c$, and the guiding layer has a thickness h. The choice of the coordinate system is critical in making the problem as simple as possible [1]. The appropriate coordinate system for this planar problem is a rectilinear cartesian system, because the three components of the field, E_x, E_y, and E_z are not coupled by reflections. For example, an electric field polarized in the y-direction, E_y, will still be an E_y directed field upon reflection at either interface; the reflection does not couple any of the vector field into the x or z directions. We place the $x = 0$ coordinate at one of the interfaces, choosing arbitrarily the top interface (between n_f and n_c).

We must consider two possible electric field polarizations, *transverse electric* or *transverse magnetic* [2]. The axis of the waveguide is oriented in the z-direction. The k-vector of the guided wave will zig-zag down the z-axis, striking the interfaces at angles greater than the critical angle. The field can be Transverse Electric (TE) or Transverse Magnetic (TM), depending on the orientation of the electric field. The TE case has no longitudinal component along the z-axis; the electric field is transverse to the plane of incidence established by the normal to the interface, and the k vector. Because of the different boundary conditions that control both fields, the TE and TM cases are distinguished in their mode characteristics as well as their polarization. In the section below, we will consider the TE case, leaving derivation of the TM case to problems at the end of the chapter.

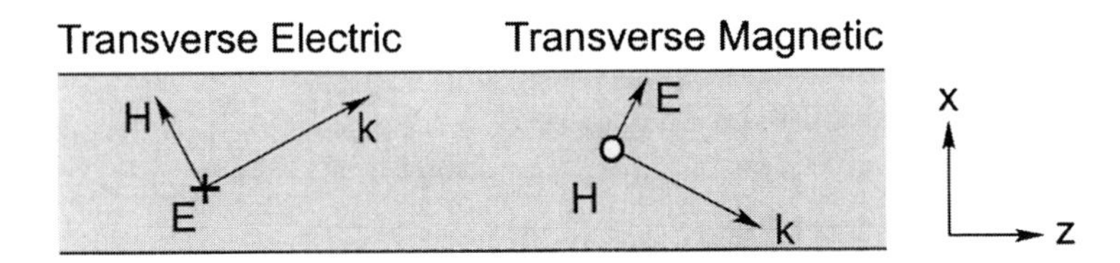

Figure 3.2. Transverse Electric (TE) and Transverse magnetic (TM) configurations. A cross indicates the field is entering the page.

In the TE case, the E field is polarized along the y-axis (into the page) of Fig. 3.2. We assume the waveguide is excited by a source with frequency ω_0 and a vacuum wavevector of magnitude $k_0 = \omega_0/c$. To find the allowed modes of the waveguide, we must first solve the wave equation in each dielectric region, and then use the boundary conditions to connect these solutions. For a sinusoidal wave with angular frequency ω_0, the wave equation (Eq. 2.30) for the electric field components in each region can be put in the scalar form

$$\nabla^2 E_y + k_0^2 n_i^2 E_y = 0 \tag{3.1}$$

where $n_i = n_f, n_s, n_c$, depending on the location. $E_y(x, z)$ is a function of both x and z, but because the slab is infinite in extent in the y-direction, E_y is independent of y. Due to the translational invariance of the structure in the z-direction, we do not expect the amplitude to vary along the z-axis, but we do expect that the phase varies. We write a trial solution to Eq. 3.1 in the form

$$E_y(x, z) = E_y(x) e^{-j\beta z} \tag{3.2}$$

β is a propagation coefficient along the z-direction, but we do not know its magnitude yet. Plugging this trial solution into Eq. 3.1, and noting that $d^2E_y/dy^2 = 0$

$$\frac{\partial^2 E_y}{\partial x^2} + (k_0^2 n_i^2 - \beta^2) E_y = 0 \tag{3.3}$$

The choice of n_i depends on the position x. For $x > 0$, we would use n_c, while for $0 > x > -h$, we would use n_f, etc. The general solution to Eq.3.3 will depend on the relative magnitude of β with respect to $k_0 n_i$. Consider the case where $\beta > k_0 n_i$. The solution to the wave equation, Eq. 3.3, will have a real exponential form

$$E_y(x) = E_0 e^{\pm\sqrt{\beta^2 - k_0^2 n_i^2}\, x} \qquad \text{for } \beta > k_0 n_i \tag{3.4}$$

where E_0 is the field amplitude at $x = 0$. To be physically reasonable, we always choose the negatively decaying branch of Eq. 3.4. This solution should remind you of the evanescent field of a total internally reflected (TIR) wave at an interface.

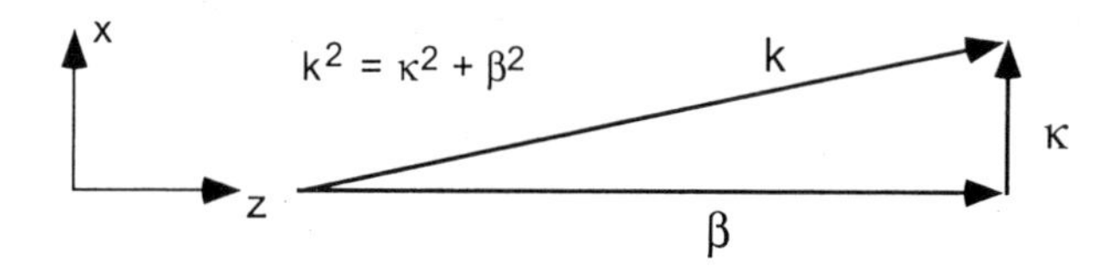

Figure 3.3. Geometric relation between β, κ, and k.

In the case where $\beta < k_o n_i$ the solution has an oscillatory form

$$E_y(x) = E_0 e^{\pm j\sqrt{k_0^2 n_i^2 - \beta^2}x} \qquad \text{for} \beta < k_0 n_i \tag{3.5}$$

So depending on the value of β, the solution can be either oscillatory or exponentially decaying. If $\beta > k_0 n_i$ we define an *attenuation coefficient*, γ,

$$\gamma = \sqrt{\beta^2 - k_0^2 n_i^2} \tag{3.6}$$

and describe the field as $E_y(x) = E_0 e^{-\gamma x}$. (Note the similarity to Eq. 2.108 for the evanescent field of a TIR wave.) If $\beta < k_0 n_i$ then we define a *transverse wavevector*, κ,

$$\kappa = \sqrt{k_0^2 n_i^2 - \beta^2} \tag{3.7}$$

so $E_y(x) = E_0 e^{\pm j\kappa x}$. From Eq. 3.7 we see that β and κ are geometrically related to the total wavevector, $k = k_o n_f$, in the guiding film.

β and κ are called the longitudinal and transverse wavevectors, respectively, inside the guiding film. These terms will be used extensively to characterize many types of waveguide mode, so become familiar with the relation shown in Fig. 3.3.

4. The Longitudinal Wavevector β

It is important to recognize that β is simply the z-component of k. Fig.3.4 plots the transverse electric field distribution in a slab waveguide for various values of β, as the angle between k and z varies from 90° to 0°. Note that the magnitude $|k|$ does not change, only its z component changes. β is the z-component. [4]

The top sketch of Fig. 3.4 shows the ray picture of the field, while the lower sketch shows the wave picture (solutions to Eqs. 3.3 and 3.4). There are three special points on the β axis: the first one is at $\beta = k_0 n_c$. For $\beta < k_0 n_c$, solutions to the wave equation in all regions of space are oscillatory (Eq. 3.5). The ray picture shows that when $\beta \approx 0$, the wave travels nearly perpendicular to the z-axis of the waveguide. Like light going through a sheet of glass, the ray refracts at the dielectric interfaces, but is not trapped. An oscillatory wave is present in the three distinct dielectric regions.

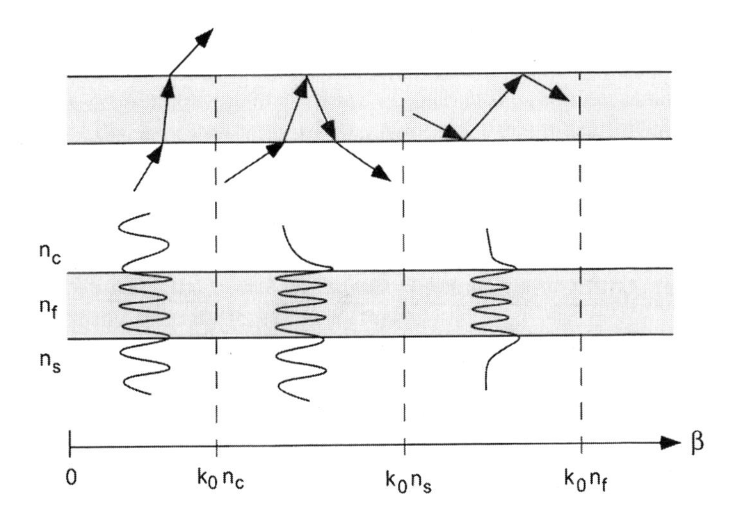

Figure 3.4. Ray and wave picture of the electromagnetic fields as a function of β.

The second special point occurs at $k_0 n_s$. For $k_0 n_c < \beta < k_0 n_s$, the ray picture shows a ray total-internally-reflecting at the film-cover interface, but refracting at the lower substrate-film interface. In the wave picture, the field becomes evanescent in the cover region. The field will still be oscillatory in the film and substrate regions. This condition is called a substrate mode.

As β increases beyond $k_0 n_s$, the evanescent conditions are satisfied in both the cover and substrate region, and oscillatory solutions are found in the film itself. Such solutions are, in fact, the guided modes of the film. The ray picture depicts a ray trapped between the two interfaces.

If β continues to increase beyond $k_0 n_f$ (physically it is not clear how this could ever be done, since β is simply the z-component of $k_0 n_f$), then Eq. 3.6 is satisfied everywhere, so the three regions must have exponential solutions. The only way to satisfy boundary conditions is to choose exponentially increasing fields in the surrounding dielectric regions, causing the field to explode toward infinity as the distance from the film increases. Satisfying this solution would require infinite energy, which is unphysical, hence it cannot occur.

We conclude from this discussion that a guided wave must satisfy the condition

$$k_0 n_s < \beta < k_0 n_f \tag{3.8}$$

where it is assumed that $n_c \leq n_s$. This is a universal condition for any dielectric waveguide, regardless of geometry.

5. Eigenvalues for the Slab Waveguide

To find the values of β that lead to allowed solutions to the wave equation, we must apply the boundary conditions to the general solutions developed in Eqs. 3.4 and 3.5. Assume that β satisfies Eq. 3.8. Then the transverse portions of the electric field amplitudes in the three regions are

$$\begin{aligned} E_y(x) &= Ae^{-\gamma_c x} \qquad 0 < x \\ E_y(x) &= B\cos(\kappa_f x) + C\sin(\kappa_f x) \qquad -h < x < 0 \\ E_y(x) &= De^{\gamma_s(x+h)} \qquad x < -h \end{aligned} \tag{3.9}$$

where $A, B, C,$ and D are amplitude coefficients to be determined from the boundary conditions, γ_c and γ_s refer to the attenuation coefficients in the cover and substrate, respectively (from Eq.3.6) and κ_f is the transverse component of k in the guiding film (from Eq. 3.7). The boundary conditions that connect the solutions at the interfaces are:

1. Tangential E is continuous
2. Tangential H is continuous

We rarely worry about continuity of the normal components of D and B, because these conditions are almost always satisfied when we satisfy the transverse conditions. Since E_y, is transverse to the interface, the first boundary condition is straightforward to apply. What about the condition for continuity of magnetic field, H? Should we write down a set of equations similar to Eq. 3.9 that describe the magnetic field as a function of position? Indeed, we could do that, but there is usually a simpler way to derive expressions for the magnetic field. If we assume that the fields are harmonic, then we can describe the magnetic intensity in terms of the electric intensity, and derive a simple boundary condition for the magnetic terms. Recall that

$$\nabla \times \mathbf{E} = -\frac{\partial \mathbf{B}}{\partial t} \tag{3.10}$$

For a sinusoidal field,

$$\mathbf{B}(t) = \mu\mathbf{H}(t) = \mu\mathbf{H}_0 e^{j\omega t} \tag{3.11}$$

so

$$\nabla \times \mathbf{E}(t) = -\mu j\omega\mathbf{H}(t) \tag{3.12}$$

We need an expression for the tangential component (the z-component in this case) of $\mathbf{H}$. Expanding the $\nabla\times$ term of Eq. 3.10 into its individual components, and taking the z-component, we get

$$\hat{z}\left(\frac{\partial E_y}{\partial x} - \frac{\partial E_x}{\partial y}\right) = -\mu j\omega H_z \tag{3.13}$$

Since there is no E_x component to the field, (it would not vary with y even if it did exist due to the infinite planar structure), we get an explicit equation for the tangential component of the magnetic field, H_z,

$$H_z = \frac{j}{\mu\omega}\frac{\partial E_y}{\partial x} \tag{3.14}$$

The tangential component of $\mathbf{H}$, H_z, is thus defined in terms of the electric field quantities. Since μ and ω are identical in all the media, the continuity of the tangential magnetic field is guaranteed if $\partial E_y/\partial x$ is made continuous across the interface. Hence we can now find the amplitude coefficients, A, B, C, and D using only the electric field description.

At the $x = 0$ interface, the condition that E_y be continuous requires that

$$Ae^{-\gamma_c 0} = B\cos(\kappa_f 0) + C\sin(\kappa_f 0) \tag{3.15}$$

which is satisfied only if $A = B$. Making the magnetic field continuous at $x = 0$ requires that the first derivative, $\partial E_y/\partial x$, be continuous at $x = 0$

$$\begin{aligned} -A\gamma_c e^{-\gamma_c 0} &= -B\kappa_f \sin(\kappa_f 0) + C\kappa_f \cos(\kappa_f 0) \\ -A\gamma_c &= +C\kappa_f \end{aligned} \tag{3.16}$$

yielding

$$C = -A\frac{\gamma_c}{\kappa_f} \tag{3.17}$$

All coefficients are written in terms of A. Using these coefficients, and applying the condition that E_y be continuous at $x = -h$ (h is a positive number) yields

$$A[\cos(-\kappa_f h) - \frac{\gamma_c}{\kappa_f}\sin(-\kappa_f h)] = De^{\gamma_s(h-h)} \tag{3.18}$$

This can be solved for D (noting $\sin(-x) = -\sin(x)$ and $\cos(-x) = \cos(x)$)

$$D = A[\cos(\kappa_f h) + \frac{\gamma_c}{\kappa_f}\sin(\kappa_f h)] \tag{3.19}$$

Putting all the terms together,

$$\begin{aligned} E_y &= Ae^{-\gamma_c x} \qquad x > 0 \\ E_y &= A\left[\cos(\kappa_f x) - \frac{\gamma_c}{\kappa_f}\sin(\kappa_f x)\right] \qquad -h < x < 0 \\ E_y &= A\left[\cos(\kappa_f h) + \frac{\gamma_c}{\kappa_f}\sin(\kappa_f h)\right] e^{\gamma_s(x+h)} \qquad x < -h \end{aligned} \tag{3.20}$$

where A is the amplitude at the $x = 0$ interface. Eq. 3.20 describes the amplitude of the electric field in all regions of the problem. Note that negative values of x must be used in the guiding and substrate layers — otherwise the formula will give nonsensical values. This equation is very handy for plotting out the mode profiles of guided modes.

Having found the amplitude coefficients in Eq. 3.20, is this description of the transverse electric field complete? Not quite! The propagation and decay constants, γ_c, γ_s, and κ_f, all depend on β, which is still undefined. The fourth, and final, boundary condition, namely the continuity of $\partial E_y/\partial x$ at $x = -h$, gives an equation for β.

$$\left.\frac{\partial E_y}{\partial x}\right|_{x=-h} = A[\kappa_f \sin(\kappa_f h) - \gamma_c \cos(\kappa_f h)] \quad \text{(film term)} \tag{3.21}$$

$$= A[\cos(\kappa_f h) + \frac{\gamma_c}{\kappa_f}\sin(\kappa_f h)]\gamma_s \quad \text{(substrate term)}$$

Divide both sides of the equation by $cos(\kappa_f h)$ to get the *eigenvalue equation*

$$\tan(h\kappa_f) = \frac{\gamma_c + \gamma_s}{\kappa_f\left[1 - \frac{\gamma_c\gamma_s}{\kappa_f^2}\right]} \tag{3.22}$$

This is a transcendental equation that must be solved numerically or graphically. All terms $(\gamma_s, \gamma_c, \kappa_f)$ depend on the value of β. Eq. 3.22 is called the *characteristic equation* for the TE modes of a slab waveguide. Solution of this equation will yield the eigenvalues, β_{TE} that correspond to allowed TE modes in the waveguide. Had we set up our initial problem with transverse magnetic fields, as opposed to transverse electric fields, we would have arrived at a different characteristic equation for the eigenvalues, β_{TM}. We leave it as an exercise (see Problem 3.1.) to confirm that for the TM case, the eigenvalue equation for β is

$$\tan(h\kappa_f) = \frac{\kappa_f\left[\frac{n_f^2}{n_s^2}\gamma_s + \frac{n_f^2}{n_c^2}\gamma_c\right]}{\kappa_f^2 - \frac{n_f^4}{n_c^2 n_s^2}\gamma_c\gamma_s} \tag{3.23}$$

Every waveguide structure, no matter what shape or symmetry, will have a characteristic equation that must be solved to find the eigenvalues of the modes.

The transcendental equation can be solved numerically on a computer, or it can be solved graphically. To provide insight into the eigenequation, the example below shows the graphical solution.

Example 3.1 Graphical and numerical solution to the β eigenvalue equation

Consider the planar dielectric structure shown in Fig. 3.5. The guiding index has value 1.50, the substrate index is 1.45, and the cover index is 1.40. This is an asymmetric waveguide. The thickness of the guiding layer is 5μm. We

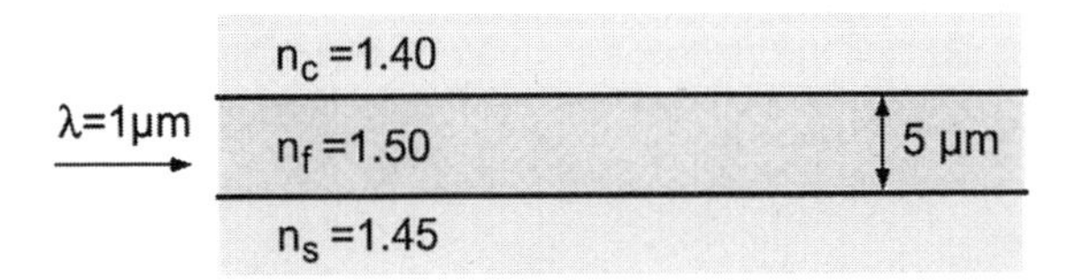

Figure 3.5. Planar slab waveguide configuration.

want to determine the allowed values of β using Eq. 3.22 for this structure. Assume that light with wavelength of $1\mu m$ is used to excite the waveguide.

Solution: We will use κ_f as the variable for plotting all the terms of the equation. This choice is arbitrary (we could have chosen β), but it makes the argument of the $\tan(\kappa_f h)$ term linear. Hence all variables must be defined in terms of κ_f

$$\begin{aligned} \beta &= \sqrt{k_0^2(1.5)^2 - \kappa_f^2} \\ \gamma_s &= \sqrt{\beta^2 - k_0^2(1.45)^2} \\ \gamma_c &= \sqrt{\beta^2 - k_0^2(1.4)^2} \end{aligned}$$

Using these values, both sides of the TE characteristic equation (Eq. 3.22) are plotted as a function of κ_f on the graph in Fig. 3.6. The variable κ_f ranges from a value of 0 (when $\beta = k_0 n_f$), to $\kappa_{max} = \sqrt{k_o^2 n_f^2 - k_o^2 n_s^2}$. The $\tan(\kappa_f h)$ term generates the typical pattern of a repeating function extending from $-\infty$ to $+\infty$. The right-hand side of Eq. 3.22 yields a slower function that diverges toward $-\infty$ around κ =20,000 cm^{-1} and then comes in from $+\infty$. At the points where the two curves cross, Eq. 3.22 is satisfied. These points represent allowed values of κ for this waveguide. From the plot we see that the allowed κ values are approximately 5,500, 12,000, 16,500, and 21,500 cm^{-1}.

This plot was generated using *Mathematica*, although there are several other suitable numeric packages that can perform these calculations and plots. To serve as a guide, the *Mathematica* code is listed below:

```
nf=1.50; ns=1.45; nc=1.40;
h=0.0005;
lambda= 10^(-4);
k=2 Pi/lambda;
beta=Sqrt[ k^2 nf^2- kappa^2];
kappamax=Sqrt[k^2 nf^2 - k^2 ns^2];
gammas=Sqrt[beta^2-k^2 ns^2];
gammac=Sqrt[beta^2-k^2 nc^2];
Plot[{Tan[kappa h],(gammas+gammac)/(kappa(1-gammas gammac/kappa^2))},
     {kappa, 1, kappamax}, PlotRange ->{-10,10}]
```

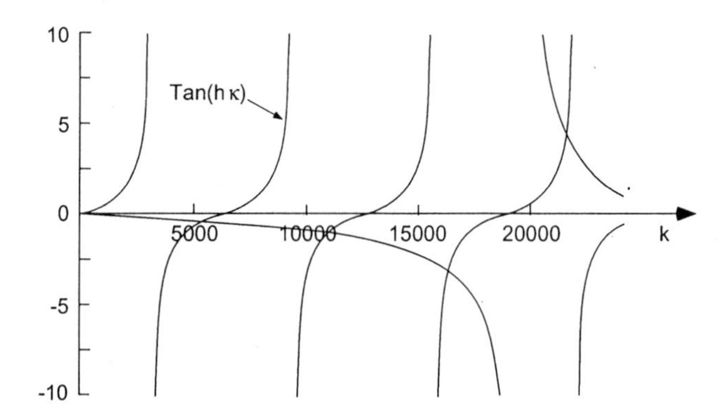

Figure 3.6. Graphical plot of Eq. 3.22 for the waveguide shown in Fig. 3.5

The transcendental characteristic equation must be solved numerically, which is a relatively straightforward action for many mathematical software packages. Again using *Mathematica* the following command was used repeatedly to find each root of the equation.

```
FindRoot[Tan[kappa h]== (gammas+gammac)/(kappa(1-
gammas gammac/kappa^2)), {kappa,5000}]
```

The last bracket of the command tells *Mathematica* to begin its search around a value of $\kappa = 5000$. The program returned the first κ value of 5497.16. To find higher roots, we used values taken from the graph as starting points, and let the computer return the more accurate value. Numerically, the eigenvalues for κ were found to be 5497.16, 10963.2, 16351, and 21545 cm^{-1}. The eigenvalues in terms of β can be found directly from the individual κ values using Eq. 3.23 to be 94087, 93608, 92819, and 91752 cm^{-1}, respectively.

This example shows some typical features of optical waveguides. First, the thickness of the guiding film need not be very thick. It is generally on the order of a few wavelengths. Second, the index difference required to achieve a guiding structure is small. In this case, $\Delta n = 0.05$ between the core and substrate. This is actually a huge difference compared to many practical devices which have index differences as small as 0.001. Finally, inspection of Fig. 3.6 shows that if the waveguide is made too thin (so that the argument κh does not extend beyond approximately $\pi/2$) it is possible that the two sets of lines will never cross, and there will be no mode allowed in the structure.

The example yielded four solutions for β, or four allowed modes. What does this mean? Each mode has the same wavelength of light, they each just travel in a slightly different direction within the waveguide. In the ray picture, the modes would be shown as four discrete rays travelling at slightly different

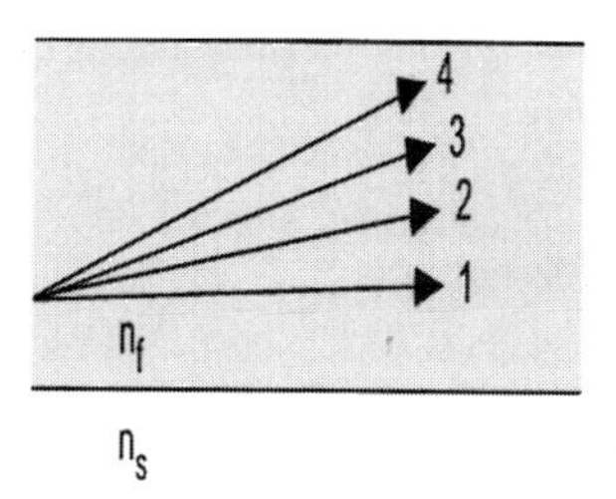

Figure 3.7. Ray depiction of the four allowed modes in the waveguide. Each ray has the same magnitude of k-vector, they are simply oriented slightly differently with respect to the z-axis.

angles, as shown in Fig. 3.7. Notice that only a few discrete rays actually propagate in the waveguide.

To those familiar with basic quantum mechanics, the problem outlined in the example above should look very familiar. This graphical technique is often used to find the allowed energy eigenvalues of a particle in a finite potential well [4]. The analogy between the particle-in-a-box and the optical waveguide problem is very strong: both situations describe *waves* which are confined between two reflecting boundaries. In both cases the waves partially tunnel into the surrounding potential barrier before turning around. Only certain allowed energies in the case of the particle, or transverse propagation coefficients (κ) in the case of the optical wave, are found to create a standing wave in the one-dimensional system.

To complete the solution, the coefficient, A, should be related to a physical parameter. In practice, A is related to the power carried in the waveguide. The power is calculated by integrating the z-component of the Poynting vector over the cross-sectional area of the guide

$$S_z = \frac{1}{2}\Re e(\mathbf{E} \times \mathbf{H} \cdot \hat{z}) \tag{3.24}$$

Note that we are using the time-averaged power. The average power in a TE mode is

$$P_z = \frac{1}{2}\int_{-\infty}^{\infty} E_y H_x dx = \left(\frac{\beta}{2\omega\mu_0}\right)\int_{-\infty}^{\infty} |E_y|^2 dx \tag{3.25}$$

where the coefficient A is contained in E_y. Since the integral spans only one direction, the integral has units of power per unit length (in the y-direction).

It is enlightening to see the actual mode solutions that correspond to each value of β. Using the following *Mathematica* commands to evaluate Eq. 3.20 with the values of β found above, and normalizing each mode using Eq. 3.25 to have 1W/unit length in each mode, we plotted the total amplitude profile for each of the allowed modes in Ex. 3.1.

```
wave[x_]:= Exp[-gammac x] /; x>0
```

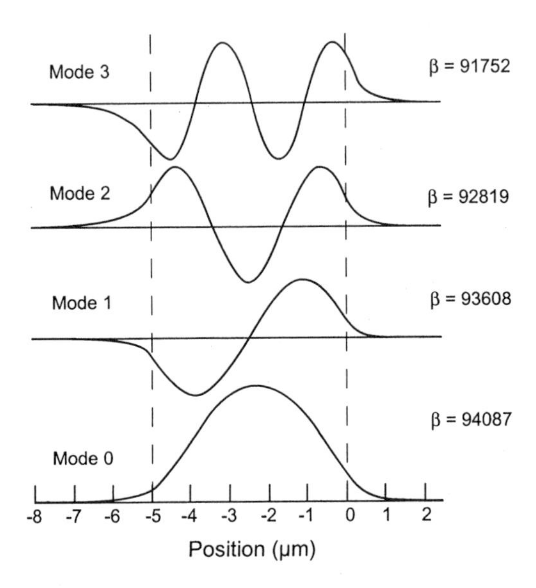

Figure 3.8. Modal field patterns for the first four TE modes of waveguide described in Ex. 2.1. The vertical lines represent the location of the dielectric interfaces.

```
wave[x_]:= Cos[kappa x]-(gammac/kappa)Sin[kappa x]/; (x<=0)&&(x>-h)
wave[x_]:= (Cos[kappa h]+(gammac/kappa)Sin[kappa h])*
Exp[gammas(x+h)]/; x<=-h;
amplitude=1/Sqrt[beta/(2omega mu) NIntegrate[(wave[x])^2,
{x,-0.001,0.0002}]]
Plot[amplitude * wave[x] ,{x, -h-0.0003, 0.0002}]
```

Fig.3.8 illustrates the amplitude solutions for the four modes of Example 3.1. Since the waveguide is asymmetric, the modes are slightly asymmetric, although it is not obvious to the casual glance. Notice that the modes have alternating even and odd symmetry, and that the evanescent tails of the higher order modes extend slightly further into the cladding than do the tails of the lowest order mode. The modes are labelled by the *number of nodes* they have. The TE_0 mode is the lowest order (which means the mode with the smallest value of κ), and it has no (0) nodes. The TE_1 mode has one node, the TE_2 mode has 2 nodes, etc. There will also be a set of TM modes with similar designations.

Power in the guiding layer is found by integrating the Poynting vector over the area of the waveguide structure. The fraction of the power contained in the core is simply

$$\frac{P_{core}}{P_{total}} = \frac{\int_{-h}^{0} E_y(x)H_x^*(x)dx}{\int_{-\infty}^{\infty} E_y(x)H_x^*(x)dx} \tag{3.26}$$

In general, higher order modes are less confined than their lower order counterparts. Application of Eq. 3.26 on mode 0 shows it has a power confinement of 99.47%, while mode 3 has only 85.9% confinement. Mode confinement is an important property for waveguide designs. A mode that is loosely confined will be more affected by bends and more strongly couple evanescently to neighboring structures than will a tightly bound mode.

6. The Symmetric Waveguide

Fig. 3.9 shows a symmetric waveguide, where a guiding film with index n_f and thickness h is surrounded on both sides by an index n_s. It is convenient to place the coordinate system in the middle of this waveguide since the fields will reflect the symmetry of the structure.

We leave it as an exercise to show that the general field description of a TE mode within this symmetric structure is

$$\begin{aligned} E_y &= Ae^{-\gamma(x-h/2)} && \text{for } x \geq h/2 \\ E_y &= A\frac{\cos\kappa x}{\cos\kappa h/2} \quad \text{or} \quad A\frac{\sin\kappa x}{\sin\kappa h/2} && \text{for } -h/2 \leq x \leq h/2 \\ E_y &= \pm Ae^{\gamma(x+h/2)} && \text{for } x \leq -h/2 \end{aligned} \tag{3.27}$$

The magnetic amplitude of the TM mode can be similarly described. There are two choices for the description of the field in the guiding layer, depending on whether a symmetric (cosine) or antisymmetric (sine) mode is excited. The fact that the modes can be uniquely characterized in terms of even or odd groups is a natural consequence of the symmetry of the index structure. The characteristic eigenvalue equation for the TE modes in a symmetric waveguide is

$$\begin{aligned} \tan\kappa h/2 &= \gamma/\kappa \quad \text{for even (cos) modes} \\ &= -\kappa/\gamma \quad \text{for odd (sin) modes.} \end{aligned} \tag{3.28}$$

The characteristic equation for the TM modes is

$$\tan\kappa h/2 = (n_f/n_s)^2\,\gamma/\kappa \quad \text{for even (cos) modes}$$

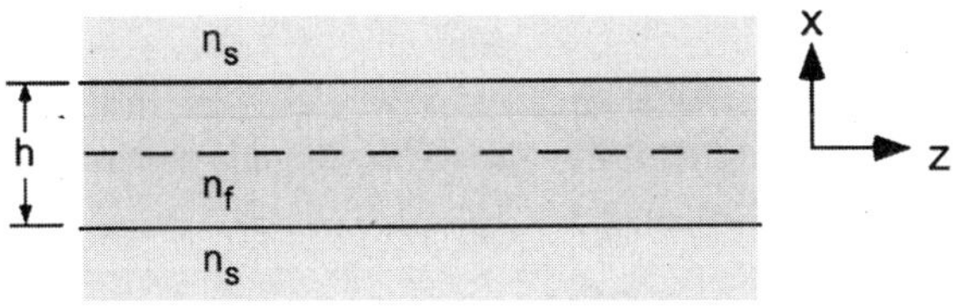

Figure 3.9. The symmetric waveguide is surrounded by material with the same index of refraction. The axis of symmetry is usually chosen to be the x=0 axis.

$$= \quad -\left(n_f/n_s\right)^2 \ \kappa/\gamma \quad \text{for odd (sin) modes.} \qquad (3.29)$$

A unique feature of the symmetric waveguide is that it will *always* support at least one mode. Consider the graphical solution for a symmetric waveguide described in Example 2 below.

Example 3.2 Modes of he symmetric waveguide

Suppose the waveguide shown in Fig. 3.9 has a film index of $n_f = 1.49$, and cladding index equal to $n_s = 1.485$. The difference in index between the two layers is very small. Let the wavelength be 0.8μm. We will use the graphical solution to find β, as it illustrates demonstrates why the symmetric waveguide will always support at least one mode. Two thicknesses will be examined; $h = 3\mu m$, and $h = 15\mu m$.

Solution: There are only two variables in this problem: γ and κ. As in the last example, we will plot functions in terms of κ.

$$\begin{aligned}\gamma_s &= \sqrt{\beta^2 - k_0^2 n_s^2} = \sqrt{k_0^2(n_f^2 - n_s^2) - \kappa^2} \\ \beta &= \sqrt{k_0^2 n_f^2 - \kappa^2}\end{aligned}$$

Evaluating these expressions, using $k_0 = 2\pi/\lambda = 7.853 \times 10^4 \text{cm}^{-1}$, yields

$$\begin{aligned}\gamma &= \sqrt{9.176 \times 10^7 - \kappa^2} \\ \beta &= \sqrt{1.3694 \times 10^{10} - \kappa^2}\end{aligned}$$

To find the eigenvalues of the TE modes, we must solve Eq. 3.28. Graphically, the functions $\tan \kappa h/2$, γ/κ, and $-\kappa/\gamma$ are plotted on the same graph as a function of κ. These are plotted against κ for the case where $h = 3\mu$m in Fig. 3.10.

The top curve, which corresponds to the even mode, begins at $+\infty$, and terminates with a value of 0. Notice that the $\tan \kappa h/2$ starts at zero and increases. It is unavoidable that the two curves cross, so there *must* always be at least one mode, no matter how thin the waveguide.

As the waveguide is made thicker, more modes appear. Consider the graphical plot of the equations for the case when the waveguide slab is 15 μm thick, as shown in Fig. 3.11.

Note that as the transverse wavevector (κ) increases, the allowed modes alternate between even and odd structure. The spatial profile of these allowed modes is very similar to that shown in Fig. 3.8.

7. Intuitive Picture of the Mode

Solution of the wave equation leads to physical solutions which we can plot out on a graph. But looking beyond the math, it is straightforward to understand

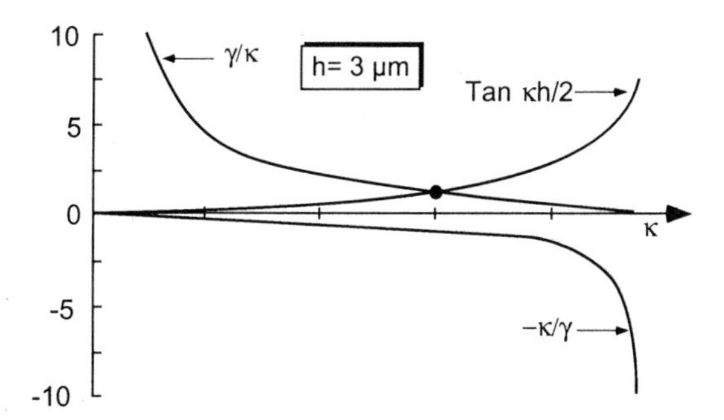

Figure 3.10. For the thin waveguide, there is only one allowed mode, which occurs near $\kappa = 6000cm^{-1}$.

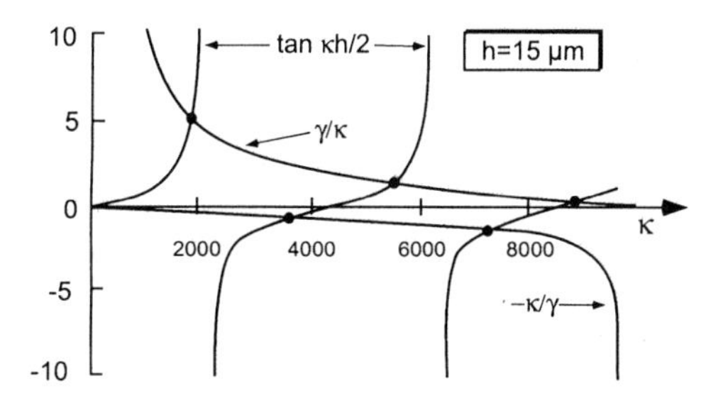

Figure 3.11. The thick waveguide supports both even and odd modes.

the mode structure. Mode structure arises from interference patterns within the waveguide between components of waves travelling in opposite directions. The field within the guiding layer of an even mode in a symmetric waveguide has the form

$$E_y(x) = A\cos\kappa x\, e^{-j\beta z} \tag{3.30}$$

Since $\cos\kappa x = (e^{j\kappa x} + e^{-j\kappa x})/2$, we can rewrite Eq. 3.30 as

$$E_y = \frac{A}{2}\left[e^{+j(\kappa x - \beta z)} + e^{-j(\kappa x + \beta z)}\right] \tag{3.31}$$

Eq. 3.31 represents the superposition of two plane waves, shown schematically in Fig. 3.12. Each plane wave has a k-vector with a transverse component, κ, and a z-component, β. One plane wave has components $\mathbf{k} = \kappa\hat{x} + \beta\hat{z}$, while the other has components $\mathbf{k} = -\kappa\hat{x} + \beta\hat{z}$. Each k-vector has a plane wave associated with it. These two plane waves zig-zag down the waveguide, continuously crossing each other's paths as they travel along. Being excited by the same source, the waves are coherent with one another at a given plane in the waveguide, and so they form stable interference patterns. The structure of the mode is a result of this interference. When the modes constructively interfere, the electric field is maximum, and where destructive interference occurs, the intensity is a minimum.

7.1 Why is β discrete?

The discrete nature of β can be found using this same model of interfering waves. To avoid decay of energy due to destructive interference as the waves travel through the waveguide, the total phase change for a point on the wavefront that travels from one interface ($x = 0$) to the next ($x = -h$), and back again, must be a multiple of 2π. For a wave incident at angle θ, a phase shift of $kn_f h \cos\theta$ is accumulated on the first transverse passage through the film, and a phase shift of $-2\Phi_c$ occurs at the film-cover interface. Another $kn_f h \cos\theta$ of phase is accumulated travelling back down, and finally there is a $-2\Phi_s$ phase shift at the film-substrate interface. The transverse resonance condition requires that

$$2kn_f h \cos\theta - 2\Phi_c - 2\Phi_s = 2\pi\nu \tag{3.32}$$

where ν is an integer. This expression is effectively a dispersion equation for the waveguide. We will use it in the next chapter to develop a generalized dispersion relation for slab waveguides of any construction.

8. Properties of Modes

Once β is determined for a waveguide, the field amplitudes can be described in all regions of the waveguide using Eqs. 3.20 or the equivalent for TM waves. We have been referring to these field distributions as *modes*. The concept of the mode is very powerful — and perhaps a little confusing to the uninitiated. Here we review some of the major properties of modes and modal analysis [5].

The general expression for the electric field solution in all space is

$$\mathbf{E}(x, y, z) = \mathbf{E}(x, y)e^{-j\beta z} \tag{3.33}$$

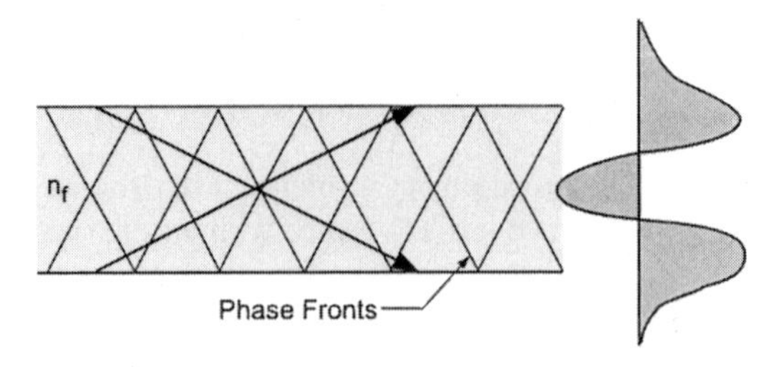

Figure 3.12. A mode can be described as having two plane waves at a slight angle to one another, forming an interference pattern. When the phase fronts cross, there is a maxima.

The term $\mathbf{E}(x, y)$ describes the transverse structure of the field, and is usually called the mode structure or mode shape. There is a corresponding magnetic field distribution for each mode, given by $\mathbf{H}(x, y, z)$. The modes have the following properties.

1. Every eigenvalue β corresponds to a distinct confined mode of the system. Every β will have a unique field distribution (shape). The amplitude of the mode is established by how much power is carried in the mode.

2. Only a finite number of modes will be guided. These are associated with the solutions to the eigenvalue equation for β. The spectrum of β for guided modes is discrete.

3. Most modes will not be guided. Most values of β will lead to unguided, or radiation modes. The spectrum of β for unguided modes is continuous, meaning there are an infinite number of unguided modes.

4. All modes are orthogonal. This is a very important point. For guided modes, orthogonality requires that

$$\int_{area} \mathbf{E}_i(x, y, z) \times \mathbf{H}_j(x, y, z)] \cdot dA = \delta_{ij} \tag{3.34}$$

   where δ_{ij} is the Kroenecker delta function, the area of the integral is the infinite xy plane at a particular value of z, and where $\mathbf{E}_i(x, y, z)$ and $\mathbf{H}_j(x, y, z)$ represent normalized modes of the system. For radiation modes, the formal relation is

$$\int_{area} [\mathbf{E}(i) \times \mathbf{H}(j)] \cdot dA = \delta(i - j)P \tag{3.35}$$

   where P is the power in the mode. Radiation modes cannot be normalized as they represent infinite plane waves. Each mode is unique, and cannot be described in terms of other modes.

5. Some modes are degenerate. Degenerate modes will share the same value of β, but will have distinguishable electric field distributions. In such degeneracies, field solutions can be found which are orthogonal, and they will satisfy Eq. 3.33. A good example of such a degeneracy is the fundamental mode in a circular dielectric fiber. The mode can have two different electric field polarizations, E_x and E_y, respectively, each of which has the same spatial energy distribution.

6. The modes of a given system form a *complete* set. Completeness means that the allowed modes span the entire space of the system. *Any* continuous distribution of electric field can be described as a superposition, or

sum, of the appropriately weighted modes of the waveguide.

$$\mathbf{E}(x,y,z) = \sum_{i}^{guided} a_i \mathbf{E}_i(x,y,z) + \int_{radiation} \hat{a}(\beta)\mathbf{E}(x,y,z,\beta)\, d\beta \tag{3.36}$$

where $\mathbf{E}_i(x,y,z)$ are the discrete modes of the system, the a_i are weighting coefficients for each mode, and the $\mathbf{E}(\beta)$ are the radiation modes of the system, with their respective weighting factor, $a(\beta)$.

The power of the mode concept lies in *completeness* and the use of *superposition*. This superposition concept is very powerful for calculating coupling between two different systems. We will explore coupling in detail in Chap.XXX. In general, a given mode in one system will be described as a superposition of modes in a second system, some of which may be guided and some may be radiative.

9. Number of Guided Modes in a Waveguide

We often use terms like "single-mode" or "multimode" to characterize a waveguide. The importance of this distinction will become apparent when we discuss information bandwidth in later chapters, and when we discuss coupling between devices. In this section, we develop some approximations of the number of guided modes in a planar waveguide.

Recall the planar waveguide described in Example 3.2. The waveguide supported a different number of modes depending on its thickness. If we had adjusted the relative indices between the layers, we would also have found that the mode number varied. The *lowest order mode* has a k vector that is nearly parallel to the z axis

$$\beta_{\text{lowest order}} \approx kn_f \tag{3.37}$$

The highest order mode will have a wavevector at nearly the critical angle

$$\beta_{\text{highest order}} \approx kn_f \cos\theta_{critical} \approx kn_s \tag{3.38}$$

The rest of the modes will have eigenvalues for β that fall between these two extremes. To get an idea of the number of modes in the waveguide, recall the general eigenvalue equation for the TE modes,

$$\tan(\kappa h) = \frac{\gamma_c + \gamma_s}{\kappa\left(1 - \frac{\gamma_c\gamma_s}{\kappa^2}\right)} \tag{3.39}$$

Graphically the two sides of the equation can be plotted against κ to create a plot such as Fig. 3.13, which is a modification of Fig. 3.6.

The right-hand-side of the equation starts at zero, slowly diverges to $-\infty$, then comes in from $+\infty$ and terminates at a value somewhere above zero at

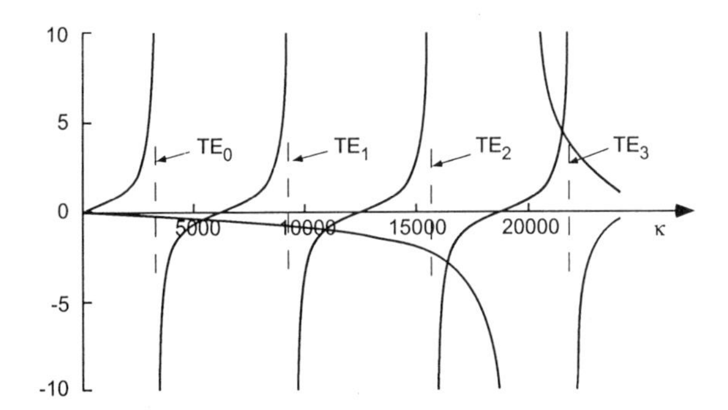

Figure 3.13. The graphical solution to the eigenmode equation for an asymmetric waveguide shows that every time the argument $\kappa_{max}h$ increases by π, another mode is allowed in the waveguide.

κ_{max}. The left-hand-side of Eq. 3.39 is a periodic function ($\tan(\kappa h)$ that goes from $-\infty$ to $+\infty$ every time κh increases by π. Notice that if the value of $\kappa_{max}h$ is greater than $\pi/2$, then we are *guaranteed* to find at least one TE mode in the waveguide. If $\kappa_{max}h > 3\pi/2$, then we are *guaranteed* to find at least two TE modes in the waveguide. These values of $\kappa_{max}h$ are known as *cut-off conditions*. Every time $\kappa_{max}h$ increases by π, another mode is allowed. The approximate number of modes, m, can be found from

$$\begin{aligned} m &= \text{Int}\left[h\kappa_{max}/\pi\right] \\ &= \text{Int}\left[hk(n_f^2 - n_s^2)^{1/2}/\pi\right] \end{aligned} \quad (3.40)$$

This approximation is most accurate when m is a large number. It is approximate because the exact location of the last crossing is not known. Note that the mode count increases with the thickness, h, of the guide, with the difference in index, $(n_f^2 - n_s^2)$, between the core and cladding, and as the wavelength of the guided light gets shorter. Also note that the point at $\kappa = 0$ is not considered to be an allowed mode, even though it appears on the graph that the two equations are crossing at that point.

We usually characterize a waveguide by its *normalized frequency*, defined as

$$V = hk(n_f^2 - n_s^2)^{1/2} \quad (3.41)$$

In terms of the normalized frequency, the approximate number of modes, m, in a waveguide is $m \approx V/\pi$. The mode cut-off conditions are usually described in terms of the normalized frequency. For example, if it is desired to build a waveguide that only carries the first three TE modes, what should the dimensions and index difference be? We can adress such issues knowing the cut-off conditions by trading thickness for index difference. The normalized frequency establishes the relationship between the parameters that influence the number of modes carried by a waveguide.

Why do we worry about the number of modes in a waveguide? The answer is pulse distortion and bandwidth. When a pulse is launched on a waveguide, such as in a digital communication system, the pulse energy will become distributed over all the allowed modes of the waveguide. If each mode travels at a slightly different velocity, as is often the case, the temporal form of the pulse will change as the pulse propagates. This lengthens the pulse, and effectively reduces the rate at which pulses can be sent.

10. Normalized Propagation Parameters

If you have calculated the mode eigenvalues of a waveguide, you already appreciate how cumbersome the process can be. We can develop normalization rules for slab waveguides which allow for simple graphical solution to the waveguide problem (an example of how this can be applied to a coupled waveguide taper is given in Example 11.4). There are five independent parameters to deal with in a slab waveguide problem: *i*) the refractive index of the guiding layer, n_f, *ii*) the refractive index of the substrate, n_s, *iii*)the refractive index of the cover, n_c,*iv*) the guiding layer thickness, h, and *v*) the vacuum wavevector, k_0. These parameters can be reduced into *normalized values* which allow a general description of the dispersion and cutoff conditions for modes. We have already defined a Normalized Frequency, V (3.41). From the five variables above, we can also define an asymmetry parameter, a, and a normalized effective index, b [4]:

$$\begin{aligned} V &= k_0 h (n_f^2 - n_s^2)^{1/2} \\ a &= (n_s^2 - n_c^2)/(n_f^2 - n_s^2) \qquad (3.42) \\ b &= (n_{eff}^2 - n_s^2)/(n_f^2 - n_s^2) \end{aligned}$$

(3.43)

where $n_{eff} = \beta/k_0$ is defined as the effective index of the guide. The normalized index, b, is zero at cutoff, and approaches unity far away from it. The asymmetry parameter ranges from 0 for a symmetric waveguide, to infinity for strong asymmetry ($n_s \gg n_c$).

These definitions are used in conjunction with the transverse resonance condition, Eq. 3.32, to define universal dispersion relations. The transverse resonance condition state that the waveguide acts like a standing wave cavity in the transverse direction. In order to be resonant, the round trip phase of a *transverse* component of k must add up to an integer number of 2π. Recall Eq. 3.32 states

$$2k_0 n_f h \cos\theta - 2\Phi_c - 2\Phi_s = 2\pi\nu,$$

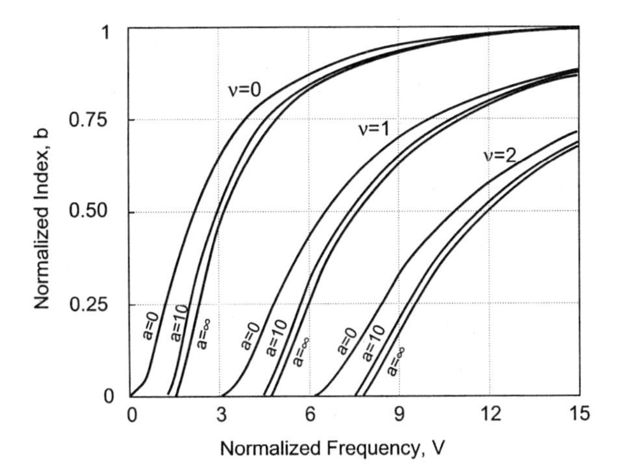

Figure 3.14. The normalized index, b, is plotted against the normalized frequency, V, for three values of the asymmetry coefficient, a, and for the first three values of ν. Values of $a = 0$, $a = 10$, and $a = \infty$ were evaluated.

where ν is an integer. Substituting the normalized parameters into this equation yields the normalized dispersion relation

$$V\sqrt{1-b} = \nu\pi + \tan^{-1}\sqrt{b/(1-b)} + \tan^{-1}\sqrt{(b+a)/(1-b)} \quad (3.44)$$

At first glance, this may appear as needless complication of straightforward equations, but there is a good reason for this normalization. We can numerically generate a set of curves which relate the normalized index, b, to the normalized frequency, V, using Eq. 3.44. Once the curves are generated, we can relate the calculation to any new waveguide through appropriate scaling. Fig. 3.13 shows the numerically derived relation between the normalized index, b, and the normalized frequency, V.

Example 3.3 Evaluation of a waveguide using normalized parameters

To illustrate the power of using the normalized parameters, let's reconsider the waveguide used in Example 3.1, which had a guiding index, $n_f = 1.5$, a substrate index $n_s = 1.45$, a cover index $n_c = 1.40$, a film thickness $h = 5\mu m$, and a driving wavelength of $\lambda_0 = 1\mu m$. Using a numerical solution, we found the eigenvalues for the first three modes to be $\beta = 94087$, 93608, and 92819 cm^{-1}. We can determine the propagation coefficients by inspection using the graph in Fig. 3.14. First we must normalize the waveguide parameters,

$$V = k_0 h (n_f^2 - n_s^2)^{1/2} = \frac{2\pi}{1\mu m} 5\mu m \sqrt{1.5^2 - 1.45^2} = 12.065$$

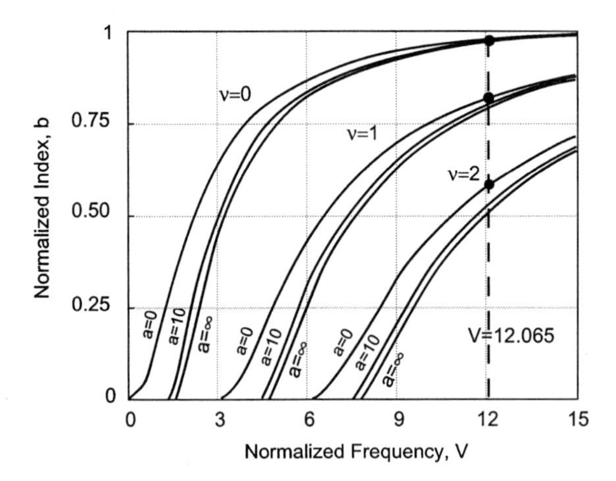

Figure 3.15. Data from Example 3.2 plotted on graph from Fig. 3.14.

$$a = (n_s^2 - n_c^2)/(n_f^2 - n_s^2) = (1.45^2 - 1.40^2)/(1.55^2 - 1.45^2) = 0.475$$

The asymmetry value $a = 0.475$ lies near the $a = 0$ value of Fig. 3.13, so we will interpolate between the two plotted lines for each value of ν in the plot. We draw a line on the graph at V=12.065, as shown in Fig. 3.14, and read the b values from the scale.

At a normalized frequency of $V = 12.065$ there are three values of b: 0.575, 0.813, and 0.965. (Note: the fourth mode that we found in Example 3.1 is not found here because the graph in Fig. 3.14 does not show a curve for the $\nu = 3$ case. The vertical line in Fig. 3.15 shows the intersections. Using the expression for b

$$b = (n_{eff}^2 - n_s^2)/(n_f^2 - n_s^2)$$

we can solve this for β noting that $n_{eff} = \beta/k_0$

$$\beta = k_0\sqrt{(n_f^2 - n_s^2)b + n_s^2}$$

Plugging values into the equation, we get the first three allowed values of β. These are tabulated alongside the "exact" values obtained by numeric technique for comparison.

The agreement is remarkable: better than 1 part in a 1000, which can be attributed to ones ability to read the graph in Fig. 3.13. The virtue of the normalized method is that the entire eigenvalue calculation can be carried out on a hand calculator, and the accuracy is probably better than we need.

Table 3.1. Comparison of β values from normalized and exact methods.

b	β_{norm} cm^{-1}	β_{exact} cm^{-1}	"error" cm^{-1}
+0.575	92926	92819	-107
+0.813	93668	93608	-60
+0.965	94139	94087	-52

From the data plotted in Fig. 3.13, we can extract some additional information. The cutoff frequency for each mode occurs when $b = 0$. Therefore the intersection of each line corresponding to a mode number ν (e.g. TE_ν or TM_ν) and an asymmetry factor a corresponds to the normalized frequency of the longest wavelength that will be carried by the mode. We see by inspection that the lowest order ($\nu = 0$) symmetric waveguide mode reaches $b = 0$ at $V = 0$, indicating that this mode never is cut-off, in agreement with our understanding of this mode. Formally, setting $b = 0$ and solving Eq. 3.44 for V yields the cutoff conditions for all modes in the step index waveguide

$$V = \tan^{-1}\sqrt{a} + \nu\pi \tag{3.45}$$

To apply the normalization technique to TM modes, we need only adjust the asymmetry parameter,

$$a = \frac{n_f^2}{n_c^2}\frac{n_s^2 - n_c^2}{n_f^2 - n_s^2} \tag{3.46}$$

11. The Numerical Aperture

A very common parameter for characterizing waveguides is the *numerical aperture*. The concept is based on ray tracing and refraction, so technically it is only applicable to multimode waveguides, although it is sometimes used in characterizations of single mode waveguides. Consider the optical waveguide shown in Fig. 3.16, where a high index layer with index n_1 is surrounded symmetrically by a lower index medium, n_2. The thickness of the guiding layer is not critical, so long as it is many times greater than the wavelength of the light being carried. We want to explore how light (in the form of rays) can couple into the end of such a structure. A ray is shown entering the edge of the waveguide, where we assume the index of refraction corresponds to that of air (essentially n=1). The entering ray is refracted according to Snell's law, bending toward the axis of the waveguide. The ray travels until it strikes one of the dielectric interfaces. If the ray strikes the interface at an angle smaller than the critical angle, it will not be guided. The ray in Fig. 3.16 is oriented such that the refracted ray in the n_1 region does not satisfy the TIR condition at

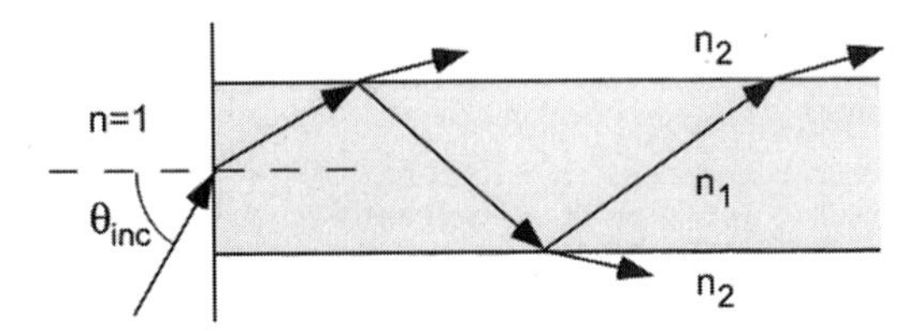

Figure 3.16. A ray incident on the waveguide at too steep an angle will not satisfy the condition for TIR inside the guide. The wave partially reflects and partially transmits at each reflection.

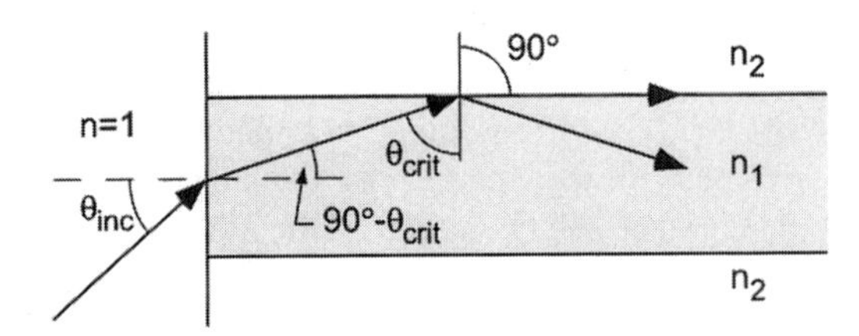

Figure 3.17. The incident ray just satisfies the condition that the refracted ray will strike the interface at the critical angle, and thus be totally internally reflected. θ_{inc} is the maximum angle for a guided ray.

the cover-film interface. This ray is partially reflected and partially transmitted at the interface. As it bounces down the waveguide, a fraction of its energy leaks out at each reflection, and the guided energy is attenuated. Such rays are unguided.

The numerical aperture is defined in terms of the maximum angle that an incident ray can have, and still be trapped by the waveguide. Consider the critical case where the ray just satisfies the TIR condition as shown in Fig. 3.17

The refracted ray strikes the interface between the guiding film, n_1, and the cladding, n_2, at and angle θ_{crit}. The angle between the ray and the axis of the waveguide is the complement of that angle, $90 - \theta_c$. Applying Snell's law to the input face of the waveguide, we can determine the maximum incident angle, θ_{max}:

$$\begin{aligned} \sin\theta_{inc} &= n_1 \sin(90 - \theta_{critical}) = n_1 \cos(\theta_{critical}) \\ &= n_1\sqrt{1 - \sin^2\theta_{critical}} = n_1\sqrt{1 - n_2^2/n_1^2} \\ &= \sqrt{n_1^2 - n_2^2} \end{aligned}$$

The numerical aperture is defined as the sin of the acceptance angle

$$NA = \sin\theta_{max} = \sqrt{n_1^2 - n_2^2} \tag{3.47}$$

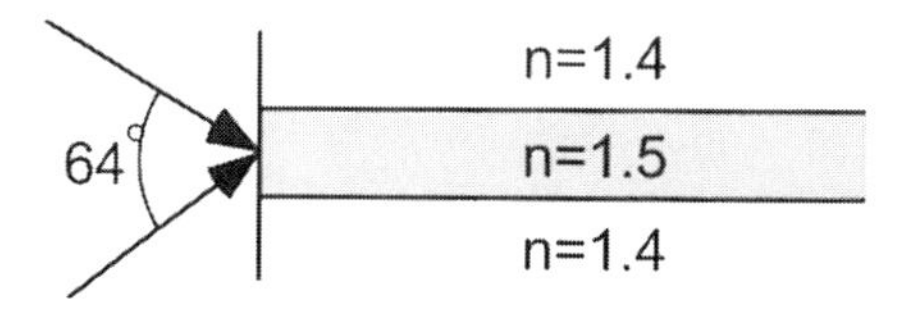

Figure 3.18. The acceptance angle of this structure is 64°. Notice that the Numerical Aperture does not depend on the dimensions of the waveguide, only the relative indices.

As we will see later, the NA is a useful parameter for large core, multimode waveguides where propagation of light is rather difficult to express in terms of electromagnetic fields. Pragmatically, to couple light into a waveguide it is essential that the light be focussed in such a way that all incident rays lie within this angle θ_{max}.

Example 3.4 The numerical aperture for a symmetric waveguide

Consider a symmetric waveguide with a guiding film index $n_1 = 1.5$, and surrounding indices $n_2 = 1.4$ as shown in Fig. 3.18. The numerical aperture is directly found from Eq. 3.46

$$NA = \sqrt{1.5^2 - 1.4^2} = 0.539$$

This corresponds to input angle,

$$\theta_{max} = \sin^{-1}(0.539) = 32°$$

so the full width of the acceptance angle is about 64°.

According to ray analysis, any ray incident on the waveguide within the numerical aperture will be guided. The NA effectively defines the cone of acceptance. Only for large structures (where the guiding film thickness is on the order of 50-100 λ), is the ray picture reasonably accurate.

12. Summary

We have introduced many important concepts of optical waveguides in this chapter. We used formal electromagnetic analysis to solve for the allowed field structure inside a dielectric waveguide. We found that boundary conditions establish the connecting formulae that define the shape of a mode. We derived a characteristic equation for both TE and TM modes. This equation is transcendental, so it requires numerical or graphical solution. We explored both forms of solution in an example. We distinguished asymmetric guides from symmetric guides, and noted that a symmetric guide will always carry at least one mode. We reviewed the mathematical details of the mode concept, stating but not proving the important properties such as orthogonality and completeness that a mode solution will satisfy. These properties will be used in later chapters when calculating coupling. Finally, we described the numerical aperture of a waveguide, based on a ray tracing analysis of the total internal reflection condition for a waveguide.

References

[1] E. Butkov, *Mathematical Physics*, Ch. 9, Addison-Wessley Publishing Co., USA (1968)

[2] J. D. Jackson, *Classical Electrodynamics, 2nd Ed.*, Ch.8., John Wiley and Sons, USA, (1975)

[3] H. Kogelnik, in *Integrated Optics, 2nd Ed., Vol. 7. Topics in Applied Physics*, edited by T. Tamir, Springer Verlag, Berlin (1982)

[4] for example, see S. Gasiorowicz, *Quantum Physics*, John Wiley and Sons, New York (1974)

[5] Allan W. Snyder and John D. Love, *Optical Waveguide Theory*, Ch. 11, Chapman and Hall, LTD, New York (1983)

Practice Problems

1. Using the boundary conditions for a TM wave, confirm Eq. 3.23 for the eigenvalue equation of an asymmetric waveguide.

2. Confirm the field expression for the TE electric field in a symmetric waveguide, Eq. 3.27.

3. Develop the eigenvalue equation for the TM modes in a symmetric waveguide.

4. For an asymmetric planar waveguide with $n_f = 1.5$, $n_s = 1.47$, and $n_c = 1.0$, determine the allowed values of β for waveguide thickness $h = 7\mu m$. Assume the excitation wavelength is $1\mu m$.

5. For the waveguide described in Problem 5., at what wavelength does a second TE mode appear?

6. Which is the lowest order mode in an asymmetric waveguide— the TE_0 mode or the TM_0 mode? Prove your case graphically.

7. Use the Fresnel reflection formula for TE waves to determine the attenuation coefficient for a leaky ray in a waveguide. Assume that the waveguide is symmetric, and is $50\mu m$ thick. Assume the the power lost per reflection is given by $\Gamma| = 1 - |r^2|$, where r is given by Eq. 2.89. Express

the answer in terms of *nepers/meter* as a function of angle. Assume the guiding layer has index 1.5, and the surrounding layers have index 1.45.

8. Evaluate the size of a confined mode as a function of the guiding film dimension. Consider a symmetric waveguide with guiding index of 3.5, and surrounding indices of 3.3.

   (a) Write an explicit description of the field for the TE_0 mode.

   (b) Define the "mode size" of the field by the distance between the points where the amplitude is reduced to $1/e$ of the peak intensity. Find the full width of the TE_0 mode for film thickness, h, ranging from 1 to 20 μm.

   (c) What film thickness leads to the smallest mode?

   (d) Calculate the mode confinement for $h = 20\mu m$, $h = 10\mu m$, and $h = 2\mu m$.

9. For the waveguide described in Ex. 3.1, at what distance into the substrate is the evanescent field equal to $0.001E_0$, where E_0 is the peak amplitude of the mode? Determine this for each mode. At what distance into the cover is the field the same value? What does this tell you about how thick to make the cladding layers on a waveguide?

10. Consider a waveguide with a guiding film $5\mu m$ thick, surrounded by thick films with index 1.45.

    (a) If the guiding film index is 1.5, calculate the numerical aperture of the waveguide.

    (b) What is the mode confinement for TE_0 mode in this waveguide?

    (c) If the guiding film is made to be $10\mu m$, what is the mode confinement of the TE_0 mode?

11. Consider a planar slab waveguide of infinite extent in the y- and z-directions. The guiding film index is 1.5, the substrate index is 1.48, and the cover index is 1.0. The thickness is $h = 2\mu$m. The waveguide is excited with a 1.3 μm source. For TE modes

    (a) What is the range of allowed β values for this waveguide?

    (b) What is the numerical aperture for this waveguide?

    (c) Numerically or graphically, determine the allowed values of β and κ for $h = 2\mu m$.

    (d) How many modes will this waveguide carry if the excitation wavelength is 0.600 μm?

12. Show that the eigenvalue equation for β for the asymmetric waveguide reduces to that of the symmetric waveguide when $n_c = n_s$.

13. Show that $H_x = \frac{\beta}{\mu\omega} E_y$ for a TE wave.

14. For a symmetric waveguide of thickness h, show that the coefficient, A, is given by

$$A^2 = \frac{4\kappa^2 \omega \mu_0 P}{\left(h + \frac{2}{\gamma}\right)(\kappa^2 + \gamma^2)}$$

where P is the power per unit length carried by the waveguide.

15. For an asymmetric planar waveguide with $n_f = 1.5$, $n_s = 1.48$, and a thickness of $8\mu m$, how many TE modes for $\lambda = 1\mu m$ will there be under the following conditions:

    (a) covered by air ($n_c = 1$)

    (b) covered by water ($n_c = 1.33$)

    (c) covered by another substrate ($n_c = 1.45$)

    (d) Explain in words why the number of modes did or did not change in these three cases.

    (e) If $n_c = 1.45$, what thickness waveguide is needed to increase the number of modes to ten?

    (f) How will the number of TM modes vary in a), b), and c)?

16. Given a symmetric waveguide with $n_f = 1.5, n_s = 1.47, \lambda = 1.0\mu m$, determine the fraction of power carried in the cladding if the guiding layer is $3\mu m$ thick.

17. Derive the mode cut-off condition for the TE_n mode in terms of V for the symmetric waveguide.

18. Repeat Ex. 3.1 for the TM mode case. Make plots of the allowed modes similar to those shown in Fig. 3.8

19. Plot the mode profiles for the TE_0 and TE_1 modes in a slab waveguide with a core index $n_f = 1.5$, $n_s = 1.49$, and $n_c = 1$. The film thickness is 10μm thick, and the guided wavelength is 1 μm.

20. Consider an asymmetric planar waveguide with a film index, $n_f = 1.50$, a substrate index $n_s = 1.495$, and a cover index $n_c = 1.40$.

    (a) If the vacuum wavelength of the guided light is 1 μm, what is the thickest that the guiding layer can be to support a TE mode? Use a computer to find the solution by making graphic plots of Eq. 3.22, and adjusting the thickness until the eigenequation cannot be satisfied.

(b) Increase the thickness of the guide by 1%, and determine the eigenvalue β of the TE mode

(c) Calculate the confinement factor for this mode which is near cut-off.

(d) Plot the mode profile for this mode. What can you generalize about modes near cut-off?

21. A planar waveguide is made with $n_f = 1.48$, $n_s = 1.46$ and $n_c = 1.44$. The thickness of the guiding film is $10\mu m$. What is the longest wavelength that can be carried in a TE mode in this waveguide?

22. Extend the development of the modal eigenequation to a four layer structure, with four potentially different indices, and two layer thickness (the cover and substrate are assumed to extend an infinite distance beyond the layers). Perhaps the simplest way to do this is to write the boundary conditions in a matrix form, and use linear algebra techniques to find the roots. Demonstrate the performance of your program with some simple structures.

Chapter 4

STEP–INDEX CIRCULAR WAVEGUIDES

1. Introduction

The circular waveguide has found extensive use in optical communications systems, especially long distance communication links. The circular waveguide has no intrinsic advantage over rectangular waveguides except in one critical area: cost. Manufacturing circular waveguides from glass is a well established technology. Industry can produce hundreds of thousands of kilometers of circular dielectric waveguide each year. The same cannot be said about planar or rectangular waveguides. In this chapter, we will develop a description of wave propagation along a circular waveguide. This chapter deals with the "step–index" fiber (Fig. 4.1). Light is guided by a high-index circular core of uniform index, surrounded by a lower-index cladding layer. The cladding layer is usually covered with a plastic coating to protect the fiber from environmental hazards and abrasion. To find the modes of the circular step-index fiber, we must solve the wave equation in cylindrical coordinates. The modes of the cylindrical structure are more abstract than those of the planar structure. Not only are they circular in symmetry which will require a more complicated solution to the wave equation, but they are two dimensional, so there will be

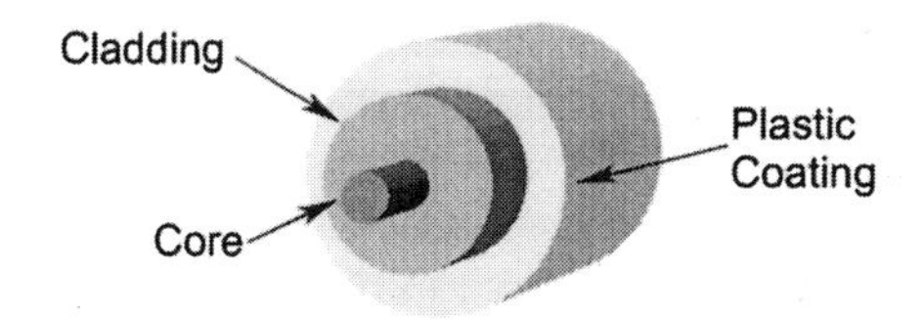

Figure 4.1. The cylindrical step waveguide consists of a high index core surrounded by a lower index cladding.

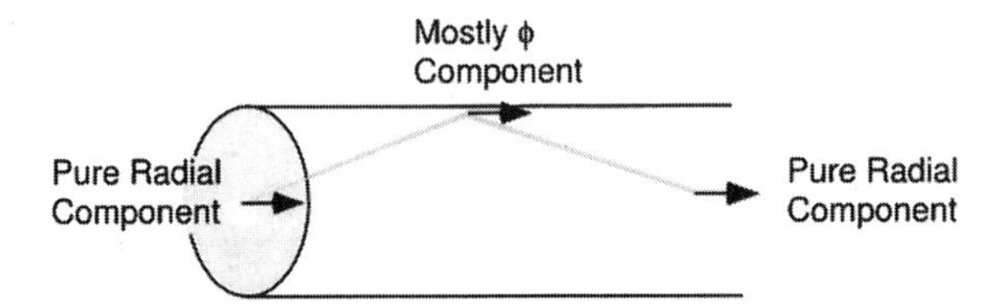

Figure 4.2. A radial field at one point in the waveguide will become an azimuthal field at another location. Notice that the field is not converted between the components by reflection, but by propagation through the coordinate system.

two mode numbers. We will see a similar effect when we discus rectangular waveguides in the next chapter.

Once the modes in the step-index fiber are established, we will develop useful formulae for mode-cutoff conditions, numerical aperture, and normalized frequency. As before, the eigenvalue equations will require graphical or numerical solution.

2. The Wave Equation in Cylindrical Coordinates

We have already derived the homogeneous wave equation for the planar waveguide structure.

$$\nabla^2 \mathbf{E} - \mu\epsilon \frac{\partial^2 \mathbf{E}}{\partial t^2} = 0 \tag{4.1}$$

To solve this equation in a cylindrical waveguide, we must write this equation in cylindrical coordinates. The electric field is a vector, and there are three components, each of which is a function of r, ϕ, and z

$$\mathbf{E}(r,\phi,z) = \hat{r}E_r(r,\phi,z) + \hat{\phi}E_\phi(r,\phi,z) + \hat{z}E_z(r,\phi,z) \tag{4.2}$$

In cylindrical coordinates, the *vector Laplacian* (∇^2) is a rather unwieldy expression (see reference [1]). The cylindrical wave equation must be evaluated in the following form:

$$\nabla(\nabla \cdot \hat{E}) - \nabla \times \nabla \times \hat{E} - \mu\epsilon \frac{\partial^2 \hat{E}}{\partial t^2} = 0 \tag{4.3}$$

Unlike the vector Laplacian in rectilinear coordinates, Eq. 4.3 can not be easily decomposed into three individual components. The transverse components of the field are tightly coupled. Imagine for example a linearly-polarized field travelling at a slight angle to the axis of a cylindrical waveguide, as shown in Fig. 4.2. At $z = 0$, the field is purely radial, but as it travels down the axis, it becomes an azimuthal (ϕ) field. It is impossible to decouple the E_r or E_ϕ components in this example.

Here is the critical point in understanding the analysis of a two-dimensional waveguide: only the $\hat{z}$-component of a field, E_z, does not couple to other components as it propagates. Even after reflection at a cylindrical surface, the E_z component remains oriented along the z-axis. Hence, we will attempt to find a solution for E_z using the wave equation. Once we have a solution for $E_z(r, \theta, \phi)$, we can use Maxwell's equations to relate E_z to E_r and E_ϕ. In this indirect fashion, all field components within a circular waveguide are derived. Fig. 4.3 shows an example of how the E_z field component remains pure.

Since E_z couples only to itself, it is possible to write the scalar wave equation for E_z directly in cylindrical coordinates,

$$\frac{1}{r}\frac{\partial}{\partial r}\left(r\frac{\partial E_z}{\partial r}\right) + \frac{1}{r^2}\frac{\partial^2 E_z}{\partial \phi^2} + \frac{\partial^2 E_z}{\partial z^2} + k_0^2 n^2 E_z = 0 \tag{4.4}$$

and to solve this equation for E_z.

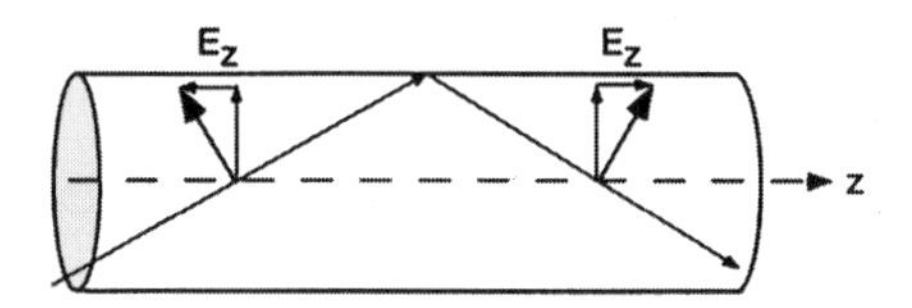

Figure 4.3. The longitudinal component of the electric field does not change through either propagation or reflection at the cylindrical surface.

3. Solution of the Wave Equation for E_z

Since E_z is a function of r, ϕ, and z, we can employ separation of variables to solve the scalar equation, Eq. 4.4. Setting $E_z(r, \phi, z) = R(r)\Phi(\phi)Z(z)$, and substituting this into Eq. 4.4 results in

$$R''\Phi Z + \frac{1}{r}R'\Phi Z + \frac{1}{r^2}R\Phi'' Z + R\Phi Z'' + k_0^2 n^2 R\Phi Z = 0 \tag{4.5}$$

Multiply Eq. 4.5 by $r^2/R\Phi Z$ to get

$$r^2\frac{R''}{R} + r\frac{R'}{R} + \frac{\Phi''}{\Phi} + r^2\frac{Z''}{Z} + k_0^2 n^2 r^2 = 0 \tag{4.6}$$

Due to the translational invariance along the z-axis, we can assume a phase term describes the z-dependence,

$$Z(z) = e^{-j\beta z} \tag{4.7}$$

where β is (again) the z-component of the wavevector, k, in the waveguide. Using Eq. 4.7, we find that $Z''/Z = -\beta^2$, which can be substituted into the wave equation

$$r^2\frac{R''}{R}+r\frac{R'}{R}+\frac{\Phi''}{\Phi}-r^2\beta^2+k_0^2n^2r^2=0 \tag{4.8}$$

Now we can use standard separation techniques to find

$$r^2\frac{R''}{R}+r\frac{R'}{R}-r^2\beta^2+k_0^2n^2r^2=-\frac{\Phi''}{\Phi}=\nu^2 \tag{4.9}$$

The term ν is called the separation constant. Eq. 4.9 can be solved directly for $\Phi(\phi)$:

$$\Phi''(\phi)=-\nu^2\Phi \tag{4.10}$$

which has the solution

$$\Phi(\phi)=Ae^{j\nu\phi}+c.c. \tag{4.11}$$

where A is a normalization constant. Since circular symmetry requires $\Phi(\phi)=\Phi(\phi+2\pi)$, we can infer that ν must be an integer.

Substituting Eq. 4.10 into Eq. 4.8 yields an equation that only contains $R(r)$:

$$r^2\frac{\partial^2 R}{\partial r^2}+r\frac{\partial R}{\partial r}+r^2\left(k_0^2n^2-\beta^2-\frac{\nu^2}{r^2}\right)R=0 \tag{4.12}$$

The solutions to this differential equation is given by Bessel functions [2]. There are many different types of Bessel function, and to the uninitiated, the choice can look formidable. Bessel functions share these properties with *sine* and *cosine* functions: *i*) the value of the function must be calculated or looked up in a table ; *ii*) the functions are orthogonal to one another; and *iii*) they are defined everywhere. It is primarily the lack of familiarity with Bessel functions that causes trepidation. *Appendix B: A Brief Synopsis of Bessel Functions* reviews useful relations and properties of relevant Bessel functions.

Two types of Bessel functions solve Eq. 4.12. Bessel Functions of the First Kind of Order ν, symbolized by $J_\nu(\kappa r)$, are the proper solution when the argument $(k_0^2n^2-\beta^2-\nu^2/r^2)$ is positive. For all cases that we will examine, ν is an integer. κ is defined through the expression

$$\kappa^2=k_0^2n^2-\beta^2 \tag{4.13}$$

Note that this is the same definition used in Chapter 3 for the transverse wavevector. The symbol, κ, has the same meaning in these cylindrical waveguide equations.

When the argument $(k_0^2n^2-\beta^2-\nu^2/r^2)$ in Eq. 4.12 is negative, Modified Bessel Functions of the Second Kind of Order ν, symbolized by $K_\nu(\gamma r)$, are the proper solution. γ is defined as

$$\gamma^2=\beta^2-k_0^2n^2 \tag{4.14}$$

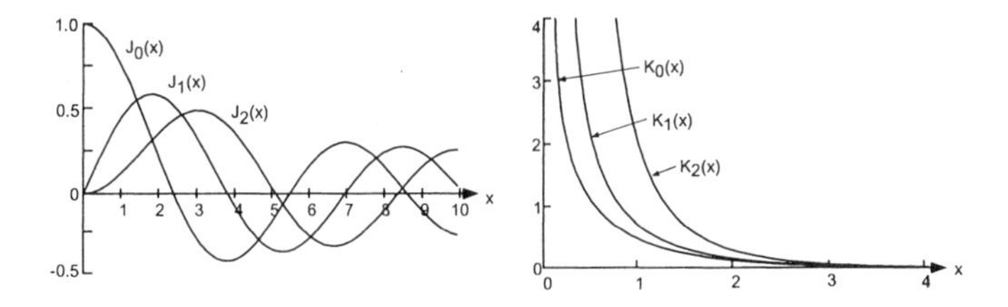

Figure 4.4. The first three Bessel functions of the first kind, $J_\nu(\kappa r)$, and of the second kind, $K_\nu(\gamma r)$.

Again, the notation is intentionally chosen to correspond to the decay parameter, γ, used in Chapter 3. As with κ, the function γ plays the same role in cylindrical waveguides that it did in planar waveguides.

Plots of both types of Bessel function are shown in Fig. 4.4. The $J_\nu(\kappa r)$ functions are periodic along the radial axis. Only $J_0(\kappa r)$ has finite value at $r = 0$; all others $J_{\nu\neq 0}(\kappa r)$ functions are zero at the origin. For large arguments, the Bessel function of the first kind can be approximated as

$$J_\nu(\kappa r) \approx \sqrt{\frac{2}{\pi\kappa r}} \cos\left(\kappa r - \frac{\nu\pi}{2} - \frac{\pi}{4}\right) \quad \text{for } \kappa r \text{ large} \tag{4.15}$$

These Bessel functions can be viewed as damped sine waves. The amplitude decreases slowly with radial distance, much like the amplitude of a spreading wave in a pond. As we shall see, the J_ν Bessel functions describe the radial standing wave in a cylindrical structure.

The modified Bessel functions, $K_\nu(\gamma r)$, display a monotonic decreasing characteristic. The higher orders of the function decrease at a slower rate, but all orders have the same functional form. In the limit of large γr, the function can be approximated as

$$K_\nu(\gamma r) \approx \frac{e^{-\gamma r}}{\sqrt{2\pi\gamma r}} \tag{4.16}$$

Again, this looks like a radially damped exponentially decreasing function. Note that at large distance, all orders of $K_\nu(\gamma r)$ look approximately the same. The $\sqrt{1/2\pi\gamma r}$ dependence is the natural decrease of a wave as it expands with radius, while the exponent represents decay due to evanescent interference. $K_\nu(\gamma r)$ functions are used to describe evanescent fields in the optical waveguide.

4. Field Distributions in the Step Index Fiber

In this and the next section, we derive expressions for the fields and the characteristic equation for the cylindrical dielectric waveguide. Consider the fiber waveguide shown in Fig. 4.5. The fiber waveguide has a core of radius

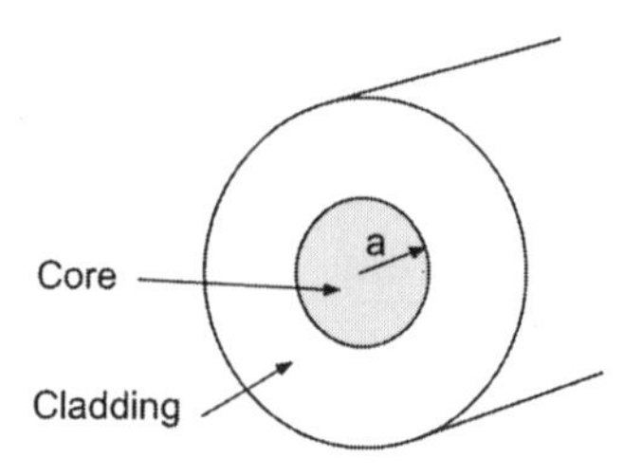

Figure 4.5. The cylindrical waveguide has a core radius of dimension a.

a surrounded by a cladding with lower index. Since we expect oscillatory solutions to the transverse wave equation in the core, $J_\nu(\kappa r)$ solutions will be sought in this region. From the analysis above we can see that an oscillatory solution only occurs when β satisfies

$$k_0 n_{core} > \beta > k_0 n_{clad} \tag{4.17}$$

Outside the higher index core, the field exponentially decays, so we choose the $K_\nu(\gamma r)$ solutions for $r > a$. The only criteria on the size of the cladding is that the evanescent field should decay to negligible values long before the outer radius of the cladding is reached. It is possible to excite "cladding modes" in which the glass core and cladding form the core of the waveguide (with a 125 μm diameter) and the plastic coating forms the cladding. Cladding modes play a role in coupling energy from fiber Bragg gratings, and causing spectral holes to appear in the reflection spectrum of some gratings.

Let's construct a solution to the wave equation. The complete longitudinal fields (E_z and H_z) in both regions can be written as

$$\begin{aligned}
\text{for } r < a \qquad E_z(r,\phi,z) &= AJ_\nu(\kappa r)e^{j\nu\phi}e^{-j\beta z} + c.c. \\
H_z(r,\phi,z) &= BJ_\nu(\kappa r)e^{j\nu\phi}e^{-j\beta z} + c.c. \\
\text{for } r > a \qquad E_z(r,\phi,z) &= CK_\nu(\gamma r)e^{j\nu\phi}e^{-j\beta z} + c.c. \\
H_z(r,\phi,z) &= DK_\nu(\gamma r)e^{j\nu\phi}e^{-j\beta z} + c.c.
\end{aligned} \tag{4.18}$$

Note that the electric and magnetic fields have the same spatial dependence. Also note that ν is a mode number, or eigenvalue. Determining the coefficients A, B, C, and D requires application of the boundary conditions, specifically, continuity of the tangential E and H fields. These steps involve a lot of mathematics, but are necessary for finding the eigenvalue equation of the step index fiber. Boundary conditions require that we know the azimuthal field components, E_ϕ and H_ϕ, in addition to the longitudinal components (E_z and H_z).

From E_z in Eq. 4.18 We can get the azimuthal components in terms of E_z from Maxwell's equations:

$$\nabla \times \mathbf{E} = -\frac{\partial \mathbf{B}}{\partial t} = -\mu \frac{\partial \mathbf{H}}{\partial t} = -\mu j \omega \mathbf{H} \tag{4.19}$$

Expanding the $\nabla \times \mathbf{E}$ term in cylindrical components, and then collecting terms, the field components H_r, H_ϕ, E_r and E_ϕ can be described [3] in terms of the longitudinal components:

$$\begin{aligned}
E_\phi &= \frac{-j}{\alpha^2}\left(\frac{\beta}{r}\frac{\partial E_z}{\partial \phi} - \omega\mu\frac{\partial H_z}{\partial r}\right) \\
E_r &= \frac{-j}{\alpha^2}\left(\frac{\mu\omega}{r}\frac{\partial H_z}{\partial \phi} + \beta\frac{\partial E_z}{\partial r}\right) \\
H_\phi &= \frac{-j}{\alpha^2}\left(\omega\epsilon\frac{\partial E_z}{\partial r} + \frac{\beta}{r}\frac{\partial H_z}{\partial \phi}\right) \\
H_r &= \frac{-j}{\alpha^2}\left(\beta\frac{\partial H_z}{\partial r} - \frac{\omega\epsilon}{r}\frac{\partial E_z}{\partial \phi}\right)
\end{aligned} \tag{4.20}$$

where α^2 stands for $k_0^2 n^2 - \beta^2$. Note that α^2 is a positive quantity in the core, and a negative quantity in the cladding for allowed values of β.

Using the longitudinal fields described in Eq. 4.18, the field components in Eqs. 4.20 can be exactly calculated. In the core region ($r < a$) we get

$$\begin{aligned}
E_r &= \frac{-j\beta}{\kappa^2}\left[A\kappa J'_\nu(\kappa r) + \frac{j\omega\mu\nu}{\beta r} B J_\nu(\kappa r)\right] e^{j\nu\phi} e^{-j\beta z} \\
E_\phi &= \frac{-j\beta}{\kappa^2}\left[\frac{j\nu}{r} A J_\nu(\kappa r) - \frac{\omega\mu}{\beta} B\kappa J'_\nu(\kappa r)\right] e^{j\nu\phi} e^{-j\beta z} \\
H_r &= \frac{-j\beta}{\kappa^2}\left[B\kappa J'_\nu(\kappa r) - \frac{j\omega\epsilon_{core}\nu}{\beta r} A J_\nu(\kappa r)\right] e^{j\nu\phi} e^{-j\beta z} \\
H_\phi &= \frac{-j\beta}{\kappa^2}\left[\frac{j\nu}{r} B J_\nu(\kappa r) + \frac{\omega\epsilon_{core}}{\beta} A\kappa J'_\nu(\kappa r)\right] e^{j\nu\phi} e^{-j\beta z}
\end{aligned} \tag{4.21}$$

where $J'_\nu(\kappa r) = dJ_\nu(\kappa r)/d(\kappa r)$. In the cladding region ($r > a$) we get

$$\begin{aligned}
E_r &= \tfrac{j\beta}{\gamma^2}\left[C\gamma K'_\nu(\gamma r) + \tfrac{j\omega\mu\nu}{\beta r} D K_\nu(\gamma r)\right] e^{j\nu\phi} e^{-j\beta z} \\
E_\phi &= \tfrac{j\beta}{\gamma^2}\left[\tfrac{j\nu}{r} C K_\nu(\gamma r) - \tfrac{\omega\mu}{\beta} D\gamma K'_\nu(\gamma r)\right] e^{j\nu\phi} e^{-j\beta z} \\
H_r &= \tfrac{j\beta}{\gamma^2}\left[D\gamma K'_\nu(\gamma r) - \tfrac{j\omega\epsilon_{clad}\nu}{\beta r} C K_\nu(\gamma r)\right] e^{j\nu\phi} e^{-j\beta z} \\
H_\phi &= \tfrac{j\beta}{\gamma^2}\left[\tfrac{j\nu}{r} D K_\nu(\gamma r) + \tfrac{\omega\epsilon_{clad}}{\beta} C\gamma K'_\nu(\gamma r)\right] e^{j\nu\phi} e^{-j\beta z}
\end{aligned} \tag{4.22}$$

where $K'_\nu(\gamma r) = dK_\nu(\gamma r)/d(\gamma r)$.

5. Boundary Conditions for the Step-Index Waveguide

To determine β, and the amplitude coefficients, A, B, C, and D of Eq. 4.18, we need to apply the boundary conditions. The boundary conditions at $r = a$ require that the four tangential components, E_z, E_ϕ, H_z, and H_ϕ be continuous at the core-cladding boundary. For example, the longitudinal electric field must satisfy $AJ_\nu(\kappa a)e^{j\nu\phi}e^{-j\beta z} = CK_\nu(\gamma a)e^{j\nu\phi}e^{-j\beta z}$ The simplest way to simultaneously satisfy all four boundary value equations is to write the four linear equations in matrix form, and then set the determinant of the matrix equal to zero.

$$\begin{pmatrix} J_\nu(\kappa a) & 0 & -K_\nu(\gamma a) & 0 \\ 0 & J_\nu(\kappa a) & 0 & -K_\nu(\gamma a) \\ \frac{\beta\nu}{a\kappa^2}J_\nu(\kappa a) & j\frac{\omega\mu}{\kappa}J'_\nu(\kappa a) & \frac{\beta\nu}{a\gamma^2}K_\nu(\gamma a) & j\frac{\omega\mu}{\gamma}K'_\nu(\gamma a) \\ -j\frac{\omega\epsilon_{core}}{\kappa}J'_\nu(\kappa a) & \frac{\beta\nu}{a\kappa^2}J_\nu(\kappa a) & -j\frac{\omega\epsilon_{clad}}{\gamma}K'_\nu(\gamma a) & \frac{\beta\nu}{a\gamma^2}K_\nu(\gamma a) \end{pmatrix} \begin{pmatrix} A \\ B \\ C \\ D \end{pmatrix} = 0 \tag{4.23}$$

For non-trivial solutions (i.e. non-zero amplitudes), the four equations will simultaneously equal zero if and only if the determinant of the matrix equals zero. Expansion of the determinant yields the "characteristic equation" for the step-index fiber.

$$\frac{\beta^2\nu^2}{a^2}\left[\frac{1}{\gamma^2}+\frac{1}{\kappa^2}\right]^2 =$$

$$\left[\frac{J'_\nu(\kappa a)}{\kappa J_\nu(\kappa a)} + \frac{K'_\nu(\gamma a)}{\gamma K_\nu(\gamma a)}\right] \cdot \left[\frac{k_0^2 n_{core}^2 J'_\nu(\kappa a)}{\kappa J_\nu(\kappa a)} + \frac{k_0^2 n_{clad}^2 K'_\nu(\gamma a)}{\gamma K_\nu(\gamma a)}\right] \tag{4.24}$$

This formidable equation requires numerical or graphical solution. There is only one unknown: β. As with the slab waveguide, the terms κ and γ are functions of β and the local index. Due to the oscillatory nature of $J_\nu(\kappa a)$, there can be several values of β for a given structure. Since there are two dimensional degrees of freedom in the cylindrical waveguide, solutions to the wave equation are labelled with two indices, ν and m. Both numbers are integers. The m value is called the *radial mode number*, and represents the number of radial nodes that exist in the field distribution. The integer ν is called the *angular mode number*, and represents the number of angular nodes that exist in the field distribution.

Once β is determined from Eq. 4.24, three of the coefficients (A, B, C, and D) can be determined in terms of the fourth by solving the individual equations of the matrix. For example, from the boundary condition for continuity of E_z at $r = a$,

$$AJ_\nu(\kappa a) = CK_\nu(\gamma a) \tag{4.25}$$

One can solve for coefficient C in terms of A

$$C = \frac{J_\nu(\kappa a)}{K_\nu(\gamma a)} A \tag{4.26}$$

Similarly D can be solved in terms of B

$$D = \frac{J_\nu(\kappa a)}{K_\nu(\gamma a)} B \tag{4.27}$$

The coefficients A and B can be related to one another using the continuity of E_ϕ or H_ϕ, and Eqs. 4.26 and 4.27. Using the electric field continuity one gets

$$B = \frac{j\nu\beta}{\omega\mu a}\left[\frac{1}{\kappa^2} + \frac{1}{\gamma^2}\right]\left[\frac{J'_\nu(\kappa a)}{\kappa J_\nu(\kappa a)} + \frac{K'_\nu(\gamma a)}{\gamma K_\nu(\gamma a)}\right]^{-1} A \tag{4.28}$$

If the magnetic field continuity is used, one gets

$$B = \frac{j\omega a}{\beta\nu}\left[\frac{n_{core}^2}{\kappa}\frac{J'_\nu(\kappa a)}{J_\nu(\kappa a)} + \frac{n_{clad}^2}{\gamma}\frac{K'_\nu(\gamma a)}{K_\nu(\gamma a)}\right]\left[\frac{1}{\kappa^2} + \frac{1}{\gamma^2}\right]^{-1} A \tag{4.29}$$

The choice of which equation to use depends on the type of mode carried in the waveguide. This is explained in the next section. Note that B/A is purely imaginary in both cases, indicating that the two longitudinal fields are $\pi/2$ out of phase. On an instantaneous basis, there is radial power flow, but due to the $\pi/2$ phase shift the power is reactive, so it averages to zero.

6. The Spatial Modes of the Step-Index Waveguide

Unlike the slab waveguide with only two possible types of mode (TE or TM), the circular waveguide has four types of mode. The quantity $|B/A|$ is of particular interest in determining the relative size of the longitudinal components of the E and H fields. These, in turn, characterize the type of mode. We will start with the simplest mode.

6.1 Transverse Electric and Transverse Magnetic Modes

Consider the characteristic equation (Eq. 4.24) for the case where $\nu = 0$. Since ν represents angular dependence of the solution, the field solutions to E_z when $\nu = 0$ will be rotationally invariant. The equation simplifies to

$$\left[\frac{J'_\nu(\kappa a)}{\kappa J_\nu(\kappa a)} + \frac{K'_\nu(\gamma a)}{\gamma K_\nu(\gamma a)}\right]\left[\frac{k_0^2 n_{core}^2 J'_\nu(\kappa a)}{\kappa J_\nu(\kappa a)} + \frac{k_0^2 n_{clad}^2 K'_\nu(\gamma a)}{\gamma K_\nu(\gamma a)}\right] = 0 \tag{4.30}$$

Either term on the left hand side can be set to zero to satisfy the equation. The two terms in Eq. 4.30 appeared individually in Eqs. 4.28 and 4.29, where the amplitude A was related to amplitude B. If the first term of Eq. 4.30 is set to zero, then A must also be zero to keep the magnitude of B in Eq. 4.28 finite. If $A = 0$, then $E_z = 0$, and the electric field will be transverse. Such modes are called TE modes.

Conversely, if the second term in Eq. 4.30 is zero, then the amplitude B will be zero (see Eq. 4.29), and the longitudinal component of the H field will be zero. The solution will therefore be a TM mode. Thus, if $\nu = 0$, the allowed modes will be either TE or TM.

The problem of finding the allowed values of the propagation vector, β, reduces to finding the roots of Eq. 4.30. These equations for the TE and TM modes can be further simplified using the Bessel function relations (see Appendix B)

$$\begin{aligned} \frac{J_\nu'}{\kappa J_\nu} &= \pm\tfrac{J_{\nu\mp1}}{\kappa J_\nu} \mp \tfrac{\nu}{\kappa^2} \\ \frac{K_\nu'}{\gamma K_\nu} &= \mp\tfrac{K_{\nu\pm1}}{\gamma K_\nu} \mp \tfrac{\nu}{\gamma^2} \end{aligned} \tag{4.31}$$

Consider first the TE mode. The first term of Eq. 4.30 should be set equal to zero. Using the relation in Eq. 4.31 the eigenvalue equation for TE modes becomes

$$-\frac{J_1(\kappa a)}{\kappa J_0(\kappa a)} - \frac{K_1(\gamma a)}{\gamma K_0(\gamma a)} = 0 \tag{4.32}$$

The other half of Eq. 4.30 is the eigenvalue equation for TM modes. These can be solved numerically or graphically. We will use *Mathematica* to do both in the following example of a TE mode.

Example 4.1: Eigenvalues for the TE modes in a step-index fiber

Let's analyze a step-index circular fiber with a core index , $n_{core} = 1.5$, a cladding index $n_{clad} = 1.45$, and with a core radius, $a = 5\mu$m. The wavelength of the light is 1.3μm. We want to determine the allowed eigenvalues for β for the TE modes. A simple *Mathematica* command evaluates and plots the two terms in Eq. 4.32

```
k=2 Pi /(1.3 10^(-4));
a=5 10^(-4);
n1=1.5;
n2=1.45;
kappamax=Sqrt[k^2(n1^2-n2^2)];
gamma = Sqrt[ kappamax^2-kappa^2];
Plot[{BesselJ[1,kappa a]/(kappa BesselJ[0,kappa a]),
  -BesselK[1,gamma a]/(gamma BesselK[0,gamma a])},{kappa,0,kappamax}]
```

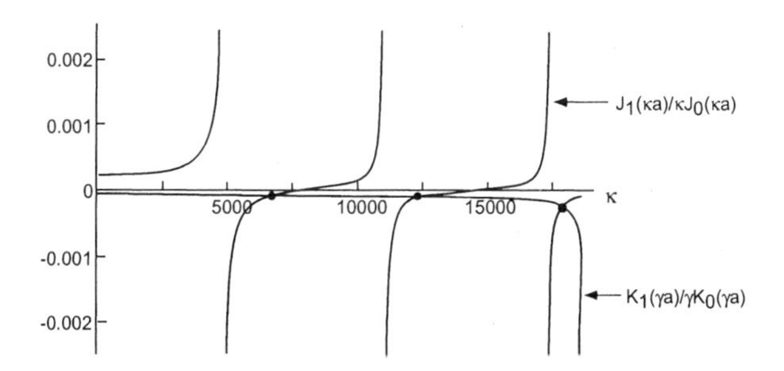

Figure 4.6. The eigenvalue equation plotted against κ for a waveguide with core index 1.5, cladding index 1.45, and wavelength 1.3 μm.

The graphical output is presented in Fig. 4.6. As in previous chapters, we chose to plot the functions against the transverse wavevector, κ, instead of against β. The plot extends from $\kappa = 0$ to κ_{max} which is given by

$$\kappa_{max} a = \sqrt{k_0^2 n_{core}^2 - k_0^2 n_{clad}^2}\, a \tag{4.33}$$

The $J_1(\kappa a)/\kappa J_0(\kappa a)$ term explodes to infinity at every root of $J_0(\kappa a)$. Since the roots of $J_0(\kappa a)$ occur (almost) periodically, the ratio J_1/J_0 regularly sweeps from $-\infty$ to $+\infty$. The K_1/K_0 term monotonically decreases as κ increases. Every time the two lines cross in Fig.4.6, there is an allowed TE mode. In this case, three TE modes are allowed, with approximate κ values of 7000, 12500, and 17500 cm^{-1}. The exact values are easily found using a root finding command. In *Mathematica* the appropriate command is

```
FindRoot[-BesselK[1, gamma a]/(gamma BesselK[0, gamma a])==
BesselJ[1,kappa a]/(kappa BesselJ[0, kappa a]), {kappa, 5200}]
```

The exact values for this example are $\kappa = 6902$, 12549, and 17795 cm^{-1}. The corresponding values of β can be determined from Eq. 4.13.

The transverse modes (TE and TM) have no azimuthal structure ($\nu = 0$). We will look at the field solutions in a later section, but in the ray picture these modes are geometrically represented by Meridional rays. As seen in Fig. 4.7, the ray associated with these modes travels through the origin, $r = 0$.

6.2 The Hybrid Modes

When $\nu \neq 0$, the characteristic equation is a little more complicated to solve. The values of β will correspond to modes which have finite components of both E_z and H_z, and are therefore neither TE nor TM modes. These modes are called EH or HE modes, depending on the relative magnitude of the longitudinal E and H components [4, 5].

If $A = 0$ then the mode is called a TE mode

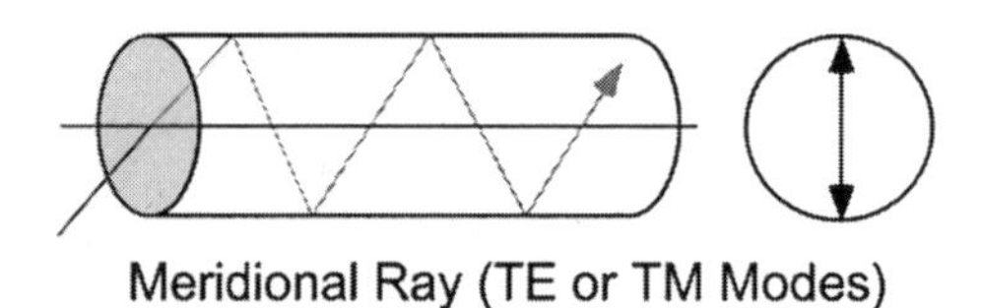

Figure 4.7. A meridional ray zig-zags down the fiber, passing through the origin. There is no angular rotation of the ray path as it propagates.

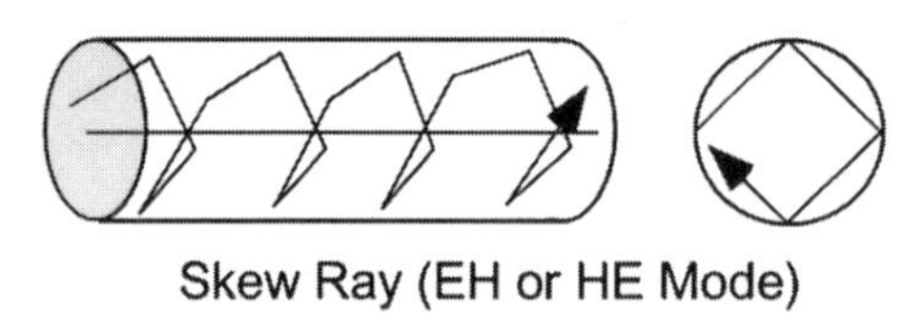

Figure 4.8. A skew ray travels in a spiral path down the fiber. The ray does not go through the origin.

If $B = 0$	then the mode is called a TM mode
If $A > B$	then the mode is called an HE mode (E_z dominates H_z)
If $A < B$	then the mode is called an EH mode (H_z dominates E_z).

The EH and HE modes are called "hybrid" modes, because they have both longitudinal H and E components in the waveguide. The EH and HE modes exist only for $\nu \geq 1$, so they have azimuthal structure. In the ray picture, these modes are called "skew" rays, because they travel down the waveguide in a screw-like pattern (Fig. 4.8), glancing off the interface as they spiral down the axis. The azimuthal structure is apparent from the cyclical path of the ray.

The EH and HE modes have complicated field patterns. These patterns are not only difficult to determine, but they are hard to visualize. Because of this, and the limited utility derived in actually graphing such distributions, we will not pursue their description. Instead, the next subsection develops a useful approximation that simplifies both the calculation and visualization of the hybrid modes.

6.3 The Linearly Polarized Modes (LP modes)

The characteristic equation for the hybrid modes is difficult to solve for β. Fortunately, a very simple and reasonable approximation makes solution straightforward [6]. Consider again the characteristic equation, Eq. 4.24

$$\frac{\beta^2\nu^2}{a^2}\left[\frac{1}{\gamma^2}+\frac{1}{\kappa^2}\right]^2 =$$

$$\left[\frac{J'_\nu(\kappa a)}{\kappa J_\nu(\kappa a)} + \frac{K'_\nu(\gamma a)}{\gamma K_\nu(\gamma a)}\right] \cdot \left[\frac{k_0^2 n_{core}^2 J'_\nu(\kappa a)}{\kappa J_\nu(\kappa a)} + \frac{k_0^2 n_{clad}^2 K'_\nu(\gamma a)}{\gamma K_\nu(\gamma a)}\right]$$

For $\nu = 1, 2, \ldots$, HE and EH modes are possible. Unfortunately, even with powerful software, finding the roots of this equation is very difficult. Dramatic simplification occurs if we make the *weakly guiding* approximation. For many practical optical fibers, the core and cladding index are nearly identical. Typical commercial fibers have $\Delta n = n_{core} - n_{clad}$ on the order of 0.001-0.005. In view of this, it is not unreasonable (at least for the purpose of finding roots) to approximate that the core and cladding index are identical, $n_{core} \approx n_{clad} = n$. This approximation will introduce an error on the order of less than 1 part per thousand in the actual value of the propagation vector, but will enable easy solution to the problem. In the weakly guiding approximation Eq. 4.24 reduces to

$$\frac{\beta^2 \nu^2}{a^2} \left[\frac{1}{\gamma^2} + \frac{1}{\kappa^2}\right]^2 = \left[\frac{J'_\nu(\kappa a)}{\kappa J_\nu(\kappa a)} + \frac{K'_\nu(\gamma a)}{\gamma K_\nu(\gamma a)}\right]^2 k_0^2 n^2 \tag{4.34}$$

This can be further simplified noting that if $n_{core} = n_{clad}$, then $\beta^2 = k_0^2 n^2$, and these terms can be cancelled from both sides. Taking advantage of some Bessel function identities

$$\begin{aligned} \frac{J'_\nu(\kappa a)}{\kappa J_\nu(\kappa a)} &= \pm \frac{J_{\nu \mp 1}}{\kappa a J_\nu(\kappa a)} \mp \frac{\nu}{\kappa^2 a} \\ \frac{K'_\nu(\gamma a)}{\gamma K_\nu(\kappa a)} &= \frac{K_{\nu \pm 1}(\gamma a)}{\gamma a K_\nu(\gamma a)} \mp \frac{\nu}{\gamma^2 a} \end{aligned} \tag{4.35}$$

simplifies Eq. 4.34, leaving only

$$\frac{J_{\nu \pm 1}(\kappa a)}{\kappa J_\nu(\kappa a)} = \mp \frac{K_{\nu \pm 1}(\gamma a)}{\gamma K_\nu(\gamma a)} \tag{4.36}$$

These are the characteristic equations for the EH (top sign) and HE (bottom sign) modes. Solution will yield the eigenvalues for the allowed modes. A little more manipulation with Bessel function identities reduces these two equations into one single equation [7]

$$\kappa \frac{J_{j-1}(\kappa a)}{J_j(\kappa a)} = -\gamma \frac{K_{j-1}(\gamma a)}{K_j(\gamma a)} \tag{4.37}$$

The indices define the mode as follows:

$j = 1$	TE, TM modes
$j = \nu + 1$	EH_ν modes
$j = \nu - 1$	HE_ν modes

Two different modes can have the same eigenvalue, or mathematically speaking, they are degenerate. In the weakly guiding approximation, the TE_{0m} is degenerate with the TM_{0m} mode (will have the same eigenvalue, β) and will propagate at the same velocity (at least to the accuracy of the weakly guiding approximation). Similarly, the $HE_{\nu+1,m}$ modes and $EH_{\nu-1,m}$ modes are degenerate.

Since degenerate modes travel at the same velocity, it is possible to define stable superpositions of different modes. Certain combinations of degenerate modes can be found which are *linearly polarized.* Furthermore the superpositions are primarily transverse modes, meaning E_z is negligible. This is best illustrated by example. We will take a "back door" approach to creating a superposition that leads to a linearly polarized mode, by initially assuming that a mode has a *transverse* field configuration, and then derive what the *longitudinal* mode structure must be.

6.4 Linearly Polarized Mode Based on a Superposition of Two Degenerate Modes

Let's start by describing an idealized transverse electric field inside a step–index fiber, polarized in the y–direction

$$E_y(r, \phi, z) = \hat{y} E_0 J_\nu(\kappa r) \cos(\nu\phi) e^{-j\beta z} \tag{4.38}$$

where E_0 is the amplitude, and the functional form is consistent with the fields we defined in Eq. 4.18 for cylindrical symmetry. We have assumed that the azimuthal dependence is in the form of a cosine term. If the electric field is travelling in the z-direction and polarized in the y-direction, then the magnetic field must also be transverse and oriented in the x-direction. Using the impedance of the medium, we can write an expression for H in terms of E

$$H_x(r, \phi, z) = \hat{x} \frac{E_y}{\eta} J_\nu(\kappa r) \cos(\nu\phi) e^{-j\beta z} \tag{4.39}$$

Since we have simply "assumed" a transverse field, it would be valuable to actually verify that the longitudinal component, E_z, is negligibly small. The longitudinal component can be found using Faraday's equation

$$\nabla \times \mathbf{H} = \frac{d\mathbf{D}}{dt} = -j\omega\epsilon\mathbf{E} \tag{4.40}$$

where we have assumed that $\mathbf{E}$ is a time harmonic field with angular frequency ω. Expanding the curl equation, and noting that only H_x has non–negligible values, E_z can be written as

$$E_z(r, \phi, z) = \frac{1}{-j\omega\epsilon} \frac{\partial H_x}{\partial y} = \frac{1}{-j\omega\epsilon} \frac{\partial}{\partial y} \left[\frac{E_y}{\eta} J_\nu(\kappa r) \cos(\nu\phi) e^{-j\beta z} \right] \tag{4.41}$$

To evaluate this derivative, the operator $\partial/\partial y$ must be written in terms of r and ϕ. Using the identities $r = \sqrt{x^2+y^2}$, and $\phi = \tan^{-1}(x/y)$, it is straightforward to show that

$$\frac{d}{dy} = \sin\phi\frac{\partial}{\partial r} + \frac{\cos\phi}{r}\frac{\partial}{\partial\phi} \tag{4.42}$$

Apply this operator to Eq. 4.41 to get

$$E_z(r,\phi.z) = \frac{-E_y}{j\eta\omega\epsilon}\left[\sin(\phi)\kappa J_\nu'(\kappa r)\cos(\nu\phi) - \frac{\nu\cos\phi}{r}J_\nu(\kappa r)\sin(\nu\phi)\right]e^{-j\beta z} \tag{4.43}$$

This can be simplified with two Bessel function identities:

$$\begin{aligned} J_\nu'(\kappa r) &= \tfrac{1}{2}\left[J_{\nu-1}(\kappa r) - J_{\nu+1}(\kappa r)\right] \\ \frac{\nu}{\kappa r}J_\nu(\kappa r) &= \tfrac{1}{2}\left[J_{\nu-1}(\kappa r) + J_{\nu+1}(\kappa r)\right] \end{aligned} \tag{4.44}$$

and the trigonometric identity

$$\sin\alpha\cos\beta = \frac{1}{2}\left[\sin(\alpha+\beta) + \sin(\alpha-\beta)\right] \tag{4.45}$$

Substituting these into Eq. 4.43, and cancelling terms yields

$$E_z(r,\phi,z) = \frac{-E_y}{j\eta\omega\epsilon}\frac{\kappa}{2}\left[J_{\nu-1}(\kappa r)\sin(\nu-1)\phi - J_{\nu+1}(\kappa r)\sin(\nu+1)\phi\right]e^{-j\beta z} \tag{4.46}$$

Recall that the general modal solution for the longitudinal field is described as

$$E_z(r,\phi,z) = AJ_\nu(\kappa r)\cos(\nu\phi)e^{-j\beta z} \tag{4.47}$$

We can see by inspection that E_z is, in fact, a superposition of two modes, one with index $\nu+1$ and the other with index $\nu-1$. Now recall that in the weakly guiding approximation, the $\text{HE}_{\nu+1,m}$ mode is degenerate with the $\text{EH}_{\nu-1,m}$ mode. This shows that it is possible to add two modes in such a way that the residual longitudinal component of the field is essentially zero. The coefficient, A, describing the amplitude term can be expressed as

$$A = E_y\frac{\kappa}{2\eta\omega\epsilon} = E_y\frac{\kappa}{2k_0 n} \tag{4.48}$$

Since the transverse wave vector, κ, is much smaller than the wavevector, k, there is little amplitude in the longitudinal field. Thus our initial assumption of a perfectly transverse field (i.e. no longitudinal components) is nearly satisfied through proper superposition of degenerate hybrid modes.

From the example, we can see how two modes can be combined to create a linearly–polarized, cartesian–coordinate referenced, electric field distribution.

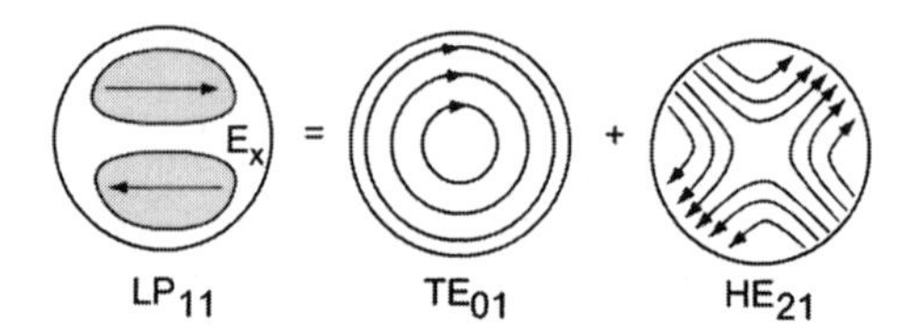

Figure 4.9. The LP_{11} mode is a superposition of the TE_{01} and HE_{21} modes. Note that the LP mode is linearly polarized, in contrast to the electric fields of the two constituent modes. E_x polarization is shown, although with appropriate superposition, an E_y polarized mode could have been created.

These superposition modes are called the $LP_{\nu m}$ modes. The designation and construction of an LP mode is as follows:

$$\begin{array}{lll} \mathrm{LP}_{1m} & \rightarrow & \text{sum of } \mathrm{TE}_{0m}, \mathrm{TM}_{0m}, \text{ and } \mathrm{HE}_{2m} \text{ modes} \\ \mathrm{LP}_{\nu m} & \rightarrow & \text{sum of } \mathrm{HE}_{\nu+1,m} \text{ and } \mathrm{EH}_{\nu-1,m} \text{ modes} \\ \mathrm{LP}_{0m} & \rightarrow & \mathrm{HE}_{1m} \text{ mode only (special case)} \end{array}$$

Fig. 4.9 shows a sketch of the mode structure of an LP_{11} mode, and a sketch of the two modes that are combined to form it. As seen from the figure, the mode profiles of the HE_{21} and TE_{01} modes are best described in cylindrical coordinates, one having purely azimuthal fields, and the other having radial fields. However, the superposition leads to a mode with two lobes that is linearly polarized. The plot shows $\hat{y}$ polarization, but the mode could also be polarized along the $\hat{x}$ axis. Also, the lobes could be rotated by 90°, making the null region lie along the $x = 0$ axis. Thus there are four degenerate LP_{11} modes (two orientations of the lobes, each with two possible polarizations).

The LP modes have many practical advantages. First, the LP modes provide an easy way to visualize the structure of the guided modes. Because most of the energy is stored in the transverse field of the LP mode, we can ignore the complications of energy stored in the longitudinal terms. Second, the LP modes represent actual energy distributions that a polarized source would excite in a fiber. For example, a polarized laser uniformly illuminating the end of a step would create a linearly polarized transverse field on the input. Finally, LP modes allow for a simplified characteristic equation that can be solved with straightforward numeric or graphical techniques.

The disadvantages of LP modes are due to the fact that they are not true modes, but are in fact a superposition of slightly nondegenerate modes. The individual EH, HE, TM, and TE modes travel at slightly different velocities, so the polarization state of the initial superposition will change as the modes propagate down the axis of the guide. The LP modes are, in summary, only an approximation of the true mode structure of the fiber. They allow a simple

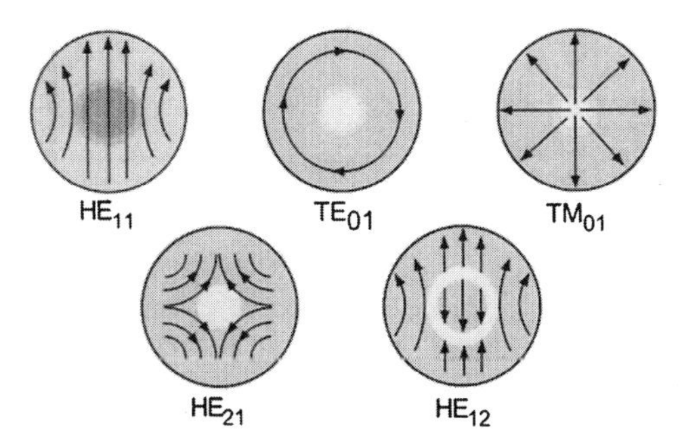

Figure 4.10. The electric field distribution of five different modes in a step index circular waveguide.

way to visualize the mode, and describe the actual field patterns excited by real sources.

6.5 Summary of Mode Shapes

As can be seen in Fig. 4.10, there are many possible orientations of the electric field within a step-index waveguide. Here we schematically represent the electric field lines for five different modes. The magnetic lines are transverse to the electric fields.

Detailed tables of the functional form of the fields in the modes of a step-index fiber can be found in the literature [5].

7. The Normalized Frequency (V-number) and Cutoff

Often we are concerned whether a given mode will propagate within a fiber. For example, we might need a single mode fiber for an experiment using a visible laser, such as the HeNe laser operating at $\lambda = 633$ nm, but all we can find is single mode fiber that is designed for operation at 1.3 μm. How can we determine if this fiber will be satisfactory? To answer this, we need to develop what are known as "cut-off" conditions, which determine under what circumstances a mode will propagate in a fiber.

The characteristic equation (Eq. 4.24) contains a term with the ratio of Bessel functions, $J_{\nu\pm1}/J_\nu$. This term explodes to infinity at each root of J_ν, as was illustrated in Fig. 4.6. To insure that there is at least one solution to the equation (i.e. one place where the lines cross), the argument κa must extend beyond the first root. Each time κa increases beyond another root of $J_{\nu\pm1}/J_\nu$, another mode will be allowed. The roots of the Bessel functions are thus the signposts for establishing mode cutoff conditions.

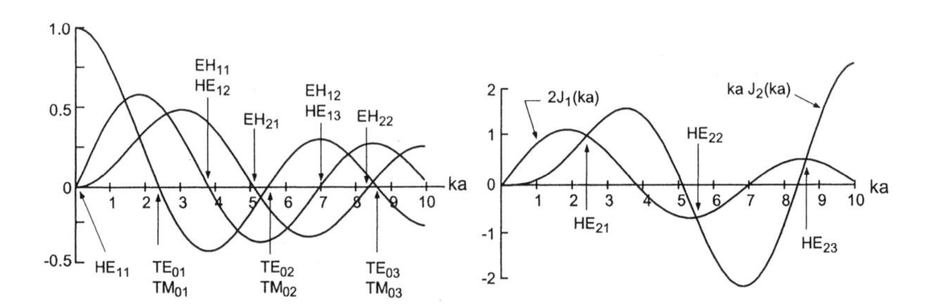

Figure 4.11. a)The first three J_ν Bessel functions are plotted, with the mode cutoff conditions of a few modes indicated at the various roots of the curves. b)The condition $2J_1(\kappa a) = \kappa a J_2(\kappa a)$ is plotted for the HE_{2m} mode cutoff conditions. Cut-off occurs where the curves cross.

We can generalize the cutoff conditions for the modes in terms of the roots of the appropriate Bessel function. For example, referring back to Fig. 4.6, it is clear that no TE mode will exist if $\kappa a < 2.405$. The TE_{01} can only exist if $\kappa a > 2.405$, so we say that the cut-off condition for the TE_{01} mode is $\kappa a = 2.405$. The cutoff condition for the TE_{02} mode occurs at the second root of the Bessel function, $J_0(\kappa a)$, which occurs 5.405. The cutoff conditions for every variety of mode can be found in a similar fashion. These cutoff conditions are

TE_{0m} modes	$\kappa a > m^{th}$ root of $J_0(\kappa a)$
HE_{1m} mode	$\kappa a > m^{th}$ root of $J_1(\kappa a)$
$\mathrm{EH}_{\nu m}$ mode	$\kappa a > m^{th}$ root of $J_\nu(\kappa a)$
	with the added constraint that the first root is not 0
$\mathrm{HE}_{\nu m}$ modes	$\left(\frac{\epsilon_{core}}{\epsilon_{clad}} + 1\right) J_{\nu-1}(\kappa a) = \frac{\kappa a}{\nu-1} J_\nu(\kappa a)$.

Fig. 4.11a shows a plot of the first three Bessel functions, with notations on the cutoff points for a few modes. For example, if κa is greater than 2.405, then the TE_{01}, TM_{01}, and HE_{21} modes will be allowed. This is in addition to the HE_{11} mode, which is always allowed. The HE_{11} mode is a special case which is described in the next section. The $\mathrm{HE}_{\nu m}$ modes have a complicated cutoff formula which requires knowledge of the refractive indices of the core and cladding. In most cases the ratio can be approximated as unity. Fig. 4.11b shows the cut-off conditions for the HE_{2m} modes.

The parameter used to characterize a waveguide is the *Normalized Frequency* or the V-number. For a cylindrical fiber, the V-number is defined as $\kappa_{max} a$.

$$V\text{-number} = a k_0 \sqrt{n_{core}^2 - n_{clad}^2} = \frac{2\pi a}{\lambda} \sqrt{n_{core}^2 - n_{clad}^2} \tag{4.49}$$

where a is the core radius. The normalized frequency provides a quick way to determine the number of modes in a waveguide, and is often used as a specification for optical fibers and devices. The cutoff conditions can all be evaluated once the V–number of a fiber is given.

Example 4.2 Number of TE Modes in a step–index fiber

Consider a step index fiber that has a core index, $n_{core} = 1.45$, a cladding index $n_{clad} = 1.44$, and a core radius of 25 μm. If the excitation wavelength is 1.5 μm, how many TE and TM modes will exist in the waveguide?

Solution: First calculate the normalized frequency for the fiber

$$\begin{aligned} V &= (2\pi\ 25\mu\text{m}/1.5\mu\text{m})\sqrt{1.45^2 - 1.44^2} \\ &= 17.802 \end{aligned} \tag{4.50}$$

The zeros of the $J_0(\kappa a) = 0$ occur at 2.504, 5.520, 8.654, 11.791, 14.931, 18.071, etc.. (See Appendix B: Bessel Functions). Clearly, V is larger than the first five roots, but is smaller than the sixth root at 18.071. So five TE modes (and five TM modes) will be allowed in this waveguide at that wavelength.

The V-number is useful for determining cutoff conditions, as well as a number of other parameters, like the total number of allowed modes and power profiles. The V-number is often specified in the purchase of optical single mode fiber. For example, the cut-off condition for a single mode fiber occurs when the V-number reaches 2.405 (the first root of the J_0 Bessel function). The term "cut-off" refers to the point where the TE_{01}, TM_{01}, and HE_{21} modes cease to propagate if V becomes smaller. The wavelength at which a single-mode fiber suddenly becomes multimode is called the "cut-off" wavelength, λ_c.

8. The Fundamental HE_{11} Mode

A mode which deserves special attention is the HE_{11} mode, sometimes called the fundamental mode, or also the LP_{01} mode. It has no cutoff condition; every step–index fiber will support at least this mode. The transverse field of the HE_{11} mode is described by the J_0 Bessel function (see Prob. 9.) in the core region, but because Bessel functions are not convenient to mathematically manipulate,the mode field distribution is often approximated by a Gaussian shape,

$$E(r) = E_0 \exp\left[-\left(\frac{r}{w}\right)^2\right] \tag{4.51}$$

The parameter, w, is adjusted to give the "best fit" between the actual Bessel function and the Gaussian approximation. For a fiber with a core diameter of a, w is chosen to be [8]

$$\frac{w}{a} = 0.65 + 1.619V^{-\frac{3}{2}} + 2.87V^{-6} \tag{4.52}$$

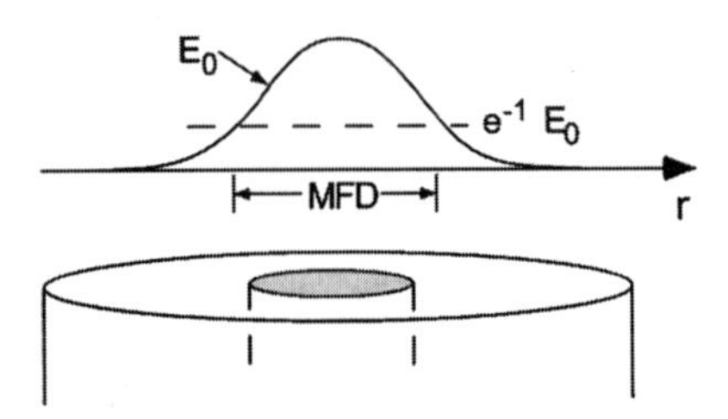

Figure 4.12. The electric field of the HE_{11} mode is transverse, and approximately Gaussian. The mode field diameter is determined by the points where the *power* is down by e^{-2}, or where the amplitude is down by e^{-1}. The MFD is not necessarily the same dimension as the core.

This approximation provides a good overlap (better than 96%) between the Bessel solution and Gaussian function over the range from $0.8\lambda_c$ to $2\lambda_c$, where λ_c is the cutoff wavelength.

The amplitude profile for the HE_{11} mode is shown in Fig. 4.12. The distance between the $1/e$ points of the amplitude profile define the *Mode Field Diameter*, MFD, which is twice the mode field radius, w. Notice that the MFD depends on the wavelength of the normalized frequency. When coupling between two single mode waveguides, matching the MFD is a critical parameter to minimize loss. When the mode is not well described by a Gaussian parameter, definition of the MFD becomes less clear. Several techniques have been proposed, and are still being considered for standards [9].

The cutoff wavelength defines the boundary between single mode and multimode operation of a fiber. Wavelengths shorter than the cutoff wavelength can excite more than one spatial mode. The cutoff wavelength is defined in terms of the cutoff parameter for the onset of the TE and TM modes, namely $V = 2.405$,

$$\lambda_c = \frac{2\pi a}{2.405}\sqrt{n_{core}^2 - n_{clad}^2} \tag{4.53}$$

The HE_{11} mode can be polarized in any arbitrary direction in the x-y plane, so it has a degeneracy of two.

9. Total Number of Modes in a Step–Index Waveguide

For large core diameter fibers with many modes, it is possible to provide an approximate formula describing the total number of modes that will propagate. Recall the characteristic equation for the LP modes (Eq. 4.36)

$$\kappa\frac{J_{j-1}(\kappa a)}{J_j(\kappa a)} = -\gamma\frac{K_{j-1}(\gamma a)}{K_j(\gamma a)}$$

For values of κa far from cutoff, the term γa will be large, and the asymptotic value of the K_ν functions can be used. Since $K_j(\gamma a) \rightarrow \sqrt{\pi/2\gamma a}e^{-\gamma a}$ for large

κa, the ratio $K_{j-1)}/K_j$ goes to unity for large arguments. The characteristic equation reduces to

$$\frac{J_{j-1}(\kappa a)}{J_j(\kappa a)} = -\frac{\gamma}{\kappa} \tag{4.54}$$

For a given value of ν, the number of allowed modes will be proportional to the number of roots of $J_j(\kappa a)$ between 0 and $\kappa a = V$. In the approximation that κa is large, the asymptotic expansion of $J_j(\kappa a)$ can be used

$$J_j(\kappa a) \approx \sqrt{\frac{2}{\pi \kappa a}} \cos(\kappa a - \frac{\nu\pi}{2} - \frac{\pi}{4}) \tag{4.55}$$

There will be one root every time the ratio goes to infinity, i.e. each time the argument increases by π. For a given value of ν, the number of roots will be approximately

$$m = (\kappa a - \frac{\nu\pi}{2} - \frac{\pi}{4})/\pi \tag{4.56}$$

Solving this in terms of the normalized frequency, $V = \kappa_{max} a$, and ignoring the $\pi/4$ term

$$V = \kappa_{max} a = (2m + \nu)\frac{\pi}{2} \tag{4.57}$$

This equation, while only an approximation, shows the general relationship between the azimuthal number, ν, and the number of radial nodes in the mode, m. As ν increases, indicating more angular lobes, the maximum value of m must decrease, implying that the radial structure becomes smoother.

Since there is an allowed mode for each value of m and ν, we can graphically plot the number of modes. The largest possible value for m, from Eq. 4.57, is V/π, when $\nu = 0$. Likewise, the maximum value for ν is $2V/\pi$. These allowed values are plotted in Fig. 4.13.

Each dot represents an allowed combination of m and ν. The total number of allowed modes is geometrically determined from the area of the triangle in Fig. 5.13, which will be $(1/2)m_{max}\nu_{max} = V^2/\pi^2$. We must recall that for each mode, there are two angular orientations (*cosine* or *sine* solution), and two possible polarizations (x or y in the LP mode approximation). The number of modes is increased by a factor of four. So the number of allowed modes in a fiber waveguide is given by the approximation

$$N = 4\frac{V^2}{\pi^2} \tag{4.58}$$

Again, we stress this formula is an approximation, and is only good when V is large.

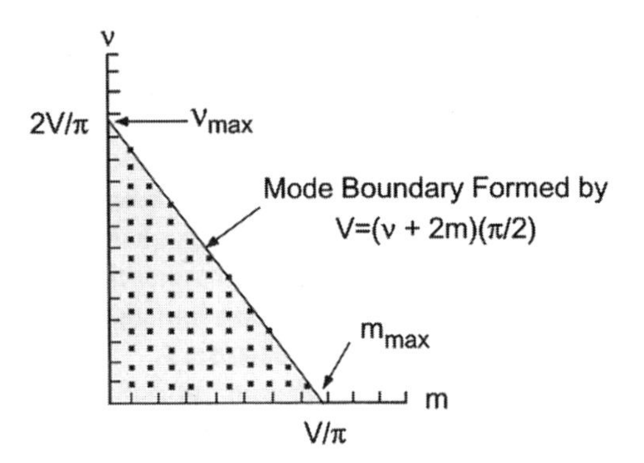

Figure 4.13. Graphical plot of the allowed values of ν and m for the step–index fiber. The boundary is determined by the condition listed in Eq. 4.58. The number of points beneath the curve is proportional to the area of the shaded region.

10. Summary

In this chapter, we developed the fundamental concepts of the circular dielectric waveguide. Solution of the wave equation in cylindrical coordinates led to mode solutions in the form of cylinder functions such as sinϕ and the Bessel functions, $J_\nu(\kappa r)$. As was found in planar waveguides, the propagation parameters, β, for the modes were found from solution of a transcendental equation, and the values of β were restricted to lie between $k_o n_{core} > \beta > k_0 n_{clad}$. We developed explicit solutions to the longitudinal electric fields of the modes, and using Maxwell's equations, we found expressions for all field components. The formal modes are complicated in terms of their field structure, so a picture based on the weakly guided mode approximation was developed which simplified both the characteristic equation for finding β, and the physical description of modes as linear superpositions which were linearly polarized.

We concluded the chapter with a number of short topics, such as the cutoff conditions, the V-parameter, the number of modes, and the power confinement of the modes. One topic that was not discussed was dispersion, which is a very important topic for any long–distance optical waveguide system. We defer a complete discussion on dispersion in waveguides until the Chapter 6, where graded index waveguides are described.

References

[1] S. Ramo, J. Whinnery, and T. Van Duzer, *Fields and Waves in Communication Electronics, 2nd ed.*, John Wiley and Sons, USA (1984)

[2] C. R. Wylie, *Advanced Engineering Mathematics, 3rd Ed.*,McGraw-Hill Book Co., New York (1966)

[3] Allen H. Cherin, *An Introduction to Optical Fibers*, McGraw-Hill Book Co., New York (1983)

[4] D. Marcuse, *Light Transmission Optics*, Ch. 8, Van Nostrand Reinhold, New York, (1982)

[5] Allan W. Snyder and John D. Love, *Optical Waveguide Theory*, Ch. 12-15, Chapman and Hall, New York, (1983)

[6] D. Gloge, "Weakly Guiding Fibers", *Applied Optics*, vol. 10, pp. 2252-2258 (1971)

[7] G. Keiser, *Optical Fiber Communications, 2nd ed.*, McGraw-Hill, Inc.,

[8] D. Marcuse, "Loss analysis of single-mode fiber splices," *Bell Sys. Tech. Jour.*, 56, pp. 703-718 (1977)

[9] T. J. Drapela, et.al., "A comparison of far–field methods for determining mode field diameter of single mode fibers using both Gaussian and Petermann definitions," *J. Lightwave Technology*, 7, pp. 1139-1152, (1989) USA, (1991)

Practice Problems

1. Confirm Eq. 4.28 (an exercise in Bessel functions).

2. Determine the cladding index of refraction of a fiber which has a core index of 1.5, a core radius of 5 μm, and $V = 2.0$ at $\lambda = 1.5 \mu m$. At what wavelength does the fiber cease to be a single mode fiber?

3. Consider a fiber with core index 1.5, cladding index 1.495, and a core radius or 5 μm.

   (a) How many modes exist for a wavelength of 1 μm?

   (b) How many modes exist for a wavelength of 0.5 μm? List each mode, and its cutoff condition.

4. Consider a step–index fiber with $n_{core} = 1.5$, $n_{clad} = 1.495$, and $a = 9 \mu$m.

   (a) What is the wavelength that corresponds to the single mode cutoff for this fiber?

   (b) At what wavelength does the HE_{33} mode cutoff?

   (c) List all the modes that propagate in the fiber at 1.5 μm.

5. For the fiber in problem 3, list all the LP modes that would exist at $\lambda = 0.8 \mu$m.

6. A step–index fiber has an $n_{core} = 1.47$ and $n_{clad} = 1.46$. The core radius is 3μm. For $\lambda = 1.3\mu$m, determine the following parameters: N.A., V-number, Mode Field Diameter of the HE_{11} mode, and the cutoff wavelength.

7. A step index fiber is operated at 1.3μm. The core radius is 25μm, $n_{core} = 1.465$ and $n_{clad} = 1.460$. Use the "weakly guided mode approximation" and plot the eigenvalue equation for the $\nu = 0$ and $\nu = 1$ cases. Find the value of β for the highest order TE mode.

8. Identify which modes exist and specify the cutoff parameter for each mode in a step–index fiber with $V = 5.5$.

9. If a single mode fiber has a longitudinal field given by

$$\begin{aligned} E_z(r,\phi,z) &= -j\frac{E}{k_0 n_{core}}\cos\phi\kappa\frac{J_1(\kappa r)}{J_0(\kappa a)} \text{ for } 0 < r < a \\ &= -j\frac{E}{k_0 n_{clad}}\cos\phi\gamma\frac{K_1(\gamma r)}{K_0(\gamma a)} \text{ for } a < r \end{aligned}$$

show that

$$\begin{aligned} E_x(r,\phi,z) &= E_0 J_0(\kappa r)/J_0(\kappa a) \text{ for } 0 < r < a \\ &= E_0 K_0(\gamma r)/K_0(\gamma a) \text{ for } a < r \end{aligned}$$

Make a sketch of the amplitude distribution inside and outside the core.

10. To learn about Bessel function identities, fill in the missing steps in the derivation of the characteristic equation for the LP modes.

11. Compare the overlap of the transverse mode for an HE_{11} mode to the gaussian approximation for that mode. Design a fiber that has a normalized frequency, $V = 2$. Explicitly describe the transverse field in all regions of space. Compare the Gaussian approximation to this by calculating the sigma squared deviation between the two normalized field patterns.

12. Make a plot for designing a single mode fiber. Use axes of core radius versus the difference in the index of refraction between the core and cladding for a single mode fiber. Plot a curve showing the boundary between single–mode and multi–mode operation of the waveguide. Do this for $\lambda = 1.5\mu$m, and 1.3μm. Assume the core index is $n = 1.5$.

13. A step–index fiber has a V-number of 10. For the LP_{11} mode, what fraction of the power is contained in the core, and what fraction is in the cladding?

14. A step–index fiber is made with a core index, $n_{core} = 1.45$ and a cladding index $n_{clad} = 1.44$. Using the Gaussian approximation, make a plot of

the core diameter versus the MFD for the HE_{11} mode over the range $V = 0.8 \rightarrow 2.2$. Assume $\lambda = 1.5\mu m$.

15. Confirm Eq. 4.32 using Eq. 4.30 and the identities listed in Eq. 4.31.

16. A step-index fiber is to be constructed using silicon for the cladding and germanium-silicon for the core. The index of the cladding is $n_{clad} = 3.5$ exactly, and the index of the core is $n_{core} = 3.503$.

    (a) What radius should the core have in order to insure that the waveguide will remain single mode until $\lambda = 1.2\mu m$?

    (b) What is the numerical aperture of this fiber?

17. Using a computer, determine the waveguide dispersion od a step-index single mode fiber. Design two single mode fibers with $V = 2$, one with a small core and large Δn, and the other with a small Δn and a large core. Choose a core index of $n = 1.50$. By calculating β at a number of wavelengths around 1.3 μm, determine the waveguide dispersion for each fiber. Which fiber has the least waveguide dispersion?

Chapter 5

RECTANGULAR DIELECTRIC WAVEGUIDES

1. Introduction

The rectangular dielectric waveguide is the most commonly used structure in integrated optics, especially in semiconductor diode lasers. Demands for new applications such as high-speed data backplanes in integrated electronics, waveguide filters, optical multiplexors, and optical switches are driving technology toward better materials and processing techniques for planar waveguide structures. The infinite slab and circular waveguides that we have already studied are not practical for use on a substrate: the slab waveguide has no lateral confinement, and the circular fiber is not compatible with the planar processing technology being used to make planar structures. The rectangular waveguide is the natural structure.

In this chapter we will study several methods for analyzing the mode structure of rectangular structures, beginning with a wave analysis based on the pioneering work of Marcatili[1]. The wave analysis provides a good description of the modes far from cut-off, but becomes less accurate for small V-number waveguides. We will then look at look at a popular analysis method called the *Effective Index Method.* One of the simplest structures to build is the ridge waveguide. An example using the effective index method will be used to illustrate these useful structures. Finally, we will review perturbative solutions to improve the results of wave analysis.

Applications for rectangular waveguides typically involve short lengths (distances of no more than a few centimeters). Unlike the optical fiber, which is primarily a way to convey optical signals over a long distance, integrated waveguides are used to make devices such as power splitters, wavelength-selective filters and drops, modulators, switches, and other devices that are useful for controlling information on an optical network. The physical mechanisms used

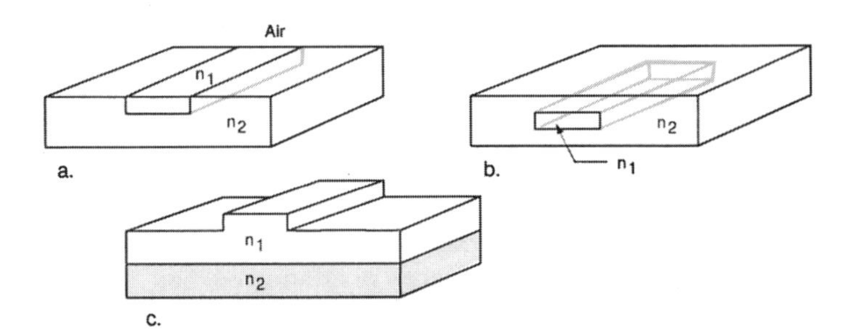

Figure 5.1. Three possible configurations for rectangular waveguides. In a) the cover index is air. In b) the guiding layer is completely surrounded by a cladding layer. In c) lateral confinement is established by the dielectric ridge on top of the substrate.

to control this information varies with the device. Power splitters and interferometers might use Y-junctions, while filters might rely on evanescent coupling between adjacent waveguides. Design and analysis of these problems requires knowing the exact mode structure of the field in order that coupling can be accurately predicted. We will see that unlike the planar slab waveguide or the circularly symmetric fiber waveguide, it is generally impossible to find exact analytical solutions to these structures. Most work is now done using numerical simulations, which are described in a following chapter.

2. Wave Equation Analysis of a Rectangular Waveguide

Fig. 5.1 shows three types of rectangular waveguide that can be employed in an integrated optical circuit. They illustrate the surface waveguide, the buried waveguide, and the ridge waveguide. These geometries are relatively simple to create using standard lithographic and overgrowth techniques. As usual, the index of the rectangular region must be slightly larger than the surrounding medium for the structure to guide light. Our goal is to determine the mode structure of these waveguides. To begin the analysis we will develop a wave equation expression that is accurate well above cutoff.

A cross-section of a generalized embedded waveguide is shown in Fig. 5.2. There are nine distinct dielectric regions in this structure. Analysis is difficult because it is impossible to simultaneously satisfy all the boundary conditions in this structure.

The difficulty in analyzing this structure originates in the four shaded regions. These regions act as the coupling zones for the x and y solutions of the field. Well above cut-off, the electromagnetic mode is tightly confined within the core, and the amount of energy in the corner regions is negligible, so the wave equation can be solved using standard separation of variables. Near cut-off, however, the mode will have a significant amount of power in the corner

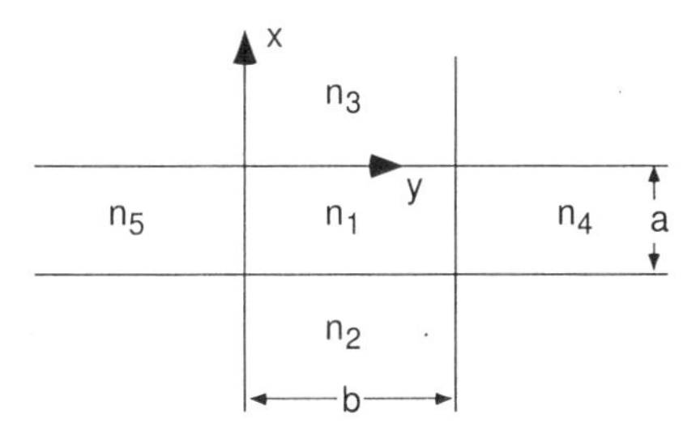

Figure 5.2. A general dielectric structure. The core (n_1) is surrounded on every side by a lower index material. The index in the shaded regions is neglected in the first-order approximation to solutions to the waveguide. The regions are numbered for identification only; there is no implicit relationship between the various indices, except that n_1 is larger than all others.

regions. The x and y dependent solutions will be strongly coupled through the boundary conditions in the corner regions, making them mathematically inseparable. Therefore, neglecting the field in the corner regions will be the substance of our approximation. As we shall show in a subsequent section, perturbation techniques can be used to "clean-up" the solutions near cut-off.

2.1 Mode Designation

The mode designation commonly used for rectangular structures is slightly different than the terminology we have used for circular or slab waveguides. Since in rectangular dielectric waveguides the field is neither purely TE or TM (there is generally a skew in the mode), a different designation is called for. In the limit of small index differences, the guided optical fields are essentially transverse, and the transverse component of the electric field will be aligned either with the x or y axis of the structure. Modes are designated E^y_{nm} if in the limit of total confinement the electric field is parallel to the y-axis, and as E^x_{nm} modes if the electric field is parallel to the x axis. As in the microwave notation, the nm subscripts designate the number of maxima in the x and y directions.

2.2 Formulation of the Boundary Value Problem

To determine the mode field configuration, we must find the eigenvalue β for each mode. If we are far from cut-off, or if the index difference between the guiding region and cladding is small, the fields are effectively transverse. This condition is similar to the LP modes described in Chapter 4. Since the boundaries of the waveguide are rectangular there will be no conversion of E_x into E_y or E_z upon incidence at an interface. Therefor vector wave equation can therefore be converted into a scalar equation for each component. We assume that the longitudinal field dependence follows the form $Z(z) = exp(-j\beta z) +$

c.c., thus the scalar wave equation becomes

$$\frac{\partial^2 E}{\partial x^2} + \frac{\partial^2 E}{\partial y^2} + \left[k_0^2 n^2(x,y) - \beta^2\right] E = 0 \tag{5.1}$$

where k_0 is the free space wavevector, β is the unknown propagation constant, and $n(x, y)$ is the index of refraction for the structure. Notice that this is similar to the wave equation we solved in Chapter 3 for the slab waveguide, except now there is a possibility of change in the y-direction.

We use separation of variables to find a solution $E(x, y) = X(x)Y(y)$, where $X(x)$ is the amplitude distribution function along x, and likewise $Y(y)$ is the amplitude distribution in the y direction. Ignoring the four corner regions, there will be five solutions, one for each region of the waveguide. The guided mode solutions in the core (region 1) should vary sinusoidally along the x and y direction. Boundary conditions require that the transverse component of the fields must be continuous across each interface. In regions 2 and 3, the y-dependence must therefore have the same sinusoidal structure as in the core, but the x dependence should decrease exponentially away from the core. Similarly, in regions 4 and 5, the field will display the same sinusoidal dependence in the x direction as the field in the core, but should exponentially decay with y. The general form of the solution is

$$E(x,y) = E_0 e^{-j(k_{xi}x + k_{yi}y)} e^{-j\beta z} + c.c. \tag{5.2}$$

where the propagation coefficients, k_{xi}, and k_{yi}, can be real or imaginary, depending for which region, i, the solution is valid. The x-propagation constants k_{x1}, k_{x4}, and k_{x5} in regions 1, 4, and 5, must be identical and independent of y. Using separation of variables, letting $E(x, y) = X(x)Y(y)$, it is easy to show the wave equation can be written as

$$\frac{\ddot{X}}{X} + \frac{\ddot{Y}}{Y} + k_0^2 n^2(x,y) - \beta^2 = 0 \tag{5.3}$$

Rearranging, the equation becomes

$$\frac{\ddot{X}}{X} = -k_0^2 n^2(x,y) + \beta^2 - \frac{\ddot{Y}}{Y} = -\kappa_x^2 \tag{5.4}$$

where κ_x^2 is a separation constant. Using this result, we can solve for the Y function

$$\frac{\ddot{Y}}{Y} = -k_0^2 n^2(x,y) + \beta^2 + \kappa_x^2 = -\kappa_y^2 \tag{5.5}$$

We will assume a step-index structure. In the core, where $n(x, y) = n_1$, the guided mode solutions must be oscillatory, implying that κ_x and κ_y must be

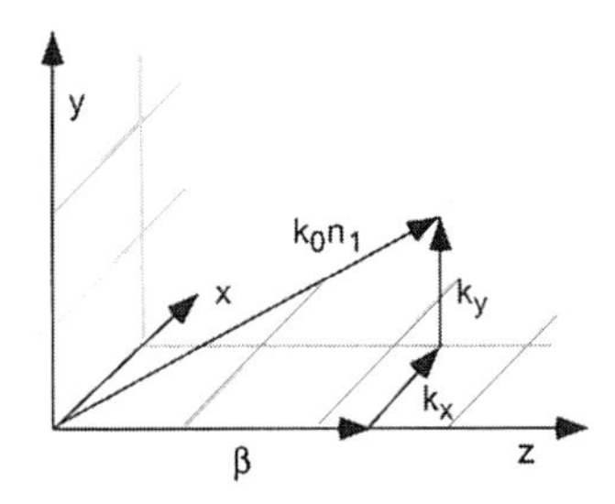

Figure 5.3. Geometric interpretation of the three propagation vectors in a rectangular waveguide.

real. The allowed core solutions will have the form

$$\begin{aligned} X(x) &= A\cos(\kappa_x x + \phi_x) \\ Y(y) &= B\cos(\kappa_y y + \phi_y) \end{aligned} \tag{5.6}$$

where ϕ_x and ϕ_y are phase constants that are adjusted to match boundary conditions. The separation constants, κ_x and κ_y, must satisfy (from Eq. 5.5)

$$\beta^2 = k_0^2 n_1^2 - \kappa_x^2 - \kappa_y^2 \tag{5.7}$$

Notice the similarity between this equation and that for the slab waveguide (Eq. 3.7). In the case of the 2-dimensional rectangular structure, there are two transverse wavevectors. Fig. 5.3 illustrates the geometric view of the relation between the three orthogonal wavevectors.

Outside the core, the guided mode solutions must have at least one component which displays exponential decay. Consider the solution in region 3 of Fig. 5.2 ($x > 0$). To match boundary conditions, the $Y(y)$ in this region must be identical to the y-solution found in the core. So the $\partial^2\Psi/\partial y^2$ term simply becomes $-\kappa^2$, and the equation in region 3 reduces to

$$\frac{\partial^2\Psi}{\partial x^2} - \kappa_y^2\Psi + (k_0^2 n_3^2 - \beta^2)\Psi = 0) \tag{5.8}$$

Substituting Eq. 5.7 into Eq. 5.8, and using the notation of separation of variables, yields

$$\begin{aligned} \frac{\ddot{X}_3}{X_3} &= -(k_0^2 n_3^2 - \beta^2 - \kappa_y^2) \\ &= (k_0^2 n_1^2 - k_0^2 n_3^2 - \kappa_x^2) \end{aligned} \tag{5.9}$$

which has the solution

$$X_3(x) = e^{-\gamma_3 x} \quad \text{for} \quad x > 0 \tag{5.10}$$

exp(-γ3 x) exp(γ5 y)	Cos(κy y + Φy) exp(-γ3 x) 3	exp(-γ3 x) exp(-γ4 (y-b))
Cos(κx x + Φx) exp(γ5 y) 5	Cos(κx x +Φx) Cos(κy y + Φy) 1	Cos(κx x + Φx) exp(-γ4 (y-b)) 4
exp(γ2 (x-a)) exp(γ5 y)	Cos(κy y + Φy) exp(γ2 (x-a)) 2	exp(-γ4 (y-b)) exp(γ2(x-a))

Figure 5.4. The rectangular waveguide can be described as nine separate regions, each with its own electromagnetic field description. (For simplicity, amplitudes are not matched across boundaries in these expressions.)

where

$$\gamma_3 = \sqrt{k_0^2(n_1^2 - n_3^2) - \kappa_x^2} \tag{5.11}$$

So the total field in region 3 can be described as

$$E(x, y) = C\cos(\kappa_y y + \phi_y)e^{-\gamma_3 x} \quad \text{for } x > 0,\ 0 < y < b \tag{5.12}$$

Through a similar set of solutions we can find the scalar solutions to the x- and y-components of the field in regions 2, 4, and 5

$$\begin{aligned} X_2(x) &= e^{\gamma_2(x+a)} \quad \text{for } x < b \\ Y_4(y) &= e^{-\gamma_4(y-b)} \quad \text{for } y > a \\ Y_5(y) &= e^{\gamma_5 y} \quad \text{for } y < 0 \end{aligned} \tag{5.13}$$

where the exponential decay constants are given by

$$\begin{aligned} \gamma_2 &= \sqrt{k_0^2(n_1^2 - n_2^2) - \kappa_x^2} \\ \gamma_4 &= \sqrt{k_0^2(n_1^2 - n_4^2) - \kappa_y^2} \\ \gamma_5 &= \sqrt{k_0^2(n_1^2 - n_5^2) - \kappa_y^2)} \end{aligned} \tag{5.14}$$

These fields are summarized in Fig. 5.4, where the appropriate product of $X(x)$ and $Y(y)$ solutions are listed in each region, and the transverse solutions for several modes are plotted in Fig. 8.5.

2.3 Solution to the Boundary Value Problems

To complete the solution, we must determine the specific values for $\kappa_{x'}$, κ_y and β. This is done by applying the boundary conditions that connect the

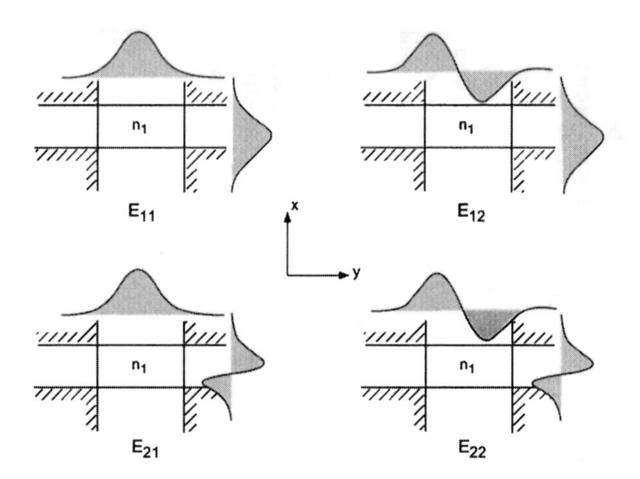

Figure 5.5. The transverse scalar field distributions for the x and y directions.

solutions between the various regions. Since there are many fields, and many interfaces, matching all the boundary conditions is a complicated and tedious process (see for example [2]). Consider the situation where an E^x mode is being guided. Fig. 5.6 shows a cross-section of the waveguide, along with the two transverse components of the field in the core.

In this waveguide, we must insure the continuity of the tangential electric field at the $y = 0$ and $y = b$ planes, and continuity of the tangential magnetic field at the $x = 0$ and $x = -a$ planes. The other boundary conditions (continuity of D_{norm} and B_{norm}) are almost automatically satisfied in the weakly guiding approximation, at least they are close enough to be insignificant. From the boundary condition point-of-view, the E^x field looks like a TE mode in a slab waveguide of thickness b, and a TM mode in a slab waveguide of thickness a. After lengthy calculation [1], the characteristic equation for κ_y can be shown

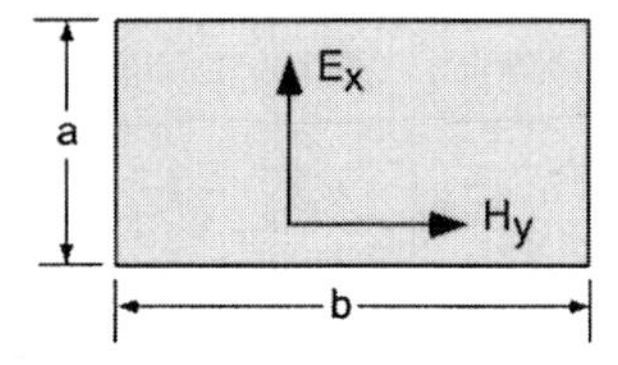

Figure 5.6. The E^x field in the core will have the electric field polarized along the x direction, and the magnetic field polarized along the y direction.

to be

$$\tan \kappa_y b = \frac{\kappa_y(\gamma_4 + \gamma_5)}{\kappa_y^2 - \gamma_4\gamma_5} \tag{5.15}$$

This is identical to the characteristic equation for the TE mode in a slab waveguide, which seems reasonable. Similarly, matching the tangential components of the H_y field only involves the interfaces at $x = 0$ and $x = -a$ between the core region and regions 2 and 3 of Fig. 5.2. After much algebra, the characteristic equation for κ_x can be shown to be [1]

$$\tan \kappa_x a = n_1^2 \frac{\kappa_x(n_2^2\gamma_3 + n_3^2\gamma_2)}{n_2^2 n_3^2 \kappa_x^2 - n_1^2\gamma_2\gamma_3} \tag{5.16}$$

which is the characteristic equation for a TM mode in a slab of thickness b. These results should seem intuitively plausible, even if the mathematical derivation is not presented in all its detail. Using these eigenvalues, we can finally determine the propagation constant β from Eq. 5.9.

The complete description of the modal fields can now be written. There are five regions, so five separate electric fields and five separate magnetic fields must be specified to completely describe the field. Reference [2]has a complete listing of these fields.

The phase terms in Eq. 5.6, are found from [2]

$$\begin{aligned} \tan \phi_x &= -\frac{n_3^2}{n_1^2}\frac{\kappa_x}{\gamma_3} \\ \tan \phi_y &= \frac{-\gamma_5}{\kappa_y} \end{aligned} \tag{5.17}$$

We can see that the rectangular waveguide mode, to first order, is simply the product of two orthogonal spatial modes, one which acts like a TE wave, and the other a TM wave. The x dependence of the mode is found by effectively solving a slab waveguide problem as if there were no structure to the waveguide in the y-direction. Similarly, the y dependence of the mode is found by treating the waveguide as a slab with infinite extent in the x-direction. The two solutions are coupled through the selection of the propagation coefficient, β, where both transverse propagation coefficients, κ_x and κ_y are subtracted quadratically from $k_0 n_1$.

The critical cut-off condition will be determined by the smaller of the two dimensions (a or b) of the waveguide. The normalized frequency for a rectangular waveguide is defined as

$$V = k_0 \frac{a}{2}\sqrt{n_1^2 - n_2^2} \tag{5.18}$$

where a is the smaller dimension, n_1 is the core index, and n_2 is the next smaller index.

The E^y_{nm} modes can be obtained in close analogy with the E^x_{nm} modes. In this case, the E_y and H_x components are dominant. Following Marcuse [2], the fields in the five regions can be derived from the longitudinal fields, E_z and H_z using Maxwell's equations.

The characteristic equations for these modes are

$$\begin{aligned} \tan \kappa_x a &= \frac{\kappa_x(\gamma_2 + \gamma_3)}{(\kappa_x^2 - \gamma_2\gamma_3)} \\ \tan \kappa_y b &= \frac{n_1^2 \kappa_y (n_5^2 \gamma_4 + n_4^2 \gamma_5)}{n_4^2 n_5^2 \kappa_y^2 - n_1^2 \gamma_4 \gamma_5} \end{aligned} \tag{5.19}$$

and the phase terms can be found from

$$\tan \eta_x = \frac{\gamma_3}{\kappa_x} \tan \eta_y = \frac{n_5^2}{n_1^2} \frac{\kappa_y}{\gamma_5}$$

Example 5.1: Analysis of a symmetric embedded waveguide

Consider the embedded waveguide shown in Fig. 5.7. The core has an index n_1, and it is surrounded by index n_2. Assume that $n_1 = 1.5, n_2 = 1.499$, $a = 5\mu$m, and $b = 10\mu$m, and the electric field is oriented in the y-direction. We will first determine the normalized propagation coefficient for the waveguide over the 0.5 to 2.0 μm region.

Developing an expression for the propagation coefficient, β, requires solving the wave equation. Because this waveguide forms a symmetric structure, we will place the origin in the center of the waveguide. We want to define the index of refraction in such a manner that it satisfies the requirement of being separable in the x and y coordinates. We define $n^2(x, y)$ as follows:

$$n^2(x, y) = n'^2(x) + n''^2(y) \tag{5.20}$$

where

$$n'^2(x) = n_1^2/2, \quad \text{for} |x| < a/2$$

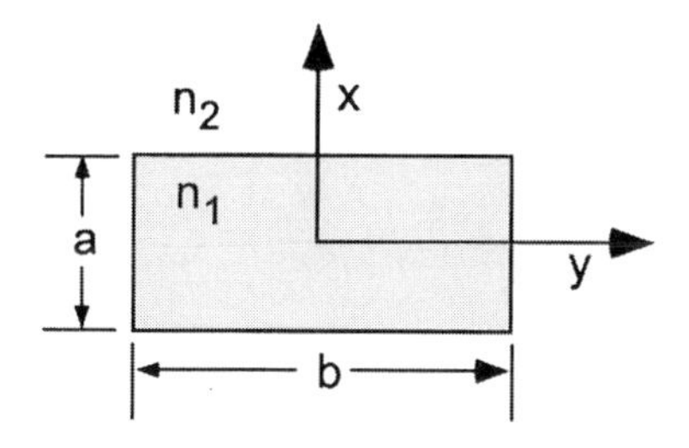

Figure 5.7. A symmetric waveguide is comprised of a rectangular dielectric of index n_1 surrounded by an index n_2. The origin of the system is situated in the center of the guiding core.

$$\begin{aligned} &= n_2^2 - n_1^2/2 \quad \text{for } |x| \geq a/2 \\ n''^2(y) &= n_1^2/2, \qquad \text{for } |y| < b/2 \\ &= n_2^2 - n_1^2/2 \quad \text{for } |y| \geq b/2 \end{aligned} \tag{5.21}$$

The definition above accurately describes the index of refraction in the five regions that we will seek a solution. The index description is not correct in the corner regions, but we are not going to solve the wave equation there, so the error is of no significance to our ability to find a solution.

The index distribution (Eq. 5.21) is symmetric, so the separable solutions to the wave equation will have the form

$$\begin{aligned} X(x) &= A\frac{\cos \kappa_x x}{\cos \kappa_x a/2} && \text{for} \quad |x| < a/2 \\ &= Ae^{-\gamma_x(|x|-a/2)} && \text{for} \quad |x| \geq a/2 \\ Y(y) &= B\frac{\cos \kappa_y y}{\cos \kappa_y b/2} && \text{for} \quad |y| < b/2 \\ &= Be^{-\gamma_y(|y|-b/2)} && \text{for} \quad |y| \geq b/2) \end{aligned}$$

for the symmetric modes. Similar expressions can be found for the antisymmetric modes (replace cosine by sine and adjust the signs). The transverse wave vectors, κ_x and κ_y, are found from the transcendental equations derived for symmetric planar waveguides in Chap. 3 (Eqs. 3.28, 3.29). If we assume an E^y mode, then the characteristic equations for the transverse wavevectors are

$$\begin{aligned} \tan \kappa_x a/2 &= \tfrac{\gamma_x}{\kappa_x} = \tfrac{\sqrt{k_0^2(n_1^2-n_2^2)-\kappa_x^2}}{\kappa_x} \\ \tan \kappa_y b/2 &= \tfrac{n_1^2}{n_2^2}\tfrac{\gamma_y}{\kappa_y} = \tfrac{n_1^2}{n_2^2}\tfrac{\sqrt{k_0^2(n_1^2-n_2^2)-\kappa_y^2}}{\kappa_y} \end{aligned}$$

From these values, we can can find the allowed value of β using Eq. 5.7

$$\beta^2 = k_0^2 n_1^2 - \kappa_x^2 - \kappa_y^2$$

Using a numeric program in *Mathematica*, we calculated the allowed transverse wavevectors and values of β over the wavelength range 0.5-2 μm. The normalized value of the propagation vector, b, defined as

$$b = \frac{\beta^2 - k_0^2 n_2^2}{k_0^2 n_1^2 - k_0^2 n_2^2} \tag{5.22}$$

was then plotted in Fig. 5.8. Please note, the normalized propagation vector, b, should not be confused with the spatial dimension, b, of the waveguide. We will rely on context to avoid any confusion between this shared symbol.

Blind application of the formula for β led to values for the normalized propagation coefficient, b, which in some regions are negative. This is clearly

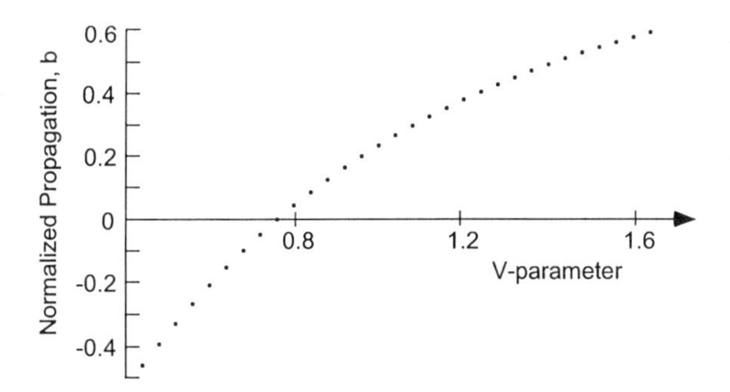

Figure 5.8. The normalized propagation coefficient for the waveguide. Note that b is less than zero for certain ranges of the normalized frequency. This is unphysical, and these modes are therefore below cut-off.

not physical; the only interpretation of these points is that they represent modes that are beyond cut-off. Unfortunately, in a symmetric waveguide we expect that there will always be at least one allowed mode, in contrast to what the numeric results are telling us. It turns out that the calculated solution is increasingly inaccurate in the region of normalized frequency $V < 2$. This is an example of where we must be careful about neglecting the corner dielectric regions. The wave equation solution is valid, however, for values of $V > 2$, as we shall show in the next section.

2.4 Solutions Near Cut-Off

The analysis presented above is an approximation based on neglect of certain regions of the waveguide. So long as the mode is well above cut-off, the solutions and expressions for the eigenvalues will be nearly indistinguishable from the exact value. For modes where $V < 2$, we can expect the exact solution to deviate from the calculated value because there will be non-negligible fields in the corner regions. To reduce this error, one must resort to numerical techniques, or to perturbation methods.

The results of Example 5.1 are disturbing: we expect that for a symmetric waveguide, there will always be at least one guided mode. In fact, this is true. So the calculations are in error. The next section deals with a first order correction to this problem.

3. Perturbation Approach to Correcting β

The major problem with the analytical approach is that it relies on the mode being tightly coupled to the core, so that relatively little field exists in the four corner regions. The source of error that arose in the last example came from neglecting the field in the corner regions surrounding the rectangular waveguide. This results in an underestimate of the propagation coefficient.

One method to find an improved solution is to apply a perturbative correction to the solutions obtained above. The true waveguide mode, Ψ, is assumed to be a solution to the scalar wave equation

$$\nabla^2\Psi + [k_0^2 n^2(x,y) - \beta^2]\Psi = 0 \tag{5.23}$$

where $n(x,y)$ is the actual index distribution for the structure. In general, $n(x,y)$ is too complicated to allow an analytical solution to the wave equation. So we seek an index distribution, $n_0(x,y)$, that is close to the actual distribution and which can be analytically solved. The wave equation for this modified index distribution,

$$\nabla^2\Psi_n + (k_0^2 n_0^2(x,y) - \beta_n^2)\Psi_n = 0 \tag{5.24}$$

has solutions Ψ_n, where n is the mode index. In general there will be an infinite number of such solutions, most of which will be unguided (radiation) modes, and a finite number will be guided modes. Based on the completeness of the modes, any reasonable distribution of electromagnetic energy can be described by an appropriate superposition of these modes. The secret to making perturbation theory work well is to select the trial index, $n_0(x,y)$, to be close enough to the actual index so that the trial and actual solutions will not be significantly different. Let's assume that a suitable index has been identified, and that one of the trial solutions, Ψ_m, closely resembles the true solution, Ψ.

When the true mode and the trial mode are very close, we can approximate the actual mode in terms of a superposition of the orthonormal calculated modes of the waveguide,

$$\Psi = \Psi_m + \sum_n a_n \Psi_n \tag{5.25}$$

where a_n is the amplitude of each of the other modes. We have separated the closest calculated mode from the superposition, so $n \neq m$ in the summation. In this way the summation represents the total perturbation on the solution. Of course, all we have done is traded an infinite series for an insolvable problem, which might strike you as poor progress. However, we can dramatically simplify this series by taking advantage of mode orthogonality. We do this in the next few steps.

If we multiply Eq. 5.24 by Ψ, and multiply Eq. 5.23 by Ψ_n, we can subtract the two equations, and then integrate over a surface transverse to the axis of the waveguide to get

$$\begin{aligned}(\beta^2 - \beta_n^2)\int_S \Psi\Psi_n dS &= k_0^2 \int_S (n^2(x,y) - n_0^2)\Psi\Psi_n dS \\ &\quad - \int_S (\Psi\nabla^2\Psi_n - \Psi_n\nabla^2\Psi) dS \end{aligned} \tag{5.26}$$

The last term vanishes due to Green's second identity, leaving

$$\beta^2 - \beta_n^2 = k_0^2 \frac{\int_S (n^2(x,y) - n_0^2)\Psi\Psi_n dS}{\int_S \Psi\Psi_n dS} \tag{5.27}$$

This gives us an expression for determining the difference between the actual value of β, and β_n for each of the modes in the simplified waveguide.

In order to get a complete description of the perturbed mode, we must evaluate the summation in Eq. 5.25. Multiplying both sides of Eq. 5.25 by one of the modes of the simplified index distribution, $\Psi_{n'}$, integrating over surface S, and invoking the orthogonality of the Ψ_n modes, we get

$$\begin{aligned} \int_S \Psi\Psi_{n'} dS &= \int_S \Psi_m \Psi_{n'} + \sum_{n \neq m} a_n \int_S \Psi_n \Psi_{n'} \\ &= \delta_{mn'} + a_n \delta_{nn'} \end{aligned} \tag{5.28}$$

where $\delta_{mn'}$ represents the Kronecker delta function, which is unity only if $m = n'$, and is equal to zero otherwise. Substituting from Eq. 5.27 the expression for $\int_S \Psi\Psi_n$, we get an expression for a_n

$$a_n = \frac{k_0^2}{(\beta^2 - \beta_n^2)} \int_S (n^2(x,y) - n_0^2)\Psi\Psi_n dS \quad m \neq n \tag{5.29}$$

Unfortunately, in order to calculate the terms a_n needed to create a superposition that resembles the true mode, Ψ, we need to know Ψ. But we do not know Ψ, and it looks like we are going in circles. Fortunately, in calculating coefficients, it is generally sufficient to retain the zeroth order solution. We can assume that the true mode Ψ closely resembles one of the modes from the modified index distribution, namely Ψ_m, and that the value of $\beta_m \approx \beta$. Replacing Ψ with Ψ_m and β with β_m, we get

$$a_n = \frac{k_0^2}{(\beta_m^2 - \beta_n^2)} \int_S (n^2(x,y) - n_0^2)\Psi_m \Psi_n dS \quad m \neq n \tag{5.30}$$

The correction to the propagation coefficient with the same approximations is

$$\delta\beta^2 = \beta^2 - \beta_m^2 = k_0^2 \frac{\int_S (n^2(x,y) - n_0^2)\Psi_m^2 dS}{\int_S \Psi_m^2 dS} \tag{5.31}$$

These are general equations that can be applied to any waveguide. The correction to β is proportional to the index difference between the actual and modified profiles, weighted by the intensity of the field in the region of the index difference. In the case of the rectangular structure, we would need to integrate only over the four corner regions where the index differs from that assumed in the solution. It should also be noted that to get a true description of the modal field distribution, the radiation modes should be included in the superposition. However, since to second order the propagation coefficient only depends on the overlap integral of the index difference with the square of the approximate mode, including the radiation modes will not significantly influence the calculation for β.

To illustrate the power of the perturbative method, let's continue Example 5.1, and see how inclusion of the field in the corner regions leads to improved physical solutions.

Example 5.2 Perturbative correction to a symmetric waveguide

In Example 5.1 we found that a solution which inaccurately predicted cut-off for the fundamental mode. We know this is physically impossible, so the solution must be in error near cut-off. In this example we apply perturbation theory. The correction to β, and thus the normalized propagation constant, is given by Eq. 5.31. In the example, the actual index in the corner regions was $n^2(x, y) = n_2^2$, while in the solution, the corner regions were given a value $n_0^2(x, y) = 2n_2^2 - n_1^2$. The difference in index between the actual and trial index is given by

$$n^2(x, y) - n_0^2 = n_2^2 - (2n_2^2 - n_1^2) = n_1^2 - n_2^2 \tag{5.32}$$

The correction to the propagation coefficient will thus be

$$\delta\beta^2 = k_0^2 \frac{4 \int_{a/2}^{\infty} \int_{b/2}^{\infty} (n_1^2 - n_2^2) \Psi_m^2 dx\, dy}{\int_{-\infty}^{\infty} \int_{-\infty}^{\infty} \Psi_m^2 dx\, dy} \tag{5.33}$$

The factor of four comes from symmetry and the fact that we must evaluate the integral over all four corners. For the trial solution, Ψ_m, we use the solution found in Example 5.1. Using the mode described in Eq. 5.21, the integral can be evaluated. This integral often will require numeric solution. Fortunately for the symmetric waveguide used in this example, a closed form solution can be found which dramatically increases calculation speed. After tedious algebra [3], the correction to the normalized propagation constant, b, can be shown to be

$$\begin{aligned} \delta b &= \left[1 + \left(\frac{k_0^2(n_1^2 - n_2^2)}{\kappa_x^2} - 1\right)^{1/2} \left(\frac{\kappa_x a \pm \sin \kappa_x a}{1 \pm \cos \kappa_x a}\right)\right]^{-1} \\ &* \left[1 + \left(\frac{k_0^2(n_1^2 - n_2^2)}{\kappa_y^2} - 1\right)^{1/2} \left(\frac{\kappa_y b \pm \sin \kappa_y b}{1 + \pm \cos \kappa_y b}\right)\right]^{-1} \end{aligned} \tag{5.34}$$

where the top sign (+) corresponds to even modes, and the bottom sign (−) corresponds to antisymmetric modes. This function can be numerically calculated and added to the original data that formed b. The results are shown in Fig.5.9.

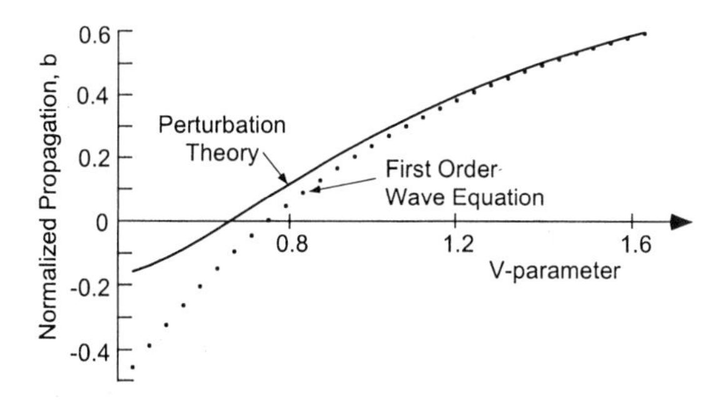

Figure 5.9. The normalized propagation constant after being corrected with the perturbation correction. The solid line represents the perturbation corrected solution, and the dotted line represents the data from Example 5.1. Notice now that the cut-off is closer to what it should be.

The two curves are superimposed on each other to illustrate the difference that the perturbation adds to the solution. While the perturbation correction does not bring the cut-off point to $V = 0$, as we know it should be, it does improve the overall curve. Comparison of this result with more accurate numerical modeling shows the perturbative correction is good to about $V = 0.7$ for the fundamental mode, and is an excellent fix for most higher order mode calculations [3]. Nevertheless, the small magnitude of the correction makes it clear that perturbation analysis is only useful for cleaning up an already reasonable calculation.

4. Effective Index Method

As we have seen in the last two sections, exact analysis of the mode structure of a dielectric waveguide can involve rather extensive calculations. In this section, a method known as the *Effective Index Method* is developed [4]. It is very similar to the first technique that we developed using the solutions to two orthogonally oriented waveguides to find the allowed value of the propagation coefficient, β, except here the direct interaction of the two waveguides is accounted for.

The effective index method converts a single two dimensional problem into two one-dimensional problems. Consider the buried rectangular waveguide shown in Fig. 5.10. To use the effective index method, we first stretch the waveguide out along its thin axis, in this case along the y-axis, forming a planar slab waveguide.

The thin one-dimensional slab waveguide can be analyzed in terms of TE or TM modes to find the allowed value of β for the wavelength and mode of interest. Once β is found, the effective index of the slab is determined through

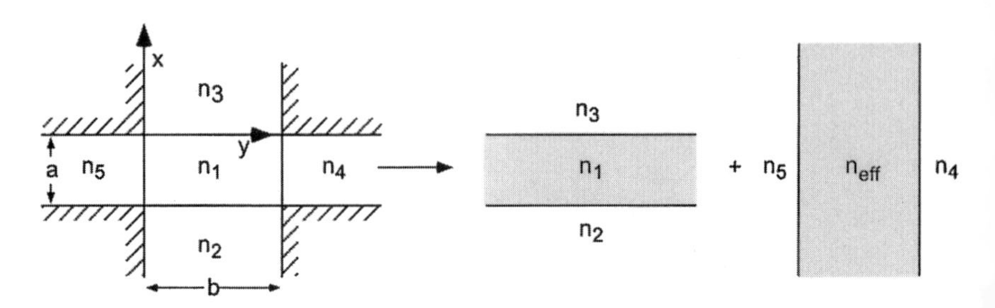

Figure 5.10. A buried dielectric waveguide can be decomposed into two spatially orthogonal waveguides: a horizontal and a vertical slab waveguide. The thin waveguide is analyzed in terms of the actual indices that form the structure. The thick waveguide is analyzed using the effective index found from the first waveguide analysis.

the expression

$$n_{eff} = \frac{\beta}{k_0} \tag{5.35}$$

where k_0 is the vacuum wavevector of the light being guided. After this effective index is determined, we return to the original structure, and stretch it along the thick axis (in this case vertical), forming a slab waveguide in the x-direction. The modes for this waveguide can now be found, only instead of using the original value of the index for the guiding film, the *effective index* found in the first step must be used. The value of β found from this last step is the actual value for the mode. A note of caution: to be accurate, the aspect ratio of the width/height aspect ratio of the waveguide must exceed a factor of three to be accurate. Thus the effective index method is not applicable to square waveguides.

As with the wave analysis, we must be careful to use the proper characteristic equations for each waveguide. For example, if in Fig. 5.10, the electric field is polarized in the x-direction, then for the thin waveguide, the field will appear to be a TM mode, and the appropriate characteristic equation must be used. When the thick slab is analyzed, the field will look like a TE mode, and so the TE characteristic equation should be used to find β. When the index difference between the guiding and cladding layers is large, using the proper characteristic equations is critical for getting a reasonable answer.

The effective index method is best illustrated with example. In Example 5.3, we present an description of a waveguide taken from published literature [5] which describes a ridge waveguide on a silicon substrate. Since silicon is the primary material for electronics, there is much interest in creating optical waveguides on silicon substrates.

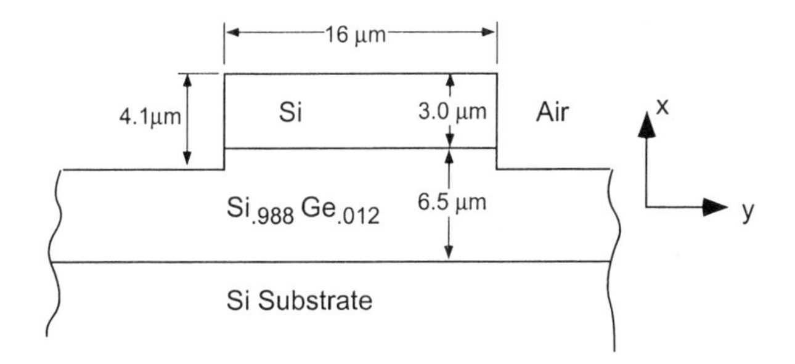

Figure 5.11. A ridge waveguide constructed from Si and SiGe alloy. The alloy is 6.5 μm thick under the Si ridge, and is 5.4 μm thick elsewhere.

Example 5.3: Silicon-Germanium Ridge Waveguide

The structure in Fig. 5.11 is to be used as a waveguide on an silicon substrate. This is an example of a ridge waveguide, where the mode is confined underneath the ridge in the Si-Ge layer. We can use the effective index method to find the eigenvalue β for the waveguide, and determine the mode size for the fundamental mode in this structure. The waveguide is to operate at $\lambda = 1.32\mu$m. Assume that the guided light is polarized in the x-direction.

The index of refraction of Si-Ge (assuming a 1.2% Ge concentration) is given by [6]

$$\begin{aligned} n_{\text{Si}} &= 3.5 \\ n_{\text{Si}_{1-x}\text{Ge}_x} &= n_{\text{Si}} + 0.104x(\text{x=0.012}) = 3.50125 \end{aligned}$$

The mode will be confined under the ridge due to the effective index created by the ridge. There are three different horizontal regions, each with a different effective index.

In order to apply the effective index method to the horizontal confining structure, we must find the effective index of all three regions. Since the field is polarized vertically (in the x-direction), the field will be a TM mode in the horizontal structures.

We will begin with the ridge. The Si cover layer can be assumed to be effectively infinite, because the evanescent waves will not penetrate very far into the layer. Therefore the structure can be considered to be symmetric waveguide, with thickness $h_1 = 6.5\mu$m. We use the characteristic equation for the TM mode in a symmetric waveguide,

$$\tan \kappa h_1/2 = \frac{n_1^2}{n_2^2}\frac{\gamma}{\kappa}$$

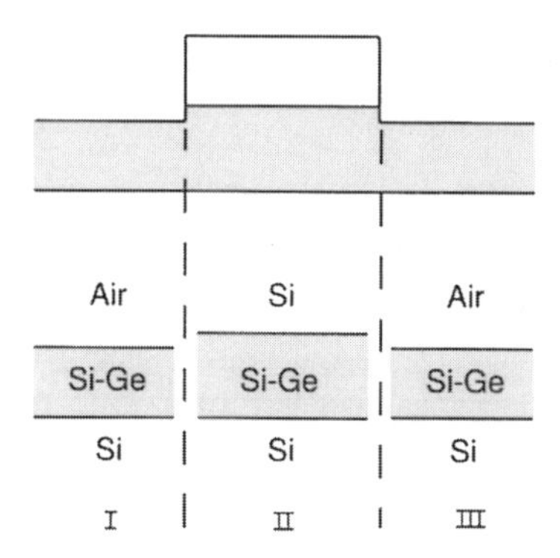

Figure 5.12. The ridge waveguide has three distinct vertical structures. Regions I and III are identical, and consist of a Si-Ge layer on a Si substrate and capped by air. Region II has a Si-Ge layer surrounded on both sides by Si.

where $n_1 = 3.50125$, $n_2 = 3.5$, and $\gamma = \sqrt{k_0^2(n_1^2 - n_2^2) - \kappa^2}$. Using numerical techniques we find

$$\begin{aligned}\kappa &= 2769.22 \text{ cm}^{-1}\\ \beta &= \sqrt{k_0^2 n_1^2 - \kappa^2} = 166636 \text{ cm}^{-1}\\ \gamma &= \sqrt{k_0^2(n_1^2 - n_2^2) - \kappa^2} = 3487 \text{ cm}^{-1}\end{aligned}$$

The effective index of the ridge section is

$$n_{eff_1} = \frac{\beta}{k_0} = 3.50077$$

Now we must repeat this process for the two side regions. These form asymmetric waveguides. The characteristic equation for the TM mode is (Eq. 3.23)

$$\tan(h_2\kappa_f) = \kappa_f \left[\frac{n_f^2}{n_s^2}\gamma_s + \frac{n_f^2}{n_c^2}\gamma_c\right]\left[\kappa_f^2 - \frac{n_f^4}{n_c^2 n_s^2}\gamma_c\gamma_s\right]^{-1}$$

Here, $h_2 = 5.4\mu\text{m}$, $\gamma_s = \sqrt{k_0^2(n_1^2 - n_s^2) - \kappa^2}$, and $\gamma_c = \sqrt{k_0^2(n_1^2 - 1) - \kappa^2}$, where $n_1 = 3.50125$, $n_c = 1$, and $n_s = 3.5$. Evaluating Eq. 3.23 numerically with these values yields

$$\begin{aligned}\kappa &= 3866.75 \text{ cm}^{-1}\\ \beta &= 166614 \text{ cm}^{-1}\\ n_{eff} &= 3.50031\end{aligned}$$

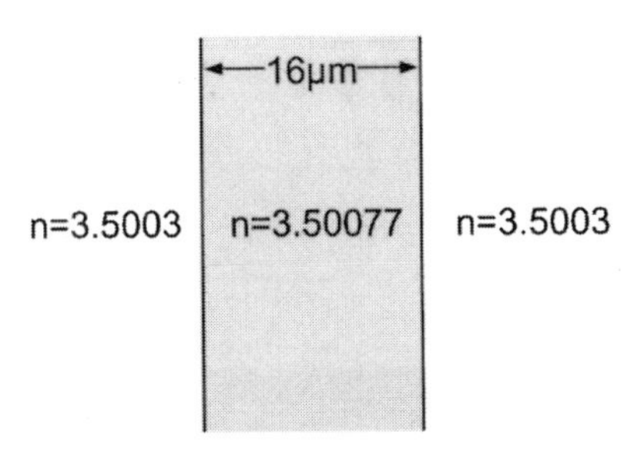

Figure 5.13. The horizontal structure of the waveguide can be modelled as a symmetric slab waveguide constructed with three layers, and the index of each layer is determined by the effective index determined by the vertical structure of the waveguide.

Armed with the three indices for regions I, II, and III, we can now describe the effective waveguide in the horizontal direction. Fig.5.13 shows the structure that must be analyzed. Note that the three regions carry the three effective indices that were evaluated in the last steps.

In this structure, the x-polarized field will appear as a TE wave, so the characteristic equation for this symmetric structure is given by Eq. 3.28, using $h_3 = 16\mu\text{m}$,

$$\tan\left(\frac{\kappa_y h_3}{2}\right) = \frac{\gamma_y}{\kappa_y} \tag{5.36}$$

Numerical solution leads to

$$\begin{aligned} \kappa_y &= 1323\ \text{cm}^{-1} \\ \gamma_y &= 2352\ \text{cm}^{-1} \\ \beta &= 166631\ \text{cm}^{-1} \end{aligned} \tag{5.37}$$

This value β is the eigenvalue for the mode of the waveguide.

Since the effective waveguide is symmetric, the lowest order spatial mode of the waveguide is described by Eq. 3.27

$$\begin{aligned} E(x) &= A\hat{x}\frac{\cos(\kappa_x x)}{\cos(\kappa_x a/2)} \quad \text{for } |x| < a/2 \\ &= A\hat{x}e^{-\gamma_x|x|} \quad \text{for} |x| > a/2 \\ E(y) &= \hat{x}\frac{\cos(\kappa_y y)}{\cos(\kappa_y b/2)} \quad \text{for } |y| < b/2 \\ &= \hat{x}e^{-\gamma_y|y|} \quad \text{for} |y| > b/2 \end{aligned}$$

Plugging numbers into these expressions and plotting the results, we find the mode field shown in Fig. 5.14.

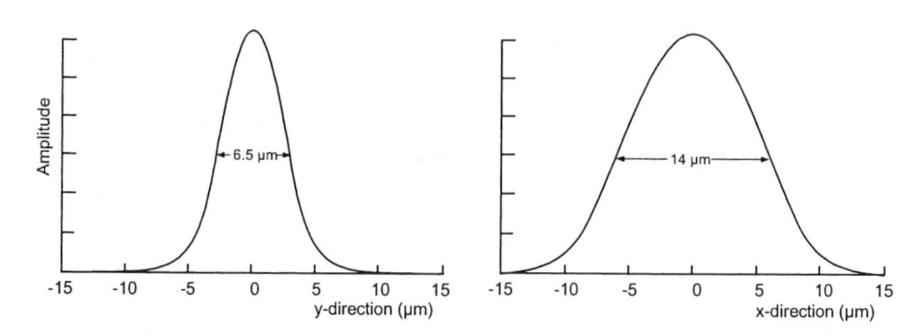

Figure 5.14. The calculated mode intensities derived from the constants given in the text. The calculated dimensions agree well with the experimentally measured values of 8.0 μm, and 13.9 μm.

5. Effective Index Method applied to Ex. 5.1

Let's complete the example we began in the wave analysis and perturbation theory sections by using the effective index method to calculate the normalized propagation coefficients for the waveguide first introduced in Ex. 5.1.

The calculation is relatively straightforward. First, the waveguide is analyzed as if it were a 5μm-thick slab waveguide. Since the electric field is oriented in the y-direction, the mode can be analyzed as a TE mode for the thin dimension. The transverse wavevector, κ_x was found for a range of wavelengths spanning $0.5 \rightarrow 2\mu$m using

$$\tan \kappa_x a/2 = \sqrt{k_0^2 n_1^2 - \kappa_x^2}/\kappa_x \tag{5.38}$$

From this data, an effective index was assigned for each k-vector

$$n_{eff} = \sqrt{k_0^2 n_1^2 - \kappa_x^2}/k_0 \tag{5.39}$$

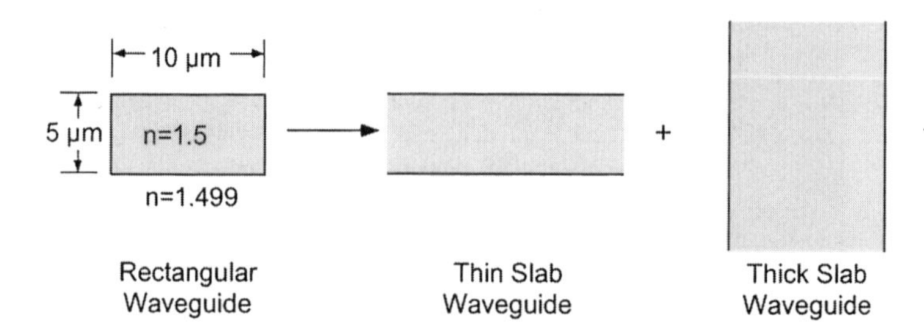

Figure 5.15. The rectangular waveguide of Example 5.1 is decomposed into two slab waveguides for analysis by the effective index method.

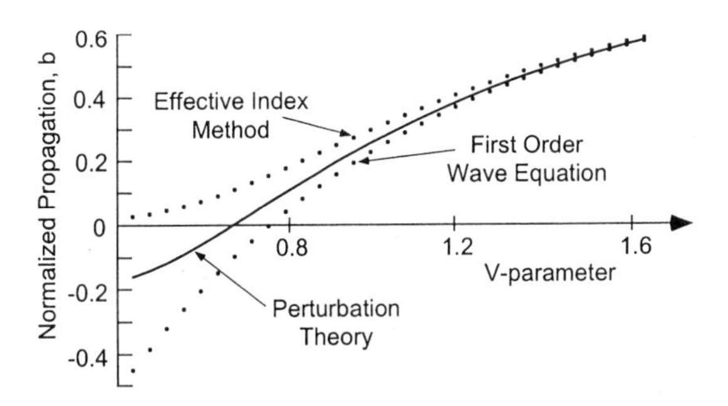

Figure 5.16. Comparison of the normalized propagation coefficient derived from the three methods described in this chapter.

Using this value of n_{eff}, the transverse wavevectors, κ_y were found using

$$\tan \kappa_y b/2 = \frac{n_eff^2}{n_2^2}\sqrt{k_0^2 n_{eff}^2 - \kappa_y^2/\kappa_y)} \tag{5.40}$$

The propagation coefficient, β was found from

$$\beta = \sqrt{k_0^2 n_{eff}^2 - \kappa_y^2} \tag{5.41}$$

Finally, to compare these results with the previous methods, the normalized propagation coefficient was calculated using the actual values of the indices, but the final value of the propagation coefficient, β

$$b = \frac{\beta^2 - k_0^2 n_2^2}{k_0^2 n_1^2 - k_0^2 n_2^2} \tag{5.42}$$

The results of the effective index method are plotted in Fig.??.

The effective index method is the only technique which predicts that there will be at least one mode. Careful comparison of the three methods discussed so-far with other more exact numerical techniques show that the perturbation solution is the most accurate around $V = 1$, while obviously the effective index technique is the only viable option in the region below about $V = 0.7$. Unfortunately, the Effective Index method is only accurate when the aspect ratio (width/height) > 3.

6. Summary

We have seen that calculation of the allowed mode field distributions and eigenvalues is an laborious task, although the techniques themselves are relatively straightforward. We explored three basic techniques: the analytical approach developed by Marcatili; the perturbation techniques which improve on the analytical solutions; and the effective index method. Achieving accurate solutions with analytical techniques is difficult because of the presence of corners. We found the solutions were accurate only far from cut-off. Perturbation techniques probably allow the greatest accuracy, although they come at the price of considerable effort. The effective index method is perhaps the most commonly employed technique for waveguide design. It is relatively straightforward to understand and apply, and its results are not too far from those generated from exact analysis. Comparison with numeric results show that near cut-off, the effective index method slightly overestimates the actual β of the mode. The effective index method also leads to ambiguous results in square waveguides, giving different results for E_{nm} and E_{mn} modes which should be degenerate.

The next chapter introduces a basic numeric technique called the Beam Propagation Method for waveguide evaluation. With the advent of stable packages that can model waveguide structures, simulation via numeric technique is becoming the standard analytical tool in waveguide design. Nevertheless, techniques such as perturbation analysis are very useful for designing certain optical devices, especially those that rely on coupling to evanescent fields.

References

[1] E. A. J. Marcatili, "Dielectric Rectangular Waveguide and Directional Coupler for Integrated Optics," *Bell System Tech. J.*, 48, 2071-2102 (1969)

[2] see for example, D. Marcuse, *Theory of Dielectric Waveguides, 2nd ed.*, p.49, Academic Press, San Diego, (1991)

[3] A. Kumar, K. Thyagarajan, and A. K. Ghatak, Optics Letters 8, 63-65 (1983)

[4] G.B. Hocker and W.K. Burns, Appl. Opt., 16, 113 (1977)

[5] S. F. Pesarcik, G. V. Treyz, S. S. Iyer, J. M. Halbout, Electron. Lett. 28, 159-160 (1992)

[6] This equation is for a particular state of strain in the SiGe layer. The "exact" formula would be $n_{SiGe_x} = n_{si} + x(0.18(1-\chi) + 0.09\chi)$ where χ describes the lattice relaxation. In this case the lattice is 84% relaxed, so $\chi = 0.84$.

Practice Problems

1. A symmetric step-index slab waveguide is shown in Fig. 5.17a. The guiding layer is 10 μm thick, and the fundamental TE mode is described

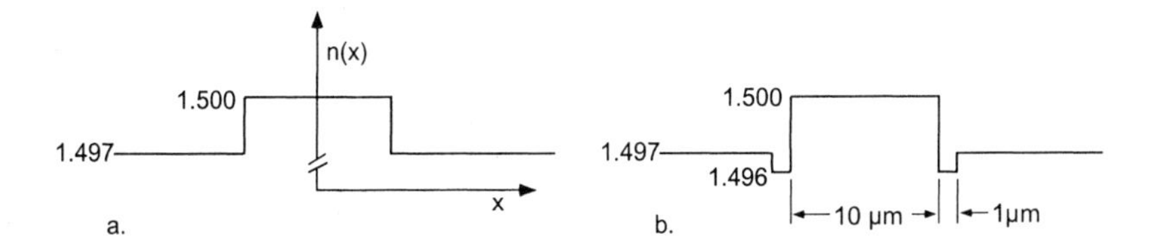

Figure 5.17. Figure for Problem 1.

as

$$
\begin{aligned}
E_y(x) &= Ae^{-\gamma(x-5\mu m)} \quad \text{for } x > 5\mu m \\
&= A\frac{\cos \kappa x}{\cos(\kappa \cdot 5\mu m)} \quad \text{for } -5\mu m < x < 5\mu m \\
&= Ae^{+\gamma(x+5\mu m)} \quad \text{for } x < -5\mu m
\end{aligned}
$$

For this waveguide, the propagation coefficient, $\beta = 94220$ cm^{-1}, and $\kappa = 2300$ cm^{-1}. The waveguide is modified by adding a region of suppressed index around the the guiding film, as shown in Fig. 5.17b. Use perturbation theory to determine the exact change, $\delta\beta^2$, for this new structure.

2. Consider a buried rectangular waveguide similar in structure to Fig. 5.1.b. The core index is $n_1 = 1.5$, while the horizontal side regions have index $n_4 = n_5 = 1.49$, and the top and bottom regions have index $n_3 = n_2 = 1.495$. The dimensions of the waveguide are $a = 10\mu$m, and $b = 5\mu$m. If an optical wave with vacuum wavelength $\lambda = 1\mu$m is carried by the guide, which direction of polarization will have the largest value of β? E^x or E^y?

3. The waveguide shown in Fig. 5.18 is to be used in an electronic package connecting two high speed computer chips. What is the cutoff wavelength for the lowest order mode? Use the first order theory described in Section 5.2 to answer this question.

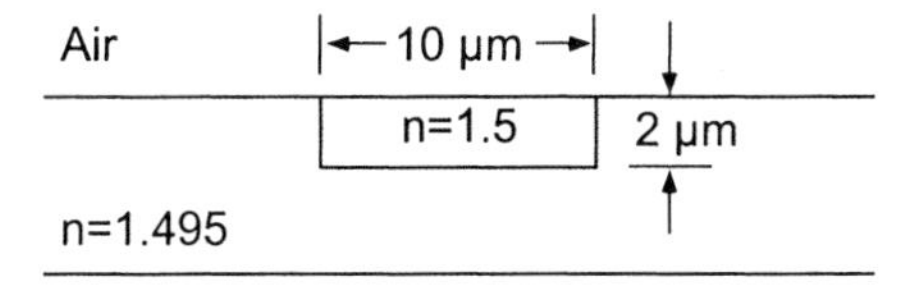

Figure 5.18. Figure for Problem 3.

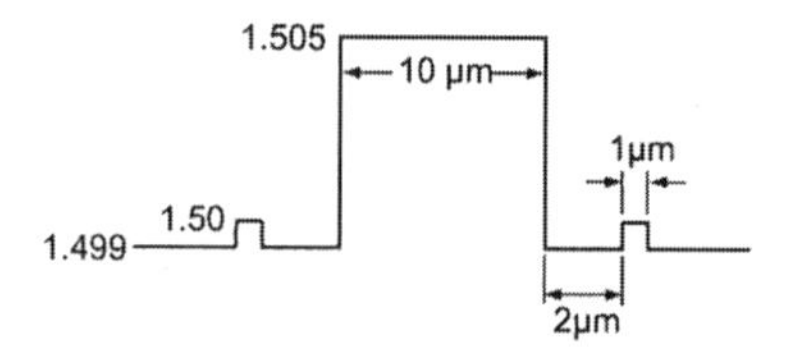

Figure 5.19. Figure for Problem 6.

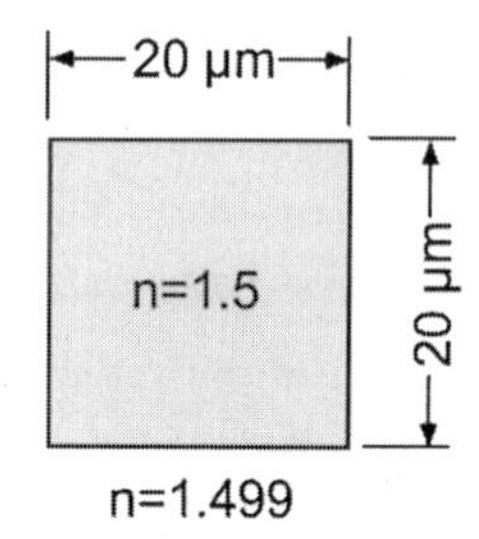

Figure 5.20. Figure for Problem 7.

4. Use the effective index method to find β in the waveguide of Fig. 5.18 for $\lambda + 0.8\mu$m.

5. For the waveguide described in Example 5.1, find the description for the E^x_{12} mode for a wavelength of 0.4 μm.

6. Use perturbation theory to find the longitudinal wavevector, β, for the waveguide structure shown in Fig. 5.19. Assume the guided wavelength is 1 μm.

7. Consider the square symmetric waveguide in Fig. 5.20, with core index $n = 1.5$, and surrounding cladding index $n = 1.499$. Since the waveguide is symmetric, it must guide at least one mode. For a wavelength of $\lambda = 1.3\mu$m, determine the amplitude distribution of the lowest order guided mode.

8. Use the effective index method to determine the mode in the ridge waveguide shown in Fig.5.21. Assume the field is polarized in the y-direction. and the wavelength is 1.55 μm.

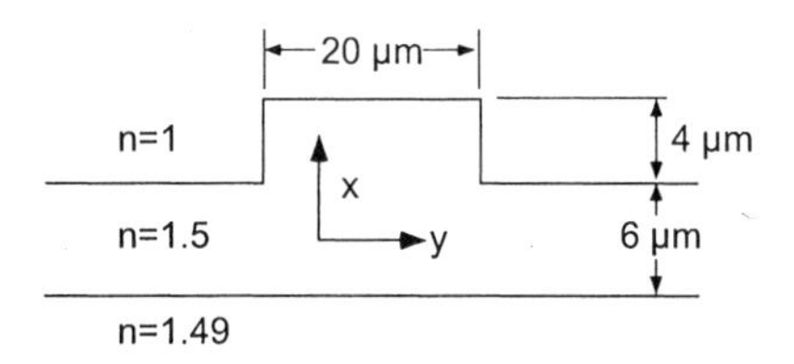

Figure 5.21. Figure for Problem 8.

9. Calculate the mode confinement factor for the mode described in Prob. 8.

10. For the waveguide described in Prob. 2., use *Mathematica* or some other numeric package to calculate the exact profile for the lowest order x-polarized mode, and make a two dimensional plot of the mode amplitude.

Chapter 6

DISPERSION IN WAVEGUIDES

1. Introduction

One of the strongest motivations for using optical waveguides is the large information capacity of an optical link. Much of this capacity comes from the high carrier frequency of the light itself (on the order of 10^{14} Hz). If even 1% of the total bandwidth available could be utilized, the information transmitted on *one optical beam* would be enough to handle over 10^8 telephone calls. While it is unlikely there will ever be demand for that many simultaneous telephone calls on one line, applications such as video require large bandwidths and can take advantage of such capacity.

The system designer must be aware of the fundamental bandwidth limitations in optical waveguides. The most prominent limitation is dispersion. Dispersion describes the spreading of a signal in time. Fig. 6.1 illustrates how dispersion limits the information capacity of a communication channel. At the input, a series of pulses (representing perhaps binary information) are launched onto an optical waveguide. Dispersion causes each of these pulses to spread in time. When they arrive at the output, the pulses have broadened to the point where they begin to seriously overlap adjacent pulses. The temporal spreading effectively establishes the maximum data rate for a communication link.

In this chapter, we examine the mechanisms which lead to dispersion in waveguides. Armed with this understanding, we can appreciate the specialized waveguides such as the graded index waveguide that are described in later chapters. There are many clever equalization techniques, both optical and electronic, which are being developed to extend the useful bandwidth of optical fibers. These topics are not covered in this chapter, but are referenced where appropriate.

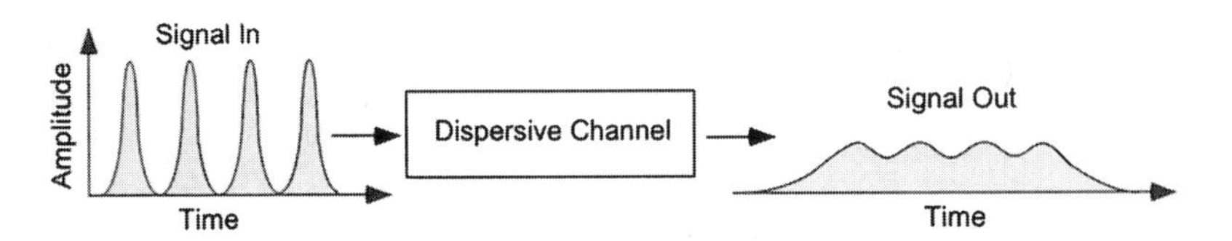

Figure 6.1. Short temporal pulses experience pulse spreading in a dispersive media. If they are not separated by enough time they will begin to overlap.

2. Three Types of Dispersion

Optical waveguides display three types of dispersion: *material dispersion*, *modal dispersion* and *waveguide dispersion*. In *material dispersion*, different wavelengths of light travel at different velocities within a given medium. Consider a pulse that has a finite spectral bandwidth, $\Delta\lambda$. If the pulse is launched in a dispersive material, each wavelength component of the pulse will travel at a different velocity. The pulse effectively spreads out (or disperses) in time and space. "Aha!", you might be thinking, "why not just make the pulse have only one wavelength?" Nice idea, but it won't work. Because of the Fourier relation between pulse duration and frequency bandwidth, *all* finite temporal pulses have a finite frequency bandwidth. Dispersion is a fundamental issue with system design.

Modal Dispersion arises in waveguides with more than one propagating mode. Unless the waveguide has been specially designed (for example, the graded index waveguide described in the next chapter), each allowed mode in the waveguide will travel with a different group velocity. The pulse energy in a waveguide will be distributed among the various allowed modes, either through the initial excitation, or through mode coupling that occurs within the waveguide. The modes arrive at the end of the waveguide slightly delayed relative to each other. This effectively spreads the temporal duration of the pulse, which again limits the bandwidth.

Waveguide dispersion is a more subtle effect. The propagation constant β depends on the wavelength, so even within a single mode different wavelengths will propagate at slightly different speeds. Compared to material and modal dispersion, waveguide dispersion is usually the smallest in magnitude. However, in the vicinity of the so-called zero dispersion point for materials, waveguide dispersion can the dominant effect in a single mode system. Waveguide dispersion can be used to cancel material dispersion, allowing the design of special "dispersion shifted" waveguides. Control of waveguide dispersion is therefore critical to many waveguide designs. In the next three sections, we will discuss these three dispersions.

3. Material Dispersion

A general understanding of the optical properties of dielectrics is essential in optoelectronics. The index of refraction, n, is the most widely used material parameter for waveguide design. In this section, we will explore the index of refraction's functional dependence on wavelength using the Lorentz model.

3.1 Frequency Dependence of the Permittivity, ϵ

We will assume that the dielectric tensor is a simple scalar, i.e., the dielectric material is isotropic. The index of refraction is defined in terms of the relative permittivity of a medium:

$$n = \sqrt{\frac{\epsilon}{\epsilon_0}} \tag{6.1}$$

The permittivity, ϵ, relates the electric flux, $\mathbf{D}$ to the electric field, $\mathbf{E}$,

$$\mathbf{D} = \epsilon \mathbf{E} \tag{6.2}$$

We want to develop a model relating the permittivity, ϵ, to an applied electric field. Consider the simple atomic model consisting of a positively charged nucleus and a surrounding negatively charged cloud. When there is no electric field present, the two charges are centered upon one another.

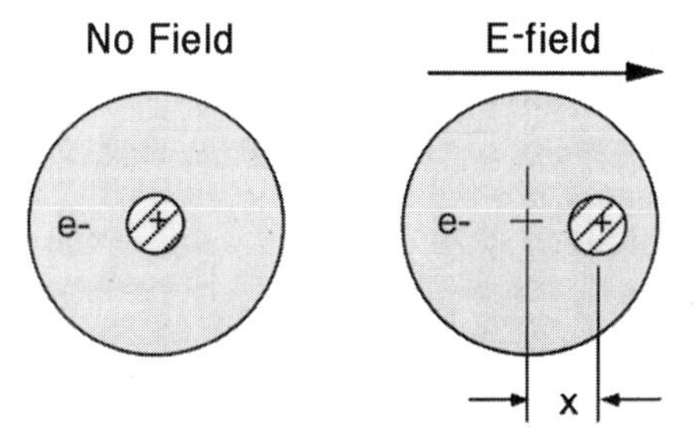

Figure 6.2. Pictorial depiction of the dipole moment induced in a neutral atom by an external field.

When an electric field is applied to the atom, the negatively-charged electrons and positively-charged nucleus experience opposite forces due to the field, and slightly separate. The charge separation forms a microscopic *dipole moment*, defined as

$$\mathbf{p} = q\mathbf{r} \tag{6.3}$$

where q is the charge, and $\mathbf{r}$ is the relative distance from equilibrium that the charges move.

The constitutive equation (6.2) can be written in terms of this dipole moment.

$$\mathbf{D} = \epsilon \mathbf{E} = \epsilon_0 \mathbf{E} + \mathbf{P} \quad (\text{C/m}^2) \tag{6.4}$$

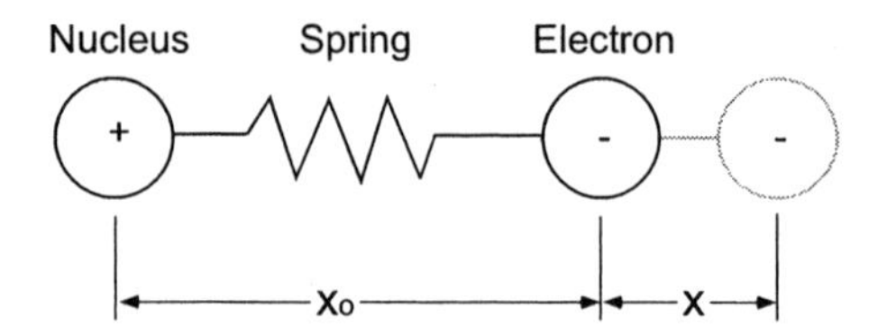

Figure 6.3. The Lorentz model of the atom consists of a heavy nucleus bound to a light electron through a spring. x_0 represents the equilibrium distance between the charges when no external forces are present. x represents the displacement from equilibrium.

where $\mathbf{P}$ is the bulk polarization of the material, defined by

$$\mathbf{P} = N\mathbf{p} = Nq\mathbf{r} \tag{6.5}$$

and N is the number of dipoles per unit volume.

To develop an analytical description of the permittivity, we will use the one dimensional Lorentz model of the atom. Lorentz [1] modelled the atom as two particles bound together by a spring. The model is shown in Fig. 6.3. The position x_0 represents the equilibrium position of the two charges when there are no external forces. x_0 could easily be zero.

Modelling the attraction between an electron and a nucleus with a spring might seem a little crude, but it is actually based on sound physical reasoning. The binding energy of an electron to a positively-charged nucleus has a general form as shown in Fig. 6.4.

The electron will reside at the minimum of a potential well, x_0. Near x_0, the potential can be approximated in a Taylor series expansion as

$$V(x) = V(x_0) + \frac{dV}{dx}(x - x_0) + \frac{1}{2}\frac{d^2V}{dx^2}(x - x_0)^2 + \frac{1}{6}\frac{d^3V}{dx^3}(x - x_0)^3 + \ldots \tag{6.6}$$

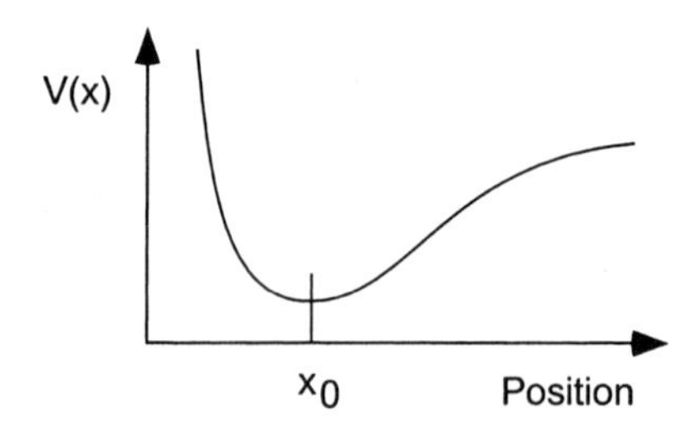

Figure 6.4. The binding potential of an electron to a positive nucleus will look roughly like this. The exact shape is generally unknown, but the potential will have a smooth minimum at some point, designated x_0.

At the minimum, x_0, the first derivative, $dV(x_0)/dx$, is zero, so it can be dropped from the expansion. To second order, the potential can be described as

$$\begin{aligned} V(x) &= V(x_0) + \frac{1}{2}\frac{d^2V}{dx^2}(x - x_0)^2 \\ &= V_0 + \frac{1}{2}kx^2 \qquad \text{when} \quad x_0 = 0 \end{aligned} \tag{6.7}$$

where k is a constant. This is the simple "Hooke's Law" potential for a spring, where k is the spring constant (not to be confused with the wavevector). This expansion is only accurate for small values of x, but if the applied electric fields are small compared to the binding potential of the electron (which is on the order of 10^{10} V/cm) the approximations are reasonable.

In the presence of an electric and magnetic fields, a charge q experiences the Lorentz force

$$\mathbf{F} = q\mathbf{E} + \hat{v} \times \mathbf{B}) \tag{6.8}$$

where $\hat{v}$ is the charge's velocity. Only at relativistic speeds or under conditions of strong DC bias, as in Hall measurements, is the magnetic term significant. We thus can ignore the magnetic force in optical interactions, simplifying the driving force on the charges to be

$$\mathbf{F} = q\mathbf{E} \tag{6.9}$$

We will use the center of mass picture, in which all the motion can be attributed to the electron with an effective mass of

$$m = \frac{m_e m_n}{m_e + m_n} \approx m_e \tag{6.10}$$

The net forces acting on the electron determine it's motion. These forces include the external force from the applied field, acceleration, friction, and spring restoring forces. Summing these forces together (recall $q = -e$)

$$m\frac{d^2\mathbf{x}}{dt^2} + \gamma m\frac{d\mathbf{x}}{dt} + k\mathbf{x} = -e\mathbf{E}(t) \tag{6.11}$$

We have introduced a decay term, $\gamma m\ dx/dt$, which acts as a friction term. Do not view this term literally: it accounts in a phenomenological manner for energy dissipation that occurs due to radiation or phonon emission.

Dividing both sides of the equation by mass, m, and recalling that for a spring-and-mass system, the resonant frequency of oscillation is given by $\omega_0 = \sqrt{k/m}$, yields

$$\frac{d^2\mathbf{x}}{dt^2} + \gamma\frac{d\mathbf{x}}{dt} + \omega_0^2\mathbf{x} = -\frac{e}{m}\mathbf{E}(t) \tag{6.12}$$

To solve for a particular solution, we must specify the driving term. Assume that $E(t)$ has a harmonic dependence, $E(t) = E_0 e^{j\omega t}$ (where it is understood that the actual electric field is given by the real part of the expression). A trial solution, $x(t) = x_0 e^{j\omega t}$, leads to the following solution:

$$\mathbf{x}(t) = \frac{-e/m}{\omega_0^2 - \omega^2 + j\gamma\omega}\mathbf{E}_0 e^{j\omega t} \tag{6.13}$$

where the displacement is parallel to the applied electric field. Knowing $\mathbf{x(t)}$, we can directly relate the effect of an applied field to the polarization of the material through the relation $p = qx$. Note that $x(t)$ is a complex number, which simply means that there will be a phase shift between the applied field and the response of the medium. Using Eq.6.4, we can determine the electric displacement vector

$$\begin{aligned} \mathbf{D} = \epsilon\mathbf{E} &= \epsilon_0\mathbf{E} + Nq\mathbf{x} = \epsilon_0\mathbf{E} - Ne\mathbf{x} \\ &= \epsilon_0\mathbf{E} + \frac{Ne^2/m}{\omega_0^2 - \omega^2 + j\gamma\omega}\mathbf{E} \\ &= \left(\epsilon_0 + \frac{Ne^2/m}{\omega_0^2 - \omega^2 + j\gamma\omega}\right)\mathbf{E} \end{aligned} \tag{6.14}$$

The frequency-dependent form of the dielectric constant can be extracted from this expression

$$\begin{aligned} \frac{\epsilon(\omega)}{\epsilon_0} &= \epsilon'(\omega) + j\epsilon''(\omega) \\ &= 1 + \frac{(Ne^2/m)(\omega_0^2 - \omega^2)}{\epsilon_0[(\omega_0^2 - \omega^2)^2 + \gamma^2\omega^2]} - j\frac{(Ne^2/m)(\gamma\omega)}{\epsilon_0[(\omega_0^2 - \omega^2)^2 + \gamma^2\omega^2]} \end{aligned} \tag{6.15}$$

The index of refraction is defined as the square root of the complex dielectric constant [2]

$$n(\omega) = \sqrt{\frac{\epsilon(\omega)}{\epsilon_0}} = \sqrt{1 + \frac{(Ne^2/m)(\omega_0^2 - \omega^2)}{\epsilon_0[(\omega_0^2 - \omega^2)^2 + \gamma^2\omega^2]} - j\frac{(Ne^2/m)(\gamma\omega)}{\epsilon_0[(\omega_0^2 - \omega^2)^2 + \gamma^2\omega^2]}} \tag{6.16}$$

The index of refraction increases with the density of dipoles (atoms in this case) through N. The denser the media, the larger will be the index of refraction. This explains why air has a low index (on the order of 1.0003) while solids have indices in the range of 1.4 - 3.5.

The imaginary part of the index of refraction leads to attenuation or gain, depending on the sign. In regions of transparency, the imaginary component of the index of refraction of dielectrics is negligibly small, dramatically simplifying the expression. We will concentrate on the transparent region in this chapter, saving a technical description of attenuation for Chapter 8.

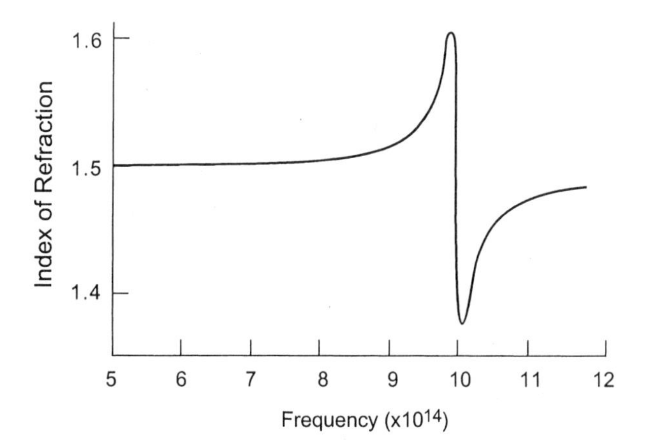

Figure 6.5. The real part of the index of refraction as predicted by the Lorentz model. In this plot, the resonant frequency, ω_0 is at 10^{15} rads/sec, and the damping constant is $\gamma = 10^{13}$ rads/sec. Note that the index n increases in magnitude, except in the immediate vicinity of the resonance.

The index of refraction slowly increases with frequency, except near a resonance, ω_0, as shown in Fig. 6.5. A resonance occurs when the frequency of the applied field is identical with a transition frequency between two energy states of the system. As the frequency passes through the resonance, the index rapidly drops and then again begins to increase. The region where the index of refraction decreases as the applied frequency increases is generally not a useful region in which to work, because the absorption losses (imaginary part of n) are highest there.

What happens when there are many electrons, each with a different resonant frequency? We simply add the effect of each resonance together

$$\epsilon = \epsilon_0 + \sum_{i=1}^{Z} \frac{f_i N e^2/m}{\omega_i^2 - \omega^2 + j\gamma\omega} \tag{6.17}$$

where f_i is called the *oscillator strength* of each resonance. This oscillator strength takes into account the possibility that each electron interacts differently with the applied field.

In optics it is more common to deal with wavelength than with frequency. Using the relation $\omega = 2\pi c/\lambda$ in Eq.6.17, and assuming that the damping terms, γ, are negligibly small, one can rewrite the expression for the index of refraction in terms of wavelength. This form, commonly called the *Sellmeier equation*, is

$$n^2 - A = \sum_k \lambda^2 - \lambda_k^2 \tag{6.18}$$

Table 6.1. Sellmeier coefficients for several dielectrics

NaCl ($A = 1.00055$)		SiO_2 ($A = 1$)		Al_2O_3($A = 1$)		CaF_2 ($A = 1$)	
$\lambda_k(\mu m)$	G_k	$\lambda_k(\mu m)$	G_k	$\lambda_k(\mu m)$	G_k	$\lambda_k(\mu m)$	G_k
0.1	0.48398	0.0684	0.69617	0.0615	1.0238	0.0503	0.56758
0.158	0.25998	0.1162	0.40794	0.1107	1.0583	0.1004	0.471091
40.5	0.08796	9.8962	0.89748	17.926	5.2808	34.649	3.84847
60.98	3.17						
0.05	0.198						
0.128	0.3869						

where the terms A, λ_k, and G_k are called the Sellmeier coefficients, and represent resonant wavelengths and oscillator strengths, respectively, for a given system. Tables of Sellmeier coefficients can be found in many physics or optics handbooks [3, 6]. As an example, Table 6.1 lists the Sellmeier coefficients for several optically transparent materials in the visible region. This data will be used in subsequent examples and homework problems.

Note that NaCl has resonances at several wavelengths, such as 50nm, 100nm, 158nm, and 40.5μm. The short wavelength resonances are due to electronic transitions within the NaCl crystal structure. The resonances at 40.5μm and 60.98μm are due to ionic vibrations. These vibrations occur at lower frequencies than the electronic resonances, because the masses of the atoms are larger than those of the electron. We will use these Sellmeier coefficients to analyze the dispersion of the materials.

3.2 Group Index and Group Delay

A useful term is the Group Delay, τ_g. Group delay is defined as the time it takes for a pulse of light to travel a unit distance. For example, it takes a light pulse about 3.336 nsec to travel one meter in vacuum. Thus the group delay for vacuum is 3.336 nsec/m. By definition, group delay is the inverse of the group velocity

$$\tau_g = 1/v_g = dk/d\omega \tag{6.19}$$

To relate this to the index of refraction, $n(\omega)$, substitute $k = n\omega/c$ into Eq.6.19

$$\begin{aligned} \tau_g &= \frac{d(n\omega/c)}{d\omega} \\ &= \frac{dn}{d\omega}\frac{\omega}{c} + \frac{n}{c}\frac{d\omega}{d\omega} \\ &= \frac{(n + \omega\frac{dn}{d\omega})}{c} \end{aligned} \tag{6.20}$$

The group delay, τ_g, depends on the index of refraction *and* the first derivative of the index with respect to frequency. Inverting τ_g, we get an expression for v_g that has a form similar to the phase velocity, $v_p = c/n$,

$$\begin{aligned} v_g &= c/(n + \omega \tfrac{dn}{d\omega}) \\ &= c/N_g \end{aligned} \tag{6.21}$$

where the term N_g is called the *Group Index*, and is defined to be

$$N_g = n + \omega \frac{dn}{d\omega} \tag{6.22}$$

We will find the Group Index more useful when defined in terms of wavelength. Simple calculation leads to

$$N_g = n - \lambda \frac{dn}{d\lambda} \tag{6.23}$$

Keep in mind the $dn/d\lambda$ is negative in most regions. The group index, N_g, is always larger than the regular index of refraction, n, except in regions of anomalous dispersion.

3.3 Group Velocity Dispersion

If a signal or pulse contains more than one wavelength, the individual components of this signal will travel at different group velocities. These components will reach the receiver at different times, effectively stretching out the time it takes for a signal to arrive. This effect is called Group Velocity Dispersion (GVD).

Consider an optical pulse with a finite spectral bandwidth, $\Delta\lambda$, travelling through a dispersive medium. The time required to travel a distance L is called the *latency*, and is the product of the group delay, τ_g with the propagation distance, L

$$\tau = \frac{L}{c} N_g(\lambda) \tag{6.24}$$

The spectral width of the pulse spans from λ_1 to λ_2 (i.e. $\Delta\lambda = |\lambda_1 - \lambda_2|$

Each wavelength component will propagate at a slightly different speed. In the time domain, the pulse spread will be

$$\begin{aligned} \Delta\tau &= \frac{L}{c}(N_g(\lambda_1) - N_g(\lambda_2)) \\ &= \frac{L}{c}\Delta N_g \\ &= \frac{L}{c}\frac{dN_g}{d\lambda}\Delta\lambda \end{aligned} \tag{6.25}$$

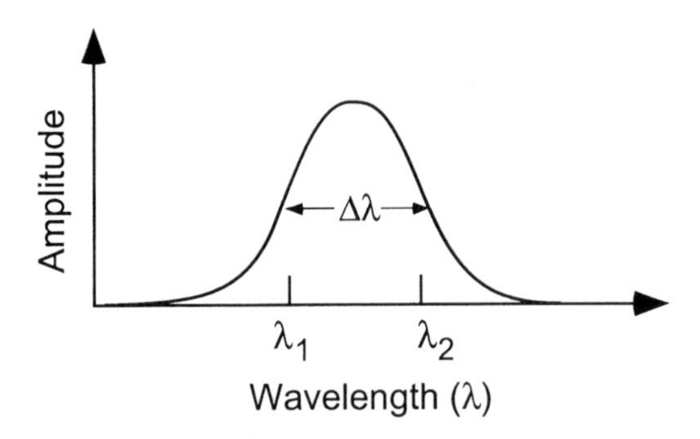

Figure 6.6. An optical pulse with a finite spread in wavelength.

So what is $dN_g/d\lambda$? We can derive an expression for this from Eq. 6.23

$$\begin{aligned} \frac{dN_g}{d\lambda} &= \frac{d}{d\lambda}(n - \lambda\frac{dn}{d\lambda}) \\ &= \frac{dn}{d\lambda} - \frac{dn}{d\lambda} - \lambda\frac{d^2n}{d\lambda^2} \\ &= -\lambda\frac{d^2n}{d\lambda^2} \end{aligned} \tag{6.26}$$

Combining Eq. 6.26 with Eq. 6.25 yields an expression for the spread in the pulse arrival time

$$\Delta\tau = -\frac{L}{c}\lambda\frac{d^2n}{d\lambda^2}\Delta\lambda \tag{6.27}$$

We see that pulse spreading depends on the *second derivative* of the material dispersion. The term

$$D = -\frac{\lambda}{c}\frac{d^2n}{d\lambda^2} \tag{6.28}$$

is called the *material dispersion*. One often finds material dispersion listed in units of psec/(nm km), i.e., the number of picoseconds the pulse will spread as it travels one kilometer per nanometer of spectral bandwidth. Also, be aware that sometimes the minus sign is carried in the material dispersion term, D, as we have done, and sometimes it is carried in the latency expression as in Eq. 6.27.

Example 6.1 Group velocity dispersion in sapphire

The Sellmeier coefficients for sapphire are listed in Table 6.1. What is the index of refraction, group velocity, and group velocity dispersion (GVD) for sapphire over the range from 0.5 to 2.5 μm?

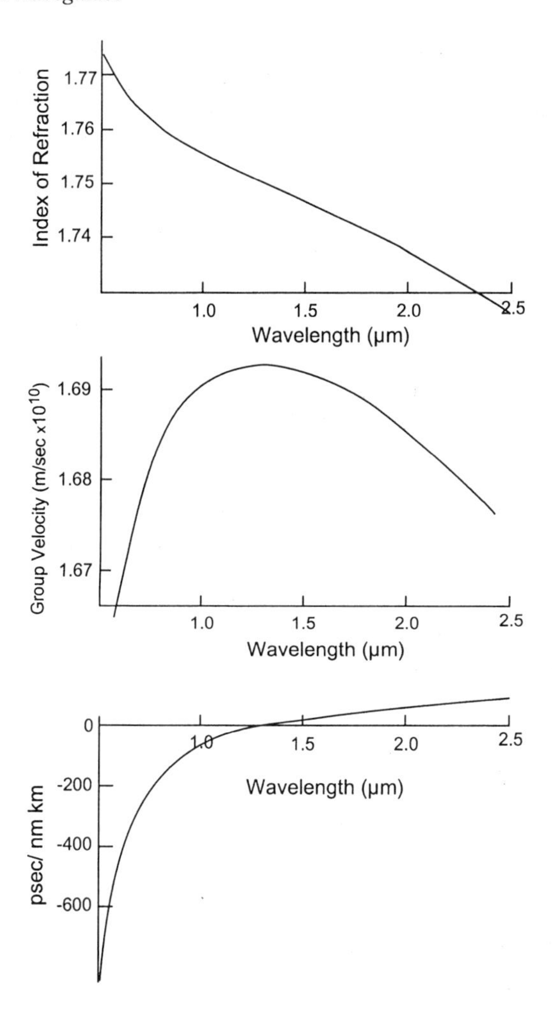

Figure 6.7. The index, group velocity, and group velocity dispersion of sapphire in the near infrared region.

Solution: The solution requires use of a computer to evaluate the Sellmeier equation for a number of wavelengths over the desired range, and to calculate the first and second derivative of the calculated index. In this example, we used *Mathematica* to evaluate first $n(\lambda)$, $dn(\lambda)/d\lambda$, and $d^2n(\lambda)/d\lambda^2$. The results from $n(\lambda)$ were plotted directly. To find the group velocity, the equation $v_g(\lambda) = c/(n - \lambda dn/d\lambda)$ was calculated and plotted. Eq. 6.28 was evaluated to find the group velocity dispersion.

Notice that $n(\lambda)$ decreases monotonically as the wavelength increases (i.e., the frequency decreases); this is normal dispersion. The group velocity reaches a maxima near $1.4\mu m$, and then decreases again. Please notice that the Group Velocity Dispersion goes through zero at the point where the group velocity is maximum. The *zero dispersion point* is the wavelength where the GVD is equal to zero.

There is a lot of information in the graphs of Fig. 6.7. First, note that the group velocity dispersion goes through zero at a particular wavelength. This wavelength is called the "zero dispersion point", and is denoted by the symbol, λ_0. For $\lambda < \lambda_0$, the material has *positive group velocity* dispersion. Positive GVD is characterized by propagation where a long wavelength pulse travels faster than a short wavelength pulse. For wavelengths longer than λ_0, *negative group velocity dispersion* is displayed.

The region near λ_0, where GVD goes through zero, is one of the desired operating points for optical communications systems, because signal distortion is minimal, however there is still some residual pulse broadening due to third order effects. The zero dispersion point can be shifted through compositional changes. Fused silica has a zero dispersion point near 1.3 μm (see problem 6.)

4. Modal Dispersion

In multimode structures, the dominant cause of pulse spreading is due to modal dispersion. The difference in velocity of the extreme modes determines the magnitude of the pulse spreading. Low-order modes ($\beta \approx kn_f$) are highly confined within the core, and effectively travel at the group velocity allowed by the guiding film. For modes near cutoff, most of the field is in the cladding layers. These modes effectively travel at the group velocity allowed by the cladding index.

The ray picture shown inFig. 6.8 provides a simple (but unfortunately incorrect!) picture of modal dispersion, where the two extreme modes travel obviously different paths, and therefore travel different path lengths as they propagate down the waveguide. Unfortunately, the ray model give opposite results, one predicting that the low order mode travels faster, while the other predicts it will travel slower. We will use the correct wave model for subsequent calculations.

If we assume that there are many modes in the waveguide, then to first approximation the difference in group delay between the fastest and slowest mode is

$$\Delta\tau = \tau_{low} - \tau_{high} \tag{6.29}$$

where τ_{low} is the group delay of the lowest order mode (i.e. the mode with the largest value of β) and τ_{high} is the group delay of the highest order mode. Recall (Eq. 6.19) that the group delay in a bulk medium is $1/v_{\text{group}} = dk/d\omega$.

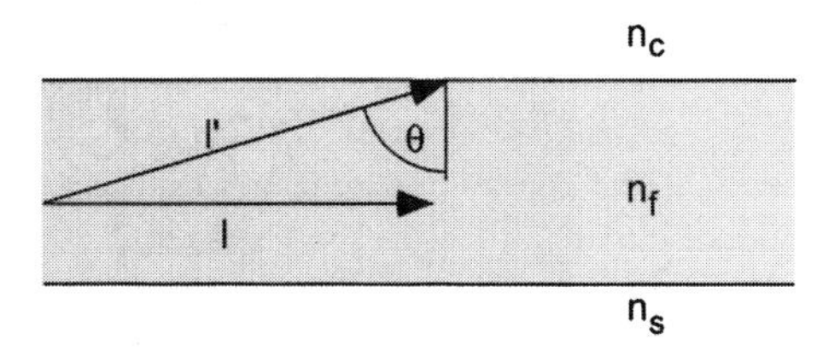

Figure 6.8. To travel a distance l down the waveguide, the ray corresponding to a low order mode can go directly, while the ray corresponding to a higher order mode must actually travel a distance $l' = l/\cos\theta$.

In a waveguide, the group delay is similarly defined in terms of the propagation coefficient

$$\tau_g = \frac{d\beta}{d\omega} \tag{6.30}$$

To evaluate Eq. 6.30, we need to express β in terms of ω. For the lowest order mode, the value of β is approximately equal to kn_f, while the highest order mode will be approximately equal to kn_s. Substituting the expression $\beta = n_f k = n_f \omega/c$ into Eq. 6.29 yields

$$\tau_{low} = \frac{d\beta}{d\omega} = \frac{d(\omega n_f/c)}{d\omega} = \frac{n_f}{c} + k_0 \frac{dn_f}{d\omega} \tag{6.31}$$

Similarly

$$\tau_{high} = \frac{n_s}{c} + k_0 \frac{dn_s}{d\omega} \tag{6.32}$$

The difference in arrival times of the two extreme modes is then

$$\Delta\tau_g = \frac{n_f - n_s}{c} + k_o \left(\frac{dn_f}{d\omega} - \frac{dn_s}{d\omega} \right) \tag{6.33}$$

The first term in Eq. 6.33 is due to the different effective z-components of the extreme modes' k-vectors. The second term is due to dispersion effects in the material that make up the waveguide. While the individual terms of this difference can be significant, the difference in material *dispersion* of the guiding and substrate layers is usually insignificant, i.e. $dn_f/d\omega \approx dn_s/d\omega$. This assumption is reasonable when the materials that make up the waveguide structure are nearly the same, such as fused silica guiding film surrounded by doped fused silica. Then the differential modal group delay is

$$\Delta\tau_g = (\frac{n_f - n_s}{c}) \tag{6.34}$$

The total pulse spreading due to modal dispersion is obtained from the group delay dispersion

$$\Delta\tau = \Delta\tau_g L = \frac{(n_f - n_s)}{c} L \tag{6.35}$$

The pulse spreading leads to a basic limitation on the information capacity of a multimode waveguide. If a temporally short pulse is used to excite a waveguide at $t = 0$, the energy in each mode will travel at a slightly different velocity. At the end of the waveguide, the energy will arrive as a series of mini-pulses each carried by an individual mode. This effectively temporally spreads the pulse.

Example 6.2 Modal dispersion in a planar waveguide

Consider an asymmetric planar waveguide where the guiding film index is $n_f = 1.48$, the substrate index is $n_s = 1.46$, and the cover index is $n_c = 1$ (air). The guiding film is 50μm thick, and the optical signal has a wavelength $\lambda = 1.3\mu$m, which corresponds to the zero dispersion wavelength of the material, so material dispersion effects are negligible.

1. What is the modal dispersion for this waveguide?
2. If a 1 nanosecond pulse is launched onto the waveguide, determine its temporal duration be after travelling 100 meters.
3. If this waveguide connects two circuit boards that are separated by 1 meter, what is the fastest digital data rate that can be sent across this waveguide without pulses running into each other?

Solution First we should confirm that this is a multimode waveguide (after all, if it is a single mode waveguide, there will be no mode dispersion). Using Eq. 2.47, the number of guided TE modes is approximately

$$m \approx \left[\frac{2 \times 50\mu m \sqrt{1.48^2 - 1.46^2}}{1.3\mu m} \right] = 18.6$$

so this is certainly a multimode waveguide. In addition there are about the same number of TM modes. Since all the modes are spread over the available range of allowed values of β, we can be confident that the extreme modes are near the limiting values, $k_0 n_f$ and $k_0 n_s$. The modal dispersion in this case is

$$\Delta\tau_g = \frac{1.48 - 1.46}{3 \times 10^8 \text{m/sec}} = 66 \text{ psec/meter}$$

Note that the cover index (air in this example) has no effect on the modal dispersion. The range and magnitude of β are determined by the film and substrate indices. The cover index influences the exact value of each β, but does not influence the magnitude limits. Also notice the units of modal dispersion: time/distance. For optical waveguides, convenient units are psec/m or psec/km depending on the situation.

After travelling 100 meters, *any* launched pulse will spread by

$$\Delta\tau = 66 \times 10^{-12} \text{ sec/m} \cdot 100\text{m} = 6.6 \text{ nsec}$$

The launched pulse was only 1 nsec in duration, so this spread is significant. The final pulsewidth is a quadratic sum of the initial width and the additional width due to pulse spreading.

$$\tau = \sqrt{1^2 + 6.6^2}\text{nsec} = 6.67\text{nsec}$$

You were not expected to know how to combine the pulse width and pulse spreading. We will discuss how pulse spreading effects are combined in one of the last sections of this chapter.

Finally, error-free transmission requires that adjacent pulses remain distinct from one another after travelling through the waveguide. Adjacent pulses must not spread into each other by the time they arrive at the output. For a 1 meter path length, a pulse will spread by 66 picoseconds. Even if the launched pulse is 1 psec in duration, it will be approximately 66 psec long by the time it travels 1 m (this neglects material dispersion, which may increase the spread even further). Subsequent pulses should be delayed by at least 66 picoseconds in order to avoid "collisions" of pulses in the receiver. The maximum data rate is then $1/66\text{psec} = 1.5 \times 10^{10}$ pulses per second. If we chose to make the waveguide longer, the maximum data rate would decrease proportionally.

There are several effective ways to counter modal dispersion. The obvious solution is to use a waveguide that only supports one mode, a *single mode* waveguide. Most high speed, long distance optical fiber used throughout the world today is single mode. The second solution is to use a *graded-index* waveguide, which is described in Chapter 7. A graded index reduces the geometric path difference between low order and high order modes.

In practice, the pulse spreading due to modal dispersion is not as great as predicted by Eq. 6.35. There are two reasons. First, the calculated delay is between the two extreme modes. If there are many modes, then each mode carries only a small fraction of the total energy. The effect of the extreme modes will be diluted. A second effect which reduces modal dispersion is due to mode coupling. We have not yet discussed mode coupling, but in regions of small dielectric perturbations, connections, etc., energy can be transferred from one mode to another. The fastest modes can only couple, and hence transfer energy, to modes which travel slower. Similarly, the slowest modes will have no choice but to couple to faster modes. The net effect is that there is an averaging of the modal velocities. This effect is difficult to quantify, because it depends on the waveguide and its coupling characteristics. But since energy is effectively *diffusing* among the modes, the length dependence of the pulse spreading goes as the $\sqrt{LL_c}$ instead of directly on L . L_c is the characteristic coupling length that depends on the coupling strength in the waveguide. Measurements of modal dispersion confirm this type of behavior.

5. Waveguide Dispersion

Waveguide dispersion has generally the smallest magnitude of the three dispersion mechanisms. Waveguide dispersion becomes significant in single mode systems operating near λ_0, the zero material dispersion point.

Consider light propagating in a single mode waveguide. Different wavelengths within this one mode will travel at slightly different velocities. The eigenvalue equation that determines β depends on the specific value of k, which is related to wavelength through $|k| = 2\pi/\lambda$. For the TE mode in a slab waveguide the eigenvalue equation is

$$\tan(h\kappa) = \frac{\gamma_c + \gamma_s}{\kappa(1 - \frac{\gamma_c \gamma_s}{\kappa^2})} \tag{6.36}$$

where γ and κ depend explicitly on the wavevector

$$\kappa = \sqrt{k_0^2 n_f^2 - \beta^2} \quad \text{and} \quad \gamma_{s,c} = \sqrt{\beta^2 - k_0^2 n_{s,c}} \tag{6.37}$$

If the wavelength changes slightly, the value of β will also change slightly. The mode will still have the same basic electric field distribution, e.g. a TE_2 mode will still be a TE_2 mode after the slight change in wavelength, but it will propagate with a slightly different speed.

The effective group delay (per unit length) of a mode is given by (see Eq. 6.30)

$$\tau_{wg} = \frac{1}{c} \frac{d\beta}{dk} \tag{6.38}$$

Using arguments similar to those used to derive the expression for material dispersion, we can develop an expression for the dispersion. Consider a pulse propagating on a single mode waveguide, with a finite spectral bandwidth expressed in terms of the wavevector, Δk. Each value of k will have a unique value of β. The temporal pulse spreading due to dispersion over a path length L will be

$$\begin{aligned} \Delta\tau_{wg} &= \frac{1}{c}\left(\frac{d\beta_1}{dk_1} - \frac{d\beta_2}{dk_2}\right) \\ &\simeq \frac{1}{c}\frac{d^2\beta}{dk^2}\Delta k \end{aligned} \tag{6.39}$$

To convert this into wavelength units (consistent with how we defined material dispersion), note that $\Delta k = -(2\pi/\lambda^2)\Delta\lambda$. Substituting this into Eq. 6.39 yields

$$\Delta\tau_{wg} = -\frac{k^2}{2\pi c}\frac{d^2\beta}{dk^2}\Delta\lambda \tag{6.40}$$

Since β can only be found through numerical solution of a transcendental equation, it is impossible to write a general expression for Eq. 6.40. The waveguide dispersion of a system is usually evaluated numerically. This process is best illustrated through example.

Consider a symmetric slab waveguide made with a guiding index, $n_{core} = 1.50$, and a surrounding index, $n_{clad} = 1.48$. The guiding region of the waveguide is 2 μm thick. Since the waveguide is symmetric, it will always support at least one mode. The effective phase velocity of this mode is determined by Eq. 1.44, expressed in terms of the waveguide mode parameters

$$v_p = \frac{\omega}{\beta}$$

Using the following set of commands with *Mathematica*, we were able to numerically analyze this waveguide. The eigenvalue β was found using the characteristic equation for a TE mode in a symmetric waveguide (Eq. 2.30)

$$\tan\frac{\kappa h}{2} = \frac{\gamma}{\kappa}$$

for values of k ranging from $k = 500\pi \rightarrow 30,000\pi$. The first and second derivative were determined through simple differencing.

```
gamma[kappa_]:=Sqrt[(1.5^2 - 1.48^2)(k i)^2 - kappa^2];
k=500 Pi;
h=0.0002;
beta=Table[Re [N [Sqrt [1.5^2 (k i)^2 - (FindRoot[Tan[h kappa /2] ==
    gamma[kappa]/kappa, {kappa,100}][[1,2]])^2 ]]], {i,1,60,1}];
ListPlot[beta]
deltabeta=Table[ (beta[[i]]-beta[[i-1]])/(500 Pi), {i,2,60,1}];
ListPlot[deltabeta, PlotRange->{1.48, 1.505}]
doubledeltabeta=Table[-((k i)^2 /(6 Pi))(deltabeta[[i]]-
    deltabeta[[i-1]])/(500 Pi), {i,2,59,1}];
ListPlot[doubledeltabeta]
```

Fig. 6.9 shows the calculated values of β as a function of k. The group velocity of a mode is determined from the expression

$$v_g = \frac{d\omega}{d\beta} \tag{6.41}$$

Relating this to the *group index*, N_g, and recalling that $\omega = kc$, we can find an expression for N_g

$$N_g = \frac{c}{v_g} = \frac{d\beta}{dk} \tag{6.42}$$

Thus the effective group index can be derived from the first derivative of the curve in Fig. 6.9. This is plotted in Fig. 6.10

Inspection of Fig.6.10 shows that the group index of the fundamental mode of the waveguide depends strongly on the magnitude of the wavevector, k. For

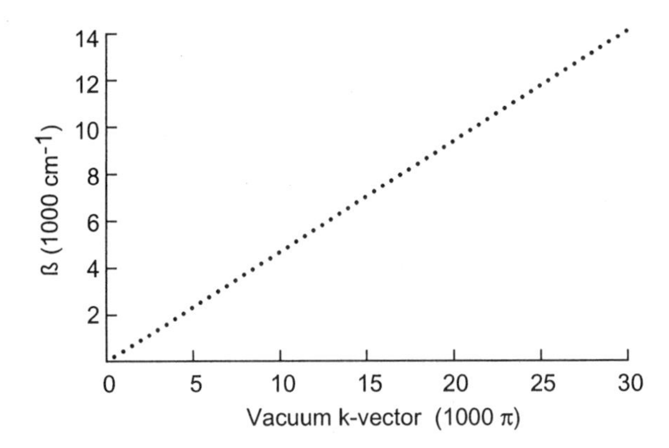

Figure 6.9. The β–k diagram for the TE_0 mode of a symmetric waveguide with a guiding film index of 1.5, and a cladding index of 1.48.

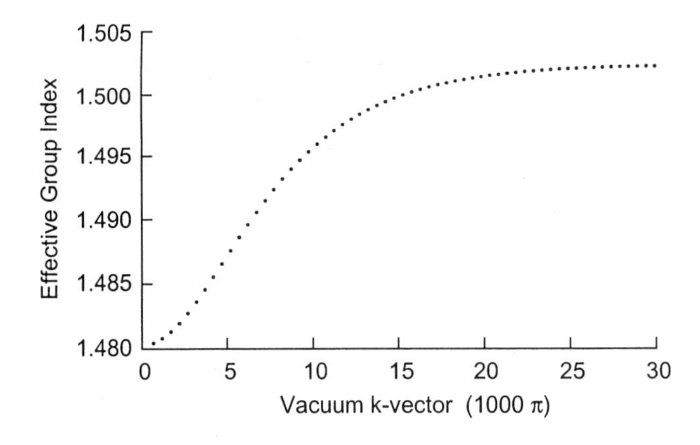

Figure 6.10. The group index experienced by a mode in the symmetric waveguide depends on the magnitude of k. For small k, the group index is approximately equal to that of the cladding, while for large k, N_g approaches the value in the guiding layer.

small values of k, the mode is weakly confined in the core, so most of the mode field travels in the lower index cladding region, and therefore sees an effective group index as determined by the cladding. As the wavevector increases, more of the mode is confined to the core. The effective group index increases from the cladding value (slightly greater than 1.48) to the core value (slightly greater than the core index, 1.5). The effective index is in fact the average index seen by the mode, and can be calculated using the mode amplitude as the weighting function. Fig. 6.11 shows the actual waveguide dispersion.

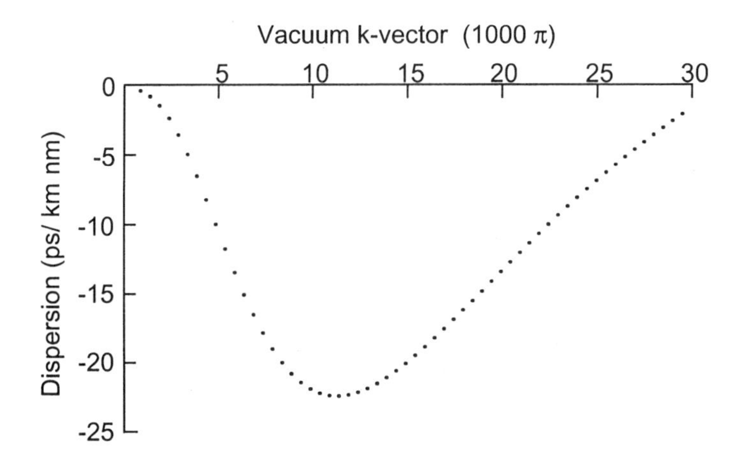

Figure 6.11. The numerically-derived waveguide dispersion. Notice it is maximum when the mode is converting from a weakly bound mode to a tightly bound mode.

The cause of waveguide dispersion is apparent from the spectral dependence of the group index (Fig. 6.10). As k increases, the effective group index of the mode increases. A pulse containing a superposition of k-vectors will spread in time because each component travels at different velocities. The second derivative of the β–k curve is proportional to the waveguide dispersion (see Eq. 6.40). In the *Mathematica* program listed above, the second derivative was calculated according the Eq. 6.40, and then multiplied by 10^{10} to convert the units (cm/sec^2) into ps/(km nm).

This example illustrates the power of numeric solution to the eigenvalue equation for a symmetric waveguide. By solving the equation repeatedly at many different values of k, we were able to map out the functional form of the dispersion. This technique forms a useful procedure for calculating such effects in new or novel structures.

6. Simultaneous Effect of Material and Modal Dispersion

Since the material and waveguide dispersion both depend on wavelength, they are highly correlated. Modal dispersion depends only on the mode structure of the waveguide, and is independent of material or waveguide dispersion. Estimating the total effective pulse broadening due to all of these effects depends in part on the pulse shape, and exact evaluation involves determining the impulse response of the waveguide [8]. We can make a simplification if we assume that the "not-unreasonable" spectral pulse shape is roughly Gaussian, $e^{-(\lambda-\lambda_0)^2/\Delta\lambda^2}$. We would expect an impulse to be broadened into a roughly Gaussian pulse due to modal dispersion after propagation through a suitable

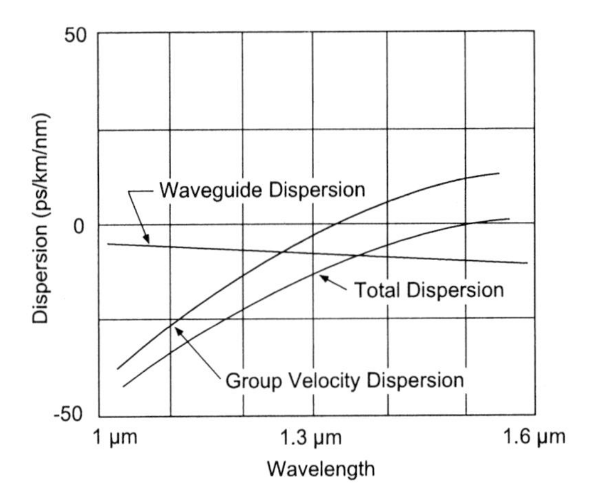

Figure 6.12. Plot of material and waveguide dispersion, simultaneously present in a waveguide. Their effects directly add.

length of fiber. We would expect similar Gaussian broadening due to material dispersion. The combined effect of convolving two Gaussian events is simply the quadratic sum of the pulse broadening

$$\tau_{total} = \sqrt{(\tau_{mat} + \tau_{wg})^2 + \tau_{modal}^2} \tag{6.43}$$

Except near the λ_0 point, the waveguide dispersion is usually negligible compared to material dispersion. However, near λ_0, the waveguide dispersion can play an important role in shifting the wavelength where the waveguide has zero effective dispersion. Consider the dispersion plot shown in Fig. 6.12 for a glass waveguide. Here the material dispersion is shown with it's characteristic curve increasing through zero near $\lambda = 1.3\mu$m, while the waveguide dispersion is a small but decreasing value in the region of the glass zero dispersion point.

The net effect of the two dispersions is to shift the point where the waveguide dispersion equals zero. Such an effect is called *dispersion shifting*, and is widely used in fiber optic waveguide design to optimize the performance of long-haul optical fiber communication links. By adjusting the zero dispersion point to match the wavelength of minimum attenuation, the maximum performance can be extracted from a waveguide.

7. Summary

The focus of this chapter was to describe the major dispersion effects in a waveguide. We began with the Lorentz model, and used it to develop an analytical expression for the index of refraction for a given material. Examples were presented showing how the Sellmeier equations could be used to determine properties such as material dispersion.

We took a look at modal dispersion, and developed an approximate formula which is adequate for extremely multimode waveguides. We mentioned the influence of mode coupling, which reduces the magnitude of this form of dispersion. Waveguide dispersion was described in general form, and illustrated through numeric example. Waveguide dispersion is most important in single-mode waveguides operating near λ_0. However, since that is where most single-mode optical communication links operate, obviously waveguide dispersion is an important topic.

In the next chapter we will explore how a graded-index profile can reduce the total modal dispersion in a multimode waveguide. Graded index waveguides do not yet match the dispersion performance of single mode waveguides, but they offer much larger areas for coupling light to and from the waveguide.

References

[1] H. A. Lorentz, *The Theory of Electrons*, p.9, Dover, New York, (1952)

[2] M. Born and E. Wolf, *Principles of Optics, 5th Ed.*, Ch. 13, Pergamon Press, Great Britain, (1975)

[3] American Institute of Physics *Physics Handbook, 3rd ed.*, Chapter 6, McGraw-Hill, Inc. USA, (1982)

[4] *OSA Handbook of Optics*, Walter Driscoll, editor, McGraw-Hill, New York, (1978)

[5] *CRC Handbook of Laser Science and Technology, Vol. III. Optical Materials:Part 1*, Marvin J. Weber, editor, CRC Press, Boca Raton, FL (1986)

[6] *Handbook of Optical Constants of Solids,* edited by Edmund Palik, Academic Press, Orlando, FL, (1985)

[7] H. Kogelnik, in *Topics in Applied Physics, Vol. 7. Integrated Optics, 2nd ed.*, T. Tamir, ed., Chap. 2, Springer-Verlag, Germany (1979)

[8] J. Gowar, *Optical Communications Systems*, Chap. 2, Prentice-Hall International, Englewood Cliffs, NJ, (1984)

[9] Luc Jeunhomme, *Single-mode fiber optics*, Marcel Decker, Inc, New York, (1983)

[10] John Senior, *Optical Fiber Communications*, Prentice-Hall International, Englewood Cliffs, NJ, (1985)

[11] G. Kaiser, *Optical Fiber Communications, 2nd ed.*, McGraw-Hill Inc., New York, (1991)

[12] A. Cherin, *Introduction to Optical Fibers*, McGraw-Hill, Inc., New York, (1983)

Practice Problems

1. A planar waveguide is made with a guiding film index of 3.5, a substrate index of 3.4, and a cover index of 3.4. The guiding film thickness is 2μm. The wavelength of interest is 900 nm. The material dispersion for this material is 400 ps/km·nm. Assume that waveguide dispersion is negligible.

   (a) Approximately how many modes exist in this structure (both TE and TM modes)

   (b) If the optical source has a 10 nm spectral width, what is the group delay dispersion due to material dispersion?

   (c) What is the modal dispersion for this waveguide?

   (d) What is the total group delay dispersion for this waveguide?

   (e) If a 1 nsec pulse is launched in one end of the waveguide, how far must it travel before it is 2 nsec long?

2. A gaussian shaped pulse, $E(t) = E_0 e^{-t^2/\tau^2}$, is launched into a dispersive waveguide. The wavelength of the pulse is chosen to be $\lambda_0 = 1.3\mu$m, the zero material dispersion wavelength. The waveguide is 10μm thick, has a guiding layer of index 1.55, and surrounding layers of index 1.52. The characteristic time constant, τ, for the pulse is 0.5 nsec.

   (a) What is the minimum possible frequency bandwidth of this pulse?

   (b) What is the wavelength bandwidth that corresponds to your answer to part (a).

   (c) What is the group delay dispersion of this structure?

   (d) What is the numerical aperture of this structure?

3. An asymmetric waveguide is 5μm thick, and has guiding index of 1.5, substrate index of 1.48, and cover index of 1.0.

   (a) Determine the cutoff wavelength for the TE_0 and TM_0 modes.

   (b) For the TE_0 mode, calculate β for 25 wavelengths spanning 1.3 μm to 1.55 μm. From these values, numerically calculate the waveguide group delay, τ_g, and the waveguide dispersion, $d\tau/d\lambda$.

   (c) Repeat the above calculation for the highest order mode in the waveguide. Which mode has the largest waveguide dispersion?

4. An LED with a spectral bandwidth of 20 nm, and a central wavelength of 850 nm, is to be used to transmit digital pulses on an optical waveguide. The symmetric waveguide is made of glass, with guiding index 1.5, and

Table 6.2. Index and Dispersion for Fused Silica Glass

$\lambda(\mu m)$	$n(\lambda)$	$\partial n(\lambda)/\partial\lambda \quad (\mu m^{-1})$
0.509	1.4619	-5.626×10^{-2}
1.014	1.4502	-1.331×10^{-2}
1.529	1.4443	-1.178×10^{-2}

surrounding layers of 1.48 μm. The material dispersion of glass at this wavelength is approximately 300 ps/km·nm.

(a) What is the maximum bit rate for a transmission length of 2 km for a single mode waveguide? (Assume that waveguide dispersion is negligible).

(b) What is the maximum bit rate for a multimode waveguide of 2 km length?

(c) What is the maximum thickness that the waveguide can be made without it becoming a multimode waveguide for TE modes?

5. The optical power that is incident on the earth's upper atmosphere from the sun is approximately 1 kW/m^2. The amount that reached the earths surface is about 30% of this value, due to scattering and absorption in the atmosphere.

   (a) If the optical field were all at one frequency, what would be the magnitude of the electric field on the surface of the earth?

   (b) The earth is approximately 150×10^6 km from the sun. If a satellite were sent into orbit around the sun, with an orbital radius of 10^6 km away from the sun, what would the optical field be on the satellite? Would non-linear optical effects be a problem in this region?

6. Table 6. lists the index of refraction and the dispersion of the index for fused silica glass at three wavelengths.

   Determine the phase velocity and group velocity for light at these three wavelengths in this material.

7. From the Sellmeier coefficients for NaCl and SiO_2, make plots of the index of refraction, phase velocity, group velocity, and group velocity dispersion for these materials over the $0.5 \rightarrow 2.5\mu$m region. Determine the zero dispersion wavelengths, λ_0, for both materials (note: The zero dispersion wavelength for NaCl lies slightly outside the specified region.).

8. Using the ground state configuration of a hydrogen atom, calculate the binding energy, E, and the average electrostatic potential, $\langle V \rangle$, between the electron and proton in the atom. What is the optical intensity (W/cm^2) that creates an equivalent electric field?

9. Near the band edge, the index of refraction of GaAs can be approximated as $n^2 = 8.950 + 2.054\lambda^2/(\lambda^2 - 0.390)$ (D. Maple, *J. Appl. Phys.* 35, p. 1241, (1964)). Plot the index of refraction and Group Velocity Dispersion for this material in the 1.0 - 2.0 μm region.

10. Calculate the FWHM frequency bandwidth, $\Delta\nu$, of

    (a) a hyperbolic secant squared pulse, $E(t) = 1/\cosh^2(t/t_0)$

    (b) an exponential pulse, $E(t) = e^{-|t|/t_0}$

    (c) a triangular pulse, with a pulse width (FWHM) of τ_0 seconds.

11. By inspection of Fig. 6.11, what *wavelength* has the largest magnitude of waveguide dispersion for the particular waveguide that is described in the example? What is the value of the waveguide dispersion in units of psec/km nm?

12. Using a computer, calculate the waveguide dispersion for the TE_0 mode in an asymmetric waveguide with the following parameters: $n_f = 1.50$, $n_c = 1$, and $n_s = 1.49$. The guiding film is 5 μm thick. Calculate the dispersion of the waveguide over a range spanning from a wavelength equal to 2/3 the cutoff wavelength to a wavelength equal to 1/3 the cutoff wavelength.

13. As a slab waveguide is made thinner, the lowest order TE_0 mode will become spatially smaller up to a certain dimension, and then will begin to grow larger as the waveguide continues to grow thinner. Is there a similar behavior in the waveguide dispersion for this mode? Is there a waveguide thickness at which the waveguide dispersion reaches a maximum or minimum? If there is, is the waveguide still single mode at this point? Assume the waveguide is comprised of two glasses with core index 1.5 and cladding index 1.495. Assume the guiding wavelength is 1 μm.

Chapter 7

GRADED INDEX WAVEGUIDES

1. Introduction

There are two ways to significantly reduce modal dispersion in a waveguide: use only single mode waveguides, or use a graded index waveguide. The first choice appears to be the simplest, but it is not always a practical solution. Single mode waveguides are much more difficult to couple light into than multimode waveguides. To help see this, consider the two planar structures shown in Fig.7.1. Both waveguides have the same indexes, but one of them has a larger guiding layer. The number of guided TE modes can be approximated from Eq.2.43,

$$m \approx \left(hk\sqrt{n_f^2 - n_s^2}\right)/\pi \tag{7.1}$$

Given identical indices of refraction, the only way to make a waveguide operate in a single mode is to reduce the thickness, h, of the guiding film. The smaller dimension of the single-mode guiding layer makes alignment between sources and other guides much more critical than with a large (multimode) structure. Connecting and aligning between multimode waveguides is easier due to the large size.

The second method for reducing modal dispersion is to use graded index waveguides. Graded index waveguides can be made with relatively large dimensions, easing the coupling and alignment problems common to single mode devices, and they can dramatically reduce modal dispersion.

Being able to analyze graded structures is essential to modern integrated optoelectronic design. Many fabrication processes such as dopant diffusion on planar structures naturally lead to graded index profiles. In this chapter, we first develop the ray picture of the graded index waveguide by using the *Eikonal* equation. Next we will develop the modal solutions to such a waveguide. Up until now, diligent application of Maxwell's equation has produced field

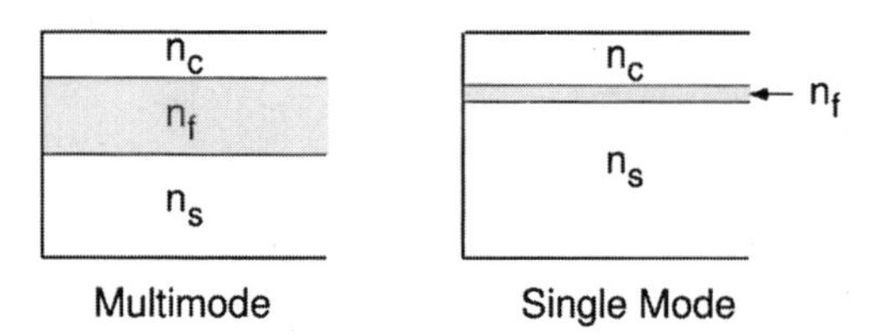

Figure 7.1. Two waveguides made with identical materials, but with different guiding film thickness. The larger film is a multimode structure.

solutions to the wave equation in the waveguide. With graded indices the equations become much more complicated. There are well known solutions to the wave equation for only a few specific index profiles. For the general index profile, one must resort to approximation or numeric techniques. Prior to the availability of numerical differential equation solvers, approximation methods such as the *WKB approximation* were applied to graded index problems. The *WKB approximation* uses a series solution to solve the wave equation in the graded index. The WKB technique is not widely used anymore, but it has been used to establish important formulae such as the number of modes and dispersion in a given waveguide. For an overview of the WKB techniques, refer to references [1],[2].

Today, with the widespread availability of powerful computers and numerical software, direct numeric solution of the wave equation is possible. This “brute force" technique allows finding precise values of the longitudinal wavevector, β, and can provide a graphical picture of the mode profile. The major limitation of direct solution is that it can not provide a general formula describing phenomena such as dispersion, group delay, or number of modes. However, due to it’s accuracy and simplicity in application, direct numeric solution of the wave equation can accurately map out dispersion, power confinement, mode shape, and most other relevant issues of a waveguide. We will provide a simple example of numeric techniques in this chapter.

In Chapter 9, we will describe two popular numerical simulation techniques called the Beam Propagation Method and the Finite Difference Finite Time technique which take a completely different numerical approach to finding the modes in a waveguide.The methods described here will provide better dispersion data, the later will prove useful in describing waveguide systems which have longitudinal structure or dimensional change.

2. Ray Tracing Model in Graded Index Material

Consider the slab waveguide shown in Fig.7.2. The index of refraction of the guiding layer is a function of position within the material. The index profile is plotted to the right of the waveguide. In this specific case, a symmetric profile is shown, with the index chosen to be highest at $x = 0$, smoothly decreasing with distance away from the central axis.

We want to examine how light propagates in such a structure. We will begin with the ray picture using the Eikonal equation.

2.1 The Eikonal Equation

Ray propagation in a graded structure is described by the *Eikonal Equation*. The term *Eikonal* comes from the Greek word for image[3]. The equation can be developed using a simple construction based on Snell's law.

Let's consider what happens when a ray is launched in a graded index material. The index gradient can be modelled as a series of microscopically thin homogeneous layers, each with an index, $n(x)$, where x is the distance from the axis ($x = 0$) to the thin layer. Consider the ray incident upon the interface between layers $n(x)$ and $n(x + \Delta x)$ at an angle θ, as shown in Fig. 7.3. This angle is the complement of the angle we normally use in applying Snell's Law.

The ray refracts at the interface between two layers. In terms of θ as we have defined it, Snell's law is

$$n(x)\cos\theta = n(x + \Delta x)\cos(\theta + \Delta\theta) \tag{7.2}$$

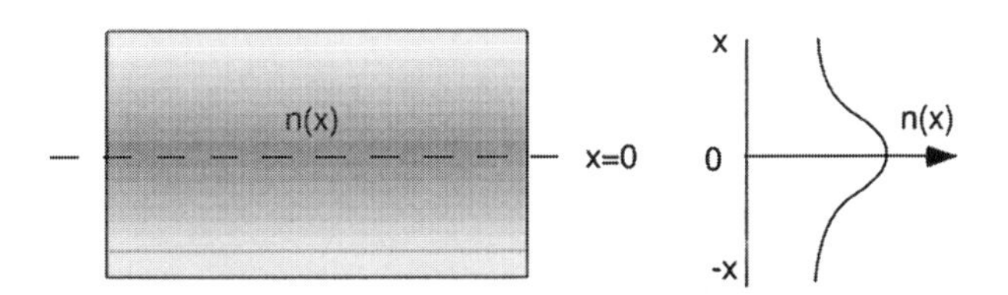

Figure 7.2. The index profile of a graded index planar waveguide.

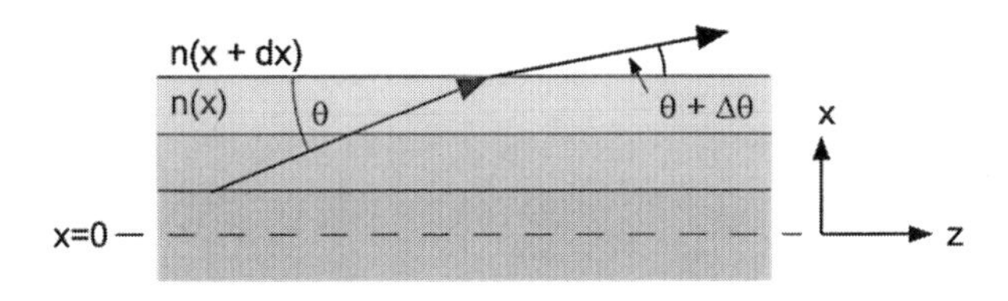

Figure 7.3. A graded index material can be modelled as a stack of thin layers, each with index n(x). Refraction occurs at the interface between two adjacent layers.

The change in direction, $\Delta\theta$, is very small. The term $n(x+\Delta x)$ can be rewritten in terms of a Taylor series expansion of $n(x)$ around x

$$n(x + \Delta x) \cong n(x) + \frac{dn}{dx}\Delta x \tag{7.3}$$

The $\cos(\theta + \Delta\theta)$ term can be expanded using the trigonometric identity

$$\begin{aligned} \cos(\theta + \Delta\theta) &= \cos\theta\cos\Delta\theta - \sin\theta\sin\Delta\theta \\ &\approx \cos\theta - \sin\theta\Delta\theta \end{aligned} \tag{7.4}$$

where we have assumed that $\cos\Delta\theta \approx 1$ and $\sin\Delta\theta \approx \Delta\theta$. Plugging Eqs. 7.3 and 7.4 into Eq. 7.2 we get

$$\begin{aligned} n(x)\cos\theta &= \left[n(x) + \frac{dn}{dx}\Delta x\right][\cos\theta - \sin\theta\Delta\theta] \\ &= n(x)\cos\theta - n(x)\sin\theta\Delta\theta + \frac{dn}{dx}\cos\theta\Delta x \\ &\quad -\frac{dn}{dx}\Delta x\sin\theta\Delta\theta \end{aligned} \tag{7.5}$$

Cancelling common terms and rearranging yields

$$n(x)\sin\theta\Delta\theta = \frac{dn}{dx}\cos\theta\Delta x - \frac{dn}{dx}\Delta x\sin\theta\Delta\theta \tag{7.6}$$

The last term in Eq. 7.6, being a product of two infinitesimals, is negligible compared to the other terms, so Eq. 7.6 can be written as

$$\frac{dn}{dx} \cong n(x)\tan\theta\frac{\Delta\theta}{\Delta x} \tag{7.7}$$

For most waveguide situations, the angle θ is going to be small (on the order of 10° or less), so small angle approximations can be used to simplify the expression:

$$\tan\theta \cong \theta \cong \frac{\Delta x}{\Delta z} \tag{7.8}$$

Substitute this into Eq. 7.7, along with $\Delta x = \Delta z\tan\theta$

$$\begin{aligned} \frac{dn}{dx} &\cong n(x)\frac{\Delta\theta}{\Delta z} \\ \lim_{\Delta z\to 0}\frac{dn}{dx} &= n(x)\frac{d}{dz}(\theta) = n(x)\frac{d}{dz}\left(\frac{dx}{dz}\right) \\ &= n(x)\frac{d^2x}{dz^2} \end{aligned} \tag{7.9}$$

Finally, solve for d^2x/dz^2

$$\frac{d^2x}{dz^2} = \frac{1}{n(x)}\frac{dn(x)}{dx} \tag{7.10}$$

This equation is called the Eikonal equation. The function $x(z)$ describes the exact ray path, and can be determined from Eq. 7.10 once $n(x)$ is known. As a practical matter, when the index of refraction is not a strong function of position, the denominator term for $n(x)$ in Eq. 7.10 is often replaced with n_0, the index of refraction at $x = 0$.

$$\frac{d^2x}{dz^2} \approx \frac{1}{n(0)}\frac{dn(x)}{dx} \tag{7.11}$$

This assumption can make an otherwise intractable differential equation manageable, while introducing negligible error so long as $n(x)$ does not change considerably over the spatial extent of the mode.

How do we interpret the Eikonal equation, Eq. 7.11? It's very simple: the Eikonal equation states that the ray always bends toward the higher index material.

Example 7.1 Ray path in a parabolic index profile

A common profile for gradient index devices is the parabolic index profile. Consider a planar waveguide which has an index profile in the $\hat{x}$-direction described by

$$n(x) = n_0\left[1 - \frac{x^2}{x_0^2}\right]$$

where x_0 is a characteristic length for the gradient. Given $n(x)$, we can evaluate dn/dx:

$$\frac{dn}{dx} = -n_0\frac{2x}{x_0^2}$$

Substitute this derivative expression into the Eikonal equation 7.10, and make the approximation that the denominator term $n(x) \approx n_0$, to get

$$\frac{d^2x}{dz^2} = -\frac{2x}{x_0^2}$$

This is a second order differential equation which has a general solution

$$x(z) = x_i \cos\left(\frac{\sqrt{2}z}{x_0}\right) + x_i' \sin\left(\frac{\sqrt{2}z}{x_0}\right)$$

The ray path from this equation is plotted in Fig.7.4.

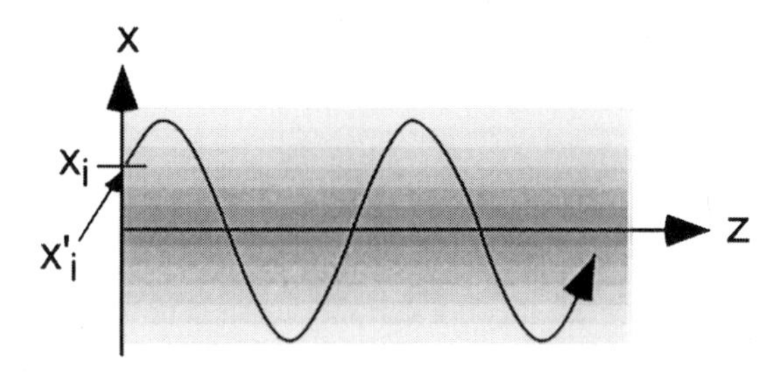

Figure 7.4. The ray path in a parabolic profile graded index.

Notice that the ray is bound to the axis, and that it has a periodic motion. The initial conditions, x_i and x_i', depend on the launch parameters for the ray. The propagating rays always *bend toward the region of higher index.* Once the ray crosses the axis of highest index, the curvature of its path changes sign in such a way as to return the ray toward the axis.

2.2 Dispersion Reduction with a Graded Index Profile

How does this graded index reduce modal dispersion? Recall that in the ray picture of the slab waveguide, modal dispersion arose due to the path differences between the high-order rays that followed a long zig-zag path down the waveguide, and low-order rays that travelled straight. Fig. 7.5 illustrates this case for two extreme modes. In the graded index structure a ray travelling near the axis will spend more time in high index material, and will travel slower than will a ray that is farther from the axis. However, rays far from the axis follow a longer sinusoidal path. Through optimal adjustment of the index gradient, it is possible to minimize the difference in group delay between the extreme rays. This will reduce modal dispersion and effectively increase the information capacity of the waveguide. The challenge in graded index waveguides is to choose an index profile that guides the light and minimizes modal dispersion.

The ray picture provides a useful illustration of how light is guided in a graded index structure. To actually perform the minimization requires use of the wave equation.

3. Modal Picture of the Graded Index Waveguide

To evaluate the properties of a graded index waveguide, we need to determine the allowed propagating modes and their dispersion characteristics. To find the modes, we must solve the wave equation for the graded index waveguide. This is usually a difficult task. Some profiles are analytically solvable,

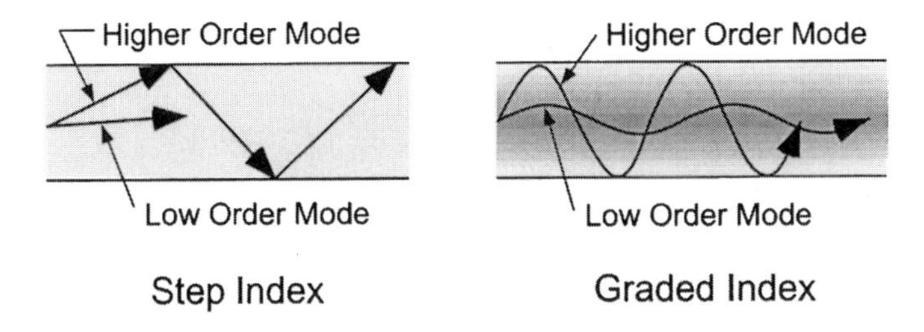

Figure 7.5. Modal dispersion arises in planar waveguides due to physical path differences between the various modes. In the graded index, the optical path length for each mode can be made the same.

however such profiles are rare and usually they do not have desirable dispersion properties to warrant their use.

For non-analytic cases approximations such as the WKB method can be used to find exact values of β, however the WKB method's real power is in the ability to generalized characteristics such as the effect of waveguide profile on dispersion. Using it to find exact values of β and the corresponding mode profile is a cumbersome task, and we will not pursue it here.

With the availability of numerical differential equation solvers, direct numeric solution of the wave equation is becoming the most widely used method for finding β in a graded index structure. Determining characteristics such as dispersion requires evaluating many different specific points, and then extracting information from the ensemble of data points. The later technique is well suited to workstation and personal computer application.

3.1 Profiles with Analytic Solutions

We will seek a TE solution to the equation

$$\frac{\partial^2 E_y}{\partial x^2} + (k_0^2 n^2(x) - \beta^2)E_y = 0 \tag{7.12}$$

in a planar waveguide structure. We assume that the index gradient extends only in the x-direction, and the electric field is polarized in the y-direction. There are a few graded profiles that have an exact analytic solution, one of which is the parabolic profile, described as

$$n^2(x) = n_0^2\left(1 - \frac{x^2}{x_0^2}\right) \qquad \text{for } x < x_0. \tag{7.13}$$

This profile is only valid where $x \ll x_0$, because $n(x)$ cannot be less than unity. The plot of Fig. 7.6 shows how the actual index eventually departs from the parabolic profile as the graded index meets the substrate index.

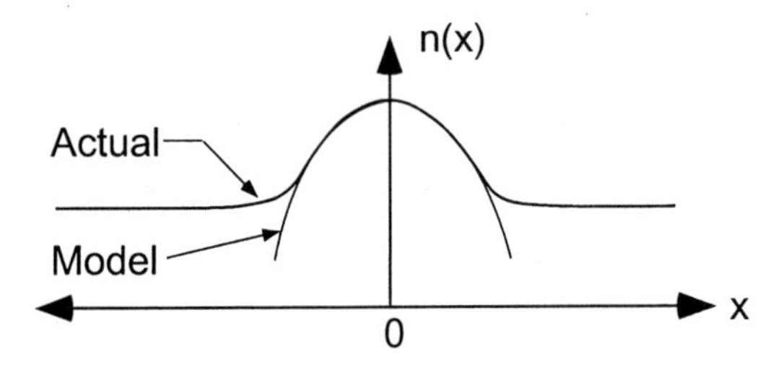

Figure 7.6. Plot of the actual index profile, compared to the plot of the model of the profile used to solve equations. The model is only accurate near $x = 0$.

Substituting Eq. 7.13 into the wave equation we get

$$\frac{\partial^2 E_y}{\partial x^2} + \left(k_0^2 n_0^2 - k_0^2 n_0^2 \frac{x^2}{x_0^2} - \beta^2 \right) E_y = 0 \tag{7.14}$$

While perhaps not obvious by inspection, Eq. 7.14 has well known solutions called the *Hermite-Gaussian* functions,

$$E_y^q(x) = H_q\left(\sqrt{2}\frac{x}{w}\right) \exp\left(\frac{-x^2}{w^2}\right) \tag{7.15}$$

where q is an integer that identifies the mode. H_q is the appropriate Hermite polynomial defined by

$$H_q(x) = (-1)^q \exp(x^2) \frac{d^q}{dx^q} \exp(-x^2) \tag{7.16}$$

The first three Hermite polynomials in x are

$$\begin{aligned} H_0(x) &= 1, \\ H_1(x) &= 2x, \\ H_2(x) &= 4x^2 - 2 \end{aligned} \tag{7.17}$$

The term w is the "beam radius," in analogy to the description of the spatial modes of a laser resonator. (The term "radius" is perhaps unfortunate in this application, since we are dealing with a planar field, not a cylindrical one). In the slab waveguide, w is defined through

$$w^2 = \left(\frac{2x_0}{k_0 n_0}\right) \tag{7.18}$$

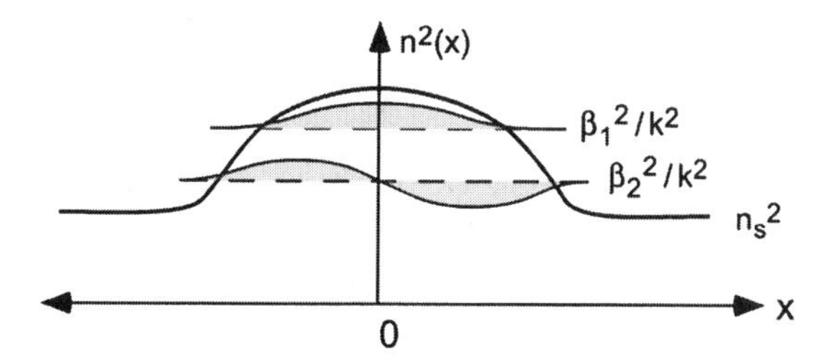

Figure 7.7. The values of β for allowed modes of the waveguide are bounded by the substrate index, $k_0 n_s$, and the maximum index in the guide, $k_0 n_0$.

The eigenvalues for the electric field propagation factor, β, are found from Eq. 7.14 to be

$$\beta_q^2 = k_0^2 n_0^2 - (2q+1)\frac{k_0 n_0}{x_o} \tag{7.19}$$

Theoretically, the parabolic profile has an infinite number of eigenvalues, β, given by Eq. 7.19. In practice, there are only a few bound modes, because the actual index profile eventually deviates from the perfect parabolic shape, as shown in Fig. 7.6. Fig. 7.7 shows two allowed values of β on the plot of the index profile, $n^2(x)$. The limits on the allowed values of β can be seen graphically. The upper limit is set by

$$\beta^2 \le k_0^2 n_0^2 \tag{7.20}$$

which is the same condition as for the slab waveguide. Similar to the slab waveguide, the lower cutoff condition requires that β remain larger than $k_0 n_s$. The solutions predicted by Eq. 7.15 begin to fail as β approaches $k_0 n_s$. This arises because the spatial structure of the individual modes extend to larger dimensions as the mode number q increases. For large q the modes get so large that they sample regions of the profile that are not parabolic, making the solutions inaccurate.

Nevertheless, the "exact" solution to a particular profile illustrates several features of graded waveguides. First, individual modes exist, and each mode has a unique field description. Second, the allowed values of the propagation coefficient are limited between $\beta_{max} = k_0 n_0$, and $\beta_{min} = k_0 n_s$. Finally, there is a limit to the spatial range over which a "solution" is accurate. This limitation is due to deviation of the mathematical description of the index profile at large values of x. There are other profiles that lead to "exact" solutions to the wave equation[4]. For most situations, however, exact solutions are not accessible, and we must use approximation techniques to describe the fields.

4. Direct Numerical Solution of the Wave Equation

The wave equation can be solved numerically using appropriate software. The numeric solution is actually quite simple to create. Given a differential equation and the necessary initial conditions, the entire solution can be generated using relatively simple recurrence relations[5] to plot the evolution of each variable. Unfortunately such techniques are not adequate for satisfying certain constraints, such as finding the eigenvalues for the equation. There are sophisticated techniques which can address this problem [4], especially when dealing with the Schrödinger equation. In this section, we want to illustrate the power of a simple *Mathematica* routine for finding the solution to the wave equation.

Let's once again consider the parabolic profile graded index so we can test any numerical solution we get against theory. To find a numeric solution, the waveguide structure must be fully specified with numeric values for all variables. As a specific example let the index of refraction be

$$n^2(x) = n_0^2 \left[1 - \left(\frac{x}{x_0}\right)^2\right] = 1.5^2 \left[1 - \left(\frac{x}{50}\right)^2\right] \tag{7.21}$$

where $n_0 = 1.5$ and $x_0 = 50\mu$m. Let the wavelength be 1 μm, so $k_0 = 2\pi$ (μm^{-1}). We will keep all dimensions in microns in this example. Plugging these values into the wave equation yields

$$\frac{\partial^2 E_y(x)}{\partial x^2} + \left((3\pi)^2(1 - \frac{x^2}{2500}) - \beta^2\right) E_y(x) = 0 \tag{7.22}$$

Notice that everything is defined except for β. Before a computer can begin to work on this problem, we must define a value of β. Here is where this technique becomes challenging. As we will see, the computer can find a solution, $E_y(x)$, for *any* value of β that we might suggest. But we know that only a few discrete values really exist. *The eigenmode solutions will be those which have a finite energy* (more precisely, we say that the solution is normalizable). Incorrect solutions will have amplitudes that tend toward infinity as the magnitude of x increases, and therefore can not represent real solutions. Thus, finding the proper solution proceeds by guessing a value of β, observing the numerically generated solution to the wave equation and seeing if it diverges toward positive or negative infinity, and then trying another value of β. The eigenvalue of β for a given equation will occur at the point where the solution does not diverge. The search for an eigenvalue is made easier by the fact that the sign of the diverging component of the profile changes as one goes past the eigenvalue. If for one trial value, β_i, the solution diverges positively, and for a different value, β_j, the solution diverges negatively, then the true value of β will lie somewhere between those two values. A series of converging guesses can rapidly zero in on the eigenvalue, β, for that mode.

To evaluate the example described above, the following commands in *Mathematica* were used.

```
n0=1.5;
x0=50;
nsquare=n0^2 (1-x^2/x0^2);
lambda=1;
k0=2 Pi/lambda;
beta=1.4920210 k0; (* trial guess*)
equation=e''[x] + (k0^2 nsquare - beta^2)e[x] ==0;
sol=NDSolve[{equation, e[0]==1, e'[0]==0}, e[x], {x, -15,15}]
Plot[e[x] /.  sol, {x, -15,15}, PlotRange->{-2,1}]
```

The key statements are the definition of the wave equation in terms of the parabolic profile `equation`, and the command `NDSolve[ ]`, which is a numerical differential equation solver. The initial conditions(`e[0]==1, e'[0]==0`) were chosen to give the field `e[x]` a unity amplitude and zero slope at $x = 0$. This ensures that we will find an even mode. To find the odd modes, we would set the amplitude to zero at $x = 0$, and defined a finite slope.

The trial value of β was manually entered each time, although it would be a simple task to put this entire process into a loop which sought the desired solution. We found very quickly that for a trial value $\beta = 1.493\ k_0$ the plot of the mode profile diverged toward negative infinity, but for $\beta = 1.492\ k_0$ the plot diverged toward positive infinity. Therefore there had to be a root somewhere between these values. After about 10 iterations the process led to a value of β accurate to 6 decimal places, which was decided to be accurate enough. Fig. 7.8 shows the plots of mode profiles for the last two iterations, taken at $\beta = 1.4920210\ k_0$ and $\beta = 1.4920211\ k_0$. The eigenvalue clearly lies somewhere between these two values. Notice that the mode profile is clearly defined if we ignore the tails extending beyond $|x| = 10\mu$m.

One solution diverges positively, while the other diverges negatively, indicating that the eigenvalue of the wave equation lies somewhere between these two values of β.

We terminated the iterations at this point, noting that the precision was probably exceeding any practical need. As a final test of this process, we can compare the numerically generated value of β to the "exact" value derived from Eq. 7.19. In this case, since there are two nodes in the waveform, this must be the $q = 2$ solution. Plugging numbers into Eq. 7.19 yields a value of $\beta = 1.49202103\ k_0$, which is consistent with our numeric result. Direct numeric solution is an extremely powerful technique for finding allowed eigenvalues of a waveguide with an arbitrary index profile. Due to this power and accuracy, it is the commonly used method by waveguide manufacturers when they are designing new optical waveguides. To develop dispersion relations, one would

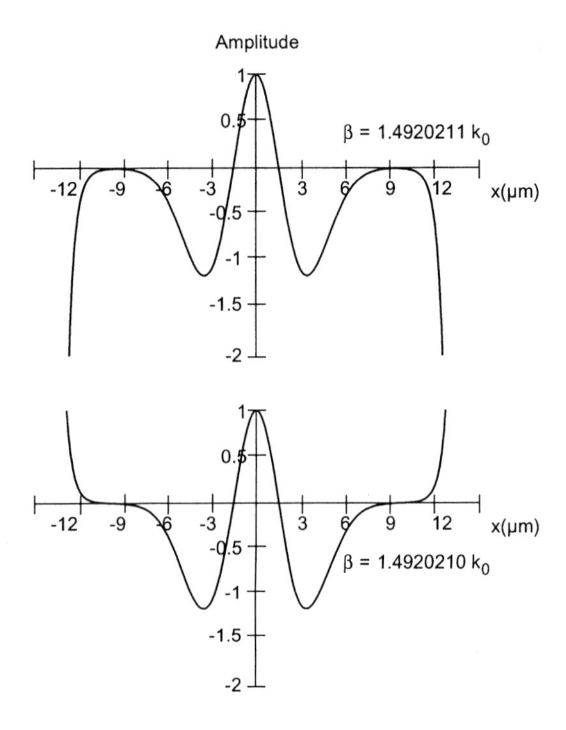

Figure 7.8. The "solution" to two trial values of β using the numerical differential equation solving routine in *Mathematica*

have to calculate a series of values, $\beta(\lambda)$, from which the appropriate derivatives could be found.

5. Summary

We developed two methods for looking at the propagation of light in a graded index waveguide. The first, based on the ray model, required the development of the Eikonal equation. Armed with the Eikonal equation, an equation of motion for rays in a graded index medium, the path of any ray can be calculated. The limitation of the ray picture is that it fails to provide information on the modal characteristics of the waveguide, including the propagation coefficient, β, the mode field size, or the dispersion properties. We then showed that direct numeric solution using package software provided a faster method for finding the eigenvalues of a graded index waveguide. In addition, a plot of the mode profile was generated.

References

[1] C. R.Pollock, *Fundamentals of Optoelectronics*, Irwin, USA, (1995)

[2] A. B. Sharma, S. J. Halme, and M. M Butusov, *Optical fiber systems and their components*, Springer Series in Optical Sciences, vol. 24, Springer, Berlin (1981)

[3] The term *Eikonal* was introduced by H. Bruns (1895)

[4] H. Kogelnik, in *Integrated Optics, Topics in Applied Physics Vol. 7, 2nd edition*, T. Tamir, ed.,Springer-Verlag, Germany, (1979)

[5] Richard Crandell, *Pascal Applications for the Sciences*, John Wiley and Sons, USA (1984)

[6] W. Press, B. Flannery, S. Teukolsky, and W. Vetterling, *Numerical Recipes, The Art of Scientific Computing*, Cambridge University Press, USA (1986)

Practice Problems

1. Is it possible to define a Numerical Aperture (N.A.) for a graded index waveguide? What would the NA of the waveguide described in Example 7.1 be?

2. Consider a graded index that forms a triangle profile

$$\begin{aligned} n(x) &= n_1 - \Delta n|x| \quad \text{for } 0 < |x| < x_0 \\ &= n_0 \quad \text{for } |x| > x_0 \end{aligned}$$

   where x_0 is the width of the triangle profile at its base. Using the Eikonal equation, develop an expression for the path of a ray through this waveguide. You will have to piece a number of solutions together, taking advantage of the translational invariance of the waveguide along the z-axis.

3. What is the modal dispersion for the parabolic index profile waveguide? Use Eq. 7.19 to determine an exact expression.

4. Use a numeric differential equation solving routine to determine the first two allowed values of β for the triangle index profile described in Problem 2. Assume that $n_1 = 1.5$, $n_0 = 1.47$, $k_0 = 2\pi\mu\text{m}{-1}$, and $x_0 = 15\mu\text{m}$. Plot the mode profiles.

5. A triangular index profile is described as

$$\begin{aligned} n(x) &= 1.50 - (0.004/\mu m)|x| \quad \text{for} \quad |x| < 5\mu m \\ &= 1.48 \quad \text{for} \quad |x| > 5\mu m \end{aligned}$$

Use a numerical differential equation solving routine to find the three allowed modes if the guided wavelength is 1 μm. Find the allowed values of β to at least 5 decimals.

6. Determine the waveguide dispersion of the triangular index profile described in Prob.5.. Using a numeric differential equation solving routine, find the eigenvalue, β, of the lowest order mode for wavelengths ranging from 0.9 to 1.2 μm. Based on this data, calculate the waveguide dispersion.

7. Use a numeric differential equation solving routine to determine the first two allowed values of β for the index profile, $n^2(x) = n_0^2 + 0.05e^{-x^2/x_0^2}$. Assume that $n_0 = 1.5$, $k_0 = 2\pi\mu\text{m}{-1}$, and $x_0 = 50\mu$m. Plot the mode profiles.

8. Calculate the waveguide dispersion of the waveguide described in Prob. 8. over the range from $\lambda = 1.5 \rightarrow 1.56\mu$m. Use the same procedure as outlined in Prob. 8..

9. Write a computer program using a suitable software package or language that numerically solves the wave equation. Add statements to the software to make it automatically iterate the trial values of β until the agreement between subsequent trial values is better than 1 part in 10^6.

10. Numerically explore several different index profiles, such as a Gaussian index, a triangle profile, a multiple step index, and an exponential profile. Which profile has the best mode confinement (defined simply by which mode is smallest between the half power points of the mode profile)? Which waveguide has the least waveguide dispersion?

11. One method for creating a slab waveguide is to bombard the surface of a substrate with high energy α-particles (ionized He accelerated to 1-2 MeV of energy). The particles cause damage to the lattice which effectively lowers the refractive index. The α-particle causes maximum damage when it has lost a good fraction of its original energy. As a result, the damage profile of irradiated glass will have a maximum a few μms below the surface. In this fashion, a buried lower-index layer can be formed on a substrate. The damage process is statistical. A simple approximation we can make to describe the resulting index profile is described as an offset Gaussian function,

$$n(x) = n_0 - \Delta n e^{-(x-h)^2/x_0^2}$$

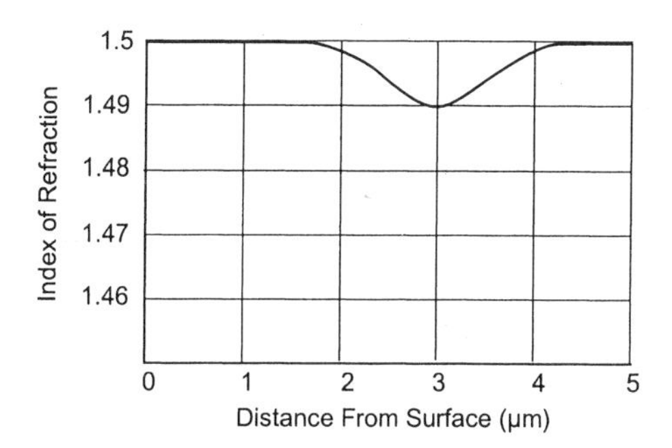

Figure 7.9. An approximation of the index profile generated by α-particle bombardment of a crystal surface.

An example of such a profile is shown below, using an average depth of 3 μm, and a Gaussian width of 0.7μm. This approximation underestimates the damage done at the surface as the α-particles penetrate into the lattice, but it provides a first-order description of the situation. Using direct numeric solution to the wave equation, determine the shape and eigenvalue of the lowest order mode for the structure shown in Fig. 7.9 ($n_0 = 1.5$, $x_0 = 0.7\mu$m, and $\Delta n = 0.01$). Determine how much energy exists in the region beyond the well. This energy will be radiated away, and represents a significant loss to the mode. Energy in the guided portion of the mode can tunnel across the well region and escape into the substrate. Since the total Δn is fixed by the optical properties of the material, the only parameter that can be varied easily is the width of the well. Explore how the mode confinement is affected as the well width, x_0, is varied.

Chapter 8

ATTENUATION AND NONLINEAR EFFECTS IN WAVEGUIDES

1. Introduction

An optical signal will be degraded by attenuation and dispersion as it propagates through a material. Dispersion can sometimes be compensated or eliminated through clever design, but attenuation simply leads to a loss of signal. Eventually the energy in the signal becomes so weak that it cannot be distinguished with sufficient reliability from the noise always present in the system. Attenuation therefore determines the maximum distance that optical links can be operated without amplification. Attenuation arises from several different physical effects. In an optical waveguide, one must consider i) intrinsic material absorptions, ii) absorptions due to impurities, iii) Rayleigh scattering, iv) surface scattering, v) bending and waveguide scattering losses, and vi) microbending loss. In terms of priority, intrinsic material absorption and Rayleigh scattering are the most serious cause of power loss for long distance optical fiber systems. Surface scattering dominates integrated waveguide losses. Impurity absorption has become less of a problem as improved material processing techniques have been developed over the years. In this chapter, we will establish the fundamental limits of attenuation, and provide a basic understanding of the attenuation processes that can be applied directly to materials such as glass or semiconductor.

You might ask, "If the signal at the end of a long link is too weak to observe, why not simply increase the input power?" We will show in this chapter that *nonlinear effects* limit the peak power that can be sent into a waveguide. Nonlinear effects play a major role in optical waveguides for two reasons. First, the small core dimensions of a typical waveguide can lead to extremely high optical intensities, even for small amounts of total power. Thus nonlinear effects can arise unintentionally. Second, the long lengths involved in certain systems, especially optical fiber links, allow small nonlinearities to add up and eventually become significant. These nonlinearities must be addressed by a system designer in order to achieve an acceptable signal-to-noise ratio for the received optical signal. There are many optical nonlinearities in optical waveguides. We will discuss three types: Stimulated Raman Scattering (SRS), Stimulated Brillioun Scattering (SBS), and Self Phase Modulation (SPM). As with all natural effects, we can either be victimized by them (and thus do our best to avoid them), or we can exploit them. Nonlinear effects can be exploited to create fascinating new devices. In this chapter we will look at the optical solitons.

2. Intrinsic Absorption Loss

In this section, we will describe attenuation due to absorption losses such as electronic and vibrational transitions in the material. Total optical attenuation is formally characterized by an expression known as Beer's Law

$$P_{out} = P_{in} e^{-\alpha z} \tag{8.1}$$

where P_{out} and P_{in} are the output and input powers of the optical wave, respectively, and α is the *attenuation coefficient*, with units of inverse length. α depends strongly on the wavelength of the light and the material system involved.

Fundamentally, absorption losses arise from the atomic or molecular resonances that we discussed in the Lorentz model of the atom in Chapter 6. An atomic transition can absorb electromagnetic energy from the applied field and store it in an excited state of the atom or solid. This energy eventually is dissipated through emission of a photon or through creation of lattice vibrations, and represents a loss to the electromagnetic field. In Chapter 6, we developed an expression for the index of refraction based on the Lorentz model. It is straightforward to show that the real (n') and imaginary (n'') parts of the index of refraction (Eq. 3.16) are:

$$\begin{aligned} n'(\omega) &= \sqrt{1 + \frac{(Ne^2/m)(\omega_0^2 - \omega^2)}{(\omega_0^2 - \omega^2)^2 + \gamma^2\omega^2}} \\ n''(\omega) &= \pm \frac{(Ne^2/m)\gamma\omega}{n'(\omega)\left[(\omega_0^2 - \omega^2)^2 + \gamma^2\omega^2\right]} \end{aligned} \tag{8.2}$$

The imaginary part of n can lead to attenuation or gain, depending on its sign. Unless special efforts are made, such as creating a population inversion, the imaginary term leads to attenuation. Consider the electric field

$$E_0 e^{-jk_0 n z} = E_0 e^{-jk_0(n' - jn'')z} = E_0 e^{-jk_0 n' z} e^{-k_0 n'' z} \tag{8.3}$$

where the sign of the imaginary term was chosen to yield a decaying amplitude. Even far from resonances, Eq. 8.2 shows that there will always be a residual absorption.

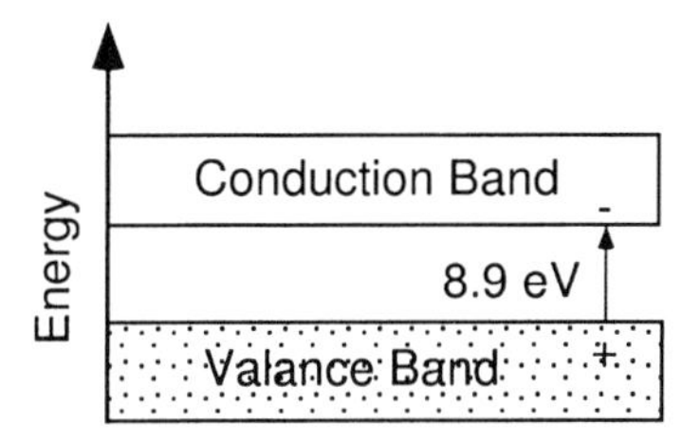

Figure 8.1. The bandgap of fused silica is about 8.9 eV. A photon with energy $\geq$ 8.9 eV is required to excite an electron from the valence band to the conduction band.

2.1 Absorption Due to Electronic Transitions

In this section we will concentrate on the optical properties of fused silica (i.e. glass), and mention analogous behavior in other systems such as semiconductors where appropriate. Fused silica has a "transmission window" in the near infrared. This transmission window lies between absorptions due to electronic transitions and Rayleigh scattering at the short wavelength side of the spectrum, and vibrational transitions on the long wavelength side of the spectrum.

Fig. 8.1 crudely illustrates the valence and conduction bands in fused silica (SiO_2). The bandgap for this important optical material is approximately 8.9eV. To optically raise an electron from the valence band to the conduction band requires a photon of energy greater than or equal to 8.9 eV. This corresponds to light with wavelength shorter than approximately 140 nm. Since the absorption process involves elevating an electron to a new state, such transitions are called *electronic transitions*, or interband transitions. In principle, photons with longer wavelengths, such as visible light, cannot excite electrons across the bandgap, so they are not absorbed by the material. This partially explains why glass is transparent in the visible region.

The absorption edges, λ_{min}, of several optical materials are tabulated below. The semiconductor materials are useful for integrated optics applications and detectors. From Table 8.1 we can see that Gallium Arsenide (GaAs) absorbs light with wavelengths shorter than approximately 0.88 μm. These interband transitions in semiconductors strongly influence the absorption and dispersion characteristics of optical waveguides made from semiconducting material. The semiconductor laser is a good example of such a waveguide system.

In practice, the sharp absorption edge predicted by the simple band model of Fig. 8.1 is not observed. The transition from absorbing to transmitting usually follows a soft curve as the wavelength changes. Fig. 8.2 shows the optical absorption coefficient of GaAs as a function of wavelength.[2] Notice that instead of changing abruptly as the photon energy drops below 1.42 eV, , the absorption decreases exponentially.

Table 8.1. Energy Bandgaps, Absorption Edges, and Refractive Index of Various Materials[1]

Material	E_g (300 K)	λ_{min}	n
PbS	0.37 eV	3.34 μm	4.1
GaAs	1.42	0.87	3.34
Si	1.12	1.10	3.5
Ge	0.67	1.85	4.1
InAs	0.35	3.54	3.1
Diamond	5.5	0.23	2.41
SiO_2	8.9	0.14	1.45

Like the GaAs example shown in Fig. 8.2, the band gap of most materials is not sharpely defined in energy. This is especially true in glass. The variety of molecular bonds and configurations in the amorphous material lead to slightly different binding energies for individual electrons, so, unlike the idealized energy picture presented in Fig.8.1, there is not a sharp edge in energy that defines the conduction or valence bands. Furthermore, thermal vibrations slightly alter the band structure of the material, further smearing out the energy distributions. As a result, the onset of optical absorption is usually a smooth function of wavelength. The empirical absorption coefficient, α from Beer's Law (Eq. 8.1, for band-edge absorptions has been found to follow a formula

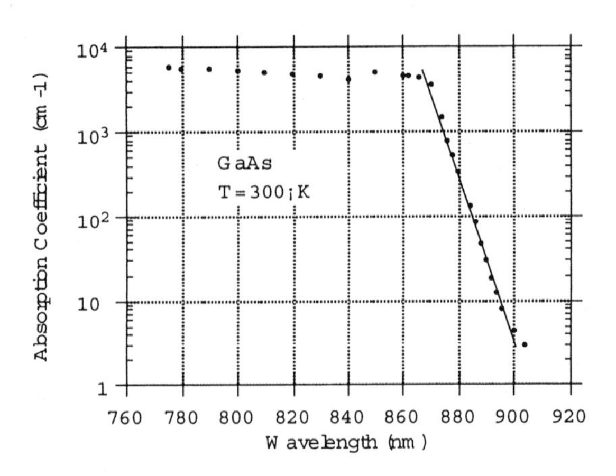

Figure 8.2. The optical absorption edge of GaAs at room temperature. Notice that the absorption does not sharply increase for photons above the bandgap energy. Rather, the absorption edge shows an exponential increase with increasing energy.

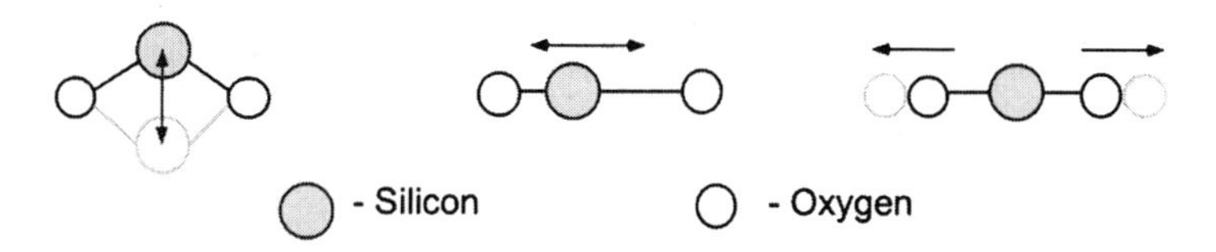

Figure 8.3. The three fundamental vibrational modes of the triatomic SiO_2 molecule.

called *Urbach's Law*

$$\alpha = Ce^{\hbar\omega/E_0} \tag{8.4}$$

where $\hbar\omega$ is the energy of a photon with with angular frequency ω ($\hbar$ is Planck's constant) , and C and E_0 are empirical constants for a given material. Using Urbach's Law in Beer's law, we see that as wavelength increases, attenuation becomes smaller, but never reaches zero — exponential functions just keep getting smaller and smaller. So even at long wavelengths there will be some residual absorption due to electronic transitions. The Urbach absorption arises from numerous weak effects such as multi-photon absorptions and combined photon-phonon interactions, as well as the Lorentzian absorption tails described in Eqs. 8.2 and 8.3.

Before we get much farther, we should note that due to the wide range of values for the attenuation coefficient, plots of the attenuation of materials are usually shown on a semilog axis. The common unit is the *decibel*, which is defined in terms of the logarithm of a power or intensity ratio

$$dB = 10 \log \frac{P_{out}}{P_{in}} \tag{8.5}$$

For example, a 20 dB attenuation would represent a reduction of power by a factor of 100, while 3 dB attenuation would represent a reduction by a factor of 2. The typical units of attenuation are in dB/km for low loss materials such as silica optical fibers, or dB/cm for lossy materials such as GaAs waveguides.

2.2 Absorption due to Vibrational Transitions

In the infrared region, the photons have typically less than 1 eV of energy, so there are few losses due to direct *electronic* transitions. However, another absorption process begins to appears. This absorption is due to *vibrational* transitions occurring in the SiO_2 structure of glass. Fig. 8.3 shows the possible vibrational modes of a SiO_2 molecule. Because it has three atoms, the SiO_2 molecule can vibrate in a bending mode, an asymmetric stretch, or a symmetric stretch.

The fundamental vibrational transitions occurs at a frequency of approximately 40 THz. This frequency corresponds to light with a wavelength of about 9 μm. The attenuation coefficient at 9μm is quite large, being over 100 cm^{-1}.

Due to nonlinear interactions in the vibrational structure of glass, there are absorptions at harmonics (technically called *overtones*) of the fundamental frequency, in particular at 3.2μm, 3.8μm, and 4.4μm. The absorption coefficients for these overtones tend to be several orders of magnitude weaker in strength than the fundamental, but they are still large enough to be observed in small thicknesses (millimeters) of glass. These absorptions have long tails, and their presence is noticeable at wavelengths far from resonance. These tails lead to a near infrared absorption that begins to become apparent around 1.5μm.

2.3 Attenuation due to Impurities

The first proposals to use optical glass fibers for long distance communication were criticized on the basis of measured attenuation coefficients of glass. Window glass had attenuations projected to be 1000 dB per kilometer, which as we mentioned in Chapter 1, is so attenuative that if all the light emitted by the sun were sent through one kilometer of such material, not a single photon would make it out the other end.

Since the intrinsic attenuation of glass could not explain this high loss, the problem of excess attenuation had to be rooted in impurities. Fused silica often contains intentional dopants, such as sodium, which modify the melting temperature or index of refraction of the glass. Glass also often contains trace amounts of unintended impurities. Transition metal impurities, such as Cr or Ti, have low lying electronic states, which can dramatically increase the absorption of visible and near infrared light in glass. At certain wavelengths, concentrations as small as *1 part per billion* of iron (Fe^{2+}) or chromium (Cr^{3+}) can increase absorption losses by 1 dB per kilometer. The pioneering work of *Kapron, Keck,* and *Mauer* et al.[3] of Corning, Inc. in the early 1970's demonstrated that the high loss was in fact due to impurities. They introduced a novel method for creating ultrapure glass fibers based on chemical vapor deposition (CVD). Based on their work, various schemes have been developed and are commercially employed in large scale operations to keep impurity concentration low.

Perhaps the most pernicious impurity is water. Water will react with many materials and form the hydroxyl ion, OH^{-1}. OH^{-1} has a vibrational stretch that corresponds to a wavelength of 2.7 μm, however important overtones occur at 1.39μm, 1.25μm, and 0.95 μm. Because of its strong optical dipole moment, only one part per million of OH^{-1} in fused silica will increase absorption losses by 30 dB per kilometer at 1.38 μm. It is important to reduce water content as much as possible during the manufacture and processing of optical fiber. Current values of OH^{-1} concentration in ultrapure glass are less than a few parts per billion. At such levels, the impurity absorptions are less than the intrinsic material losses, so there is no advantage in further reduction. However, one of the biggest concerns in optical fiber packaging and installation is ensuring that OH^- won't find a way to diffuse into the fiber. For example consider a fiber that goes under the ocean to connect two continents. Installing such a fiber is quite an expensive endeavor, so it is critical that the fiber retain its original low loss characteristics for decades. Because of the long length of such fibers, even a small additional absorption could lead to early (and expensive) failure of the link.

3. Rayleigh Scattering

Rayleigh scattering is a fundamentally different attenuation mechanism. Instead of light being absorbed and converted into stored energy within the media, light is simply scattered away from its original direction. Rayleigh scattering is responsible for giving the daytime sky its blue color, where more blue light from the sun is scattered (some of it downward to Earth) compared to the red wavelengths in the solar spectrum. We will consider the "classical" view of Rayleigh scattering, relying on electromagnetic theory rather than quantum theory to describe the physical process.

Rayleigh scattering is the scattering of light off random density fluctuations that exist in a dielectric material. When the index fluctuations occur over a dimension that is *small compared to the wavelength of light*, the density fluctuation can be viewed as a small dielectric particle which is uniformly excited by the field. The instantaneous dipole moment for such a particle is

$$\Delta p(t) = \Delta\epsilon E(t) \tag{8.6}$$

where $\Delta\epsilon$ is the excess polarizability of the random fluctuation (excess in relation to the homogeneous background). A dipole radiates power by the well known expression [4]

$$P_{rad} = \frac{\omega^4 p^2}{12\pi\epsilon_0 c^3} = \frac{\omega^4 (\Delta\epsilon)^2 E^2}{12\pi\epsilon_0 c^3} \tag{8.7}$$

If there are N independent scattering particles in a unit volume, then the total power scattered is simply N times the result of Eq. 8.7. The dipole will radiate in a plane orthogonal to the polarization of the driving field. A majority of the power scattered by the particle will be directed *away* from the original direction of the wave. This represents a loss to the wave. A key thing to note is that the radiated power increases as the fourth power of frequency. This explains why the sky is blue: statistical fluctuations in the density of air serve as the scattering points for the Rayleigh process. Some of this scattered light comes down toward the surface of the Earth, and since blue light scatters at a rate ten times that of red light, the sky looks blue. At dusk or dawn, the blue light is strongly polarized — can you explain why based on the discussion above?

So what does this have to do with waveguides? Rayleigh scattering in dielectrics arises from small density fluctuations that are frozen into the dielectric during manufacture. When an optical fiber is formed, the glass is pulled through an oven, where the molten material is stretched to become a thin fiber. After leaving the oven the glass freezes back into an amorphous solid. There is a high level of thermal agitation at the transition temperature (melting point) of glass, and this thermodynamical disorder leads to compositional and density fluctuations. These random variances are frozen in, and serve as the source for subsequent Rayleigh scattering. This is a fundamental process: there is nothing that can be done to eliminate the thermal agitation that accompanies melting material for manufacture of waveguides. In terms of material parameters, the scattering loss coefficient is [5]

$$\alpha_R = \frac{8\pi^3}{3\lambda_0^4}\left[(n^2-1)kT\beta + 2n\left(\frac{\partial n}{\partial C}\right)^2 \overline{\Delta C^2}\,\delta V\right] \tag{8.8}$$

where k is Boltzmann's constant, T is the transition temperature, β is the isothermal compressibility, n is the index of refraction, $\partial n/\partial C$ is the change of index with dopant concentration, and $\overline{\Delta C^2}$ is the mean square dopant concentration fluctuation over volume δV, which is smaller than, but on the order of, λ^3. The design of long distance communication fibers, where attenuation is critical, is influenced both by the temperature at which the fiber solidifies, and by the dopants necessary to create the desired refractive index profile. Lowering the melting point of the glass reduces subsequent Rayleigh scattering dramatically, so additional dopants are sometimes added to intentionally reduce the temperature of the melt as much as possible. Likewise, dopants having large index of refraction changes with concentration are avoided if possible, because random concentration gradients will lead to large index variations. Nevertheless, since the source of the scattering is the random fluctuations, and since these fluctuations arise from thermodynamical reasons, the losses are fundamental and cannot be compensated or eliminated.

3.1 Minimum Attenuation in Fused Silica

Fig. 8.4 shows a logarithmic plot of the near-infrared attenuation rate in a fused silica sample arising from the infrared vibrational absorption and from Rayleigh scattering. At 1μm, the intrinsic UV absorption described by Urbach's Law is insignificant compared to Rayleigh scattering. Due to its λ^{-4} behavior, Rayleigh scattering diminishes with increasing wavelength. However at about 1.6μm, the Rayleigh scattering is overwhelmed by the increasing vibrational absorption. The tails of the strong absorption at 9μm begin to appear, so losses increase dramatically as the wavelength extends above 1.6μm. The minimum absorption occurs near 1.55 μm with a value of 0.2 dB/km. The small absorption peak near 1.38 μm is due to residual OH^{-1} in the material.

The minimum attenuation value of 0.2 dB/km is a fundamental limit that cannot be further reduced in fused silica. This barrier has led some people to explore new materials. Since Rayleigh and Urbach losses decrease rapidly with

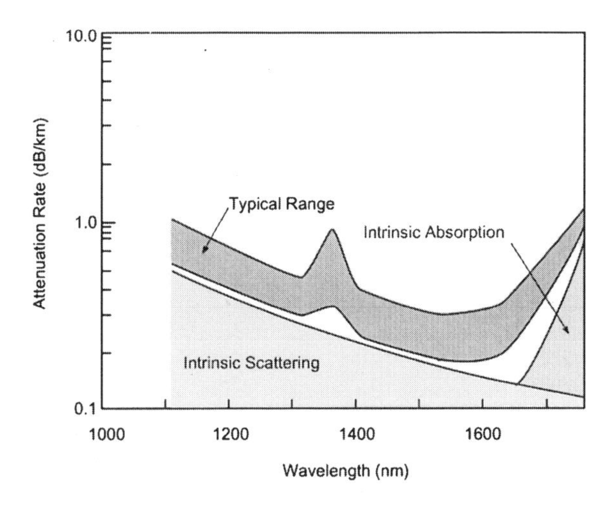

Figure 8.4. The total attenuation that is found in fused silica. The minimum occurs near 1.55 μm. (Data courtesy Corning, Inc.

increasing wavelength, attention has been focussed on moving toward materials with longer wavelength transmission windows. In SiO_2 the infrared absorption at 9 μm leads to the increase in attenuation beyond 1.6 μm. If the vibrational infrared absorption features could be moved to longer wavelengths, say 40 μm, then it is possible that the minimum attenuation could be reduced below that of SiO_2 by going toward longer wavelengths.

A simple way to reduce the frequency of the vibrational absorption ($\omega = \sqrt{k/m}$) is to use heavier atoms in the solid. Since the "spring constant", k, derived in Eq. 6.7 that describes the bonding between atoms in solids is approximately the same magnitude for most dielectric materials, our only degree of freedom for decreasing the vibrational frequency is to increase the mass. For example, if Ag, with atomic weight of 107 AMU, were substituted for Si, with atomic weight 14 AMU, we could expect to see the infrared absorption frequency reduced by approximately a factor of three. This would allow low loss operation at a longer wavelength, perhaps 2-3μm, where both the Urbach absorption and Rayleigh scattering would be significantly lower. For these reasons there has been substantial research into materials like AgCl, KRS-5, and other "heavy metal" glasses. Because of the higher mass of the atoms, these materials have potentially much lower attenuation coefficients than SiO_2; unfortunately they are difficult to manufacture. To date, they have displayed enormous losses due to waveguide imperfections such as polycrystalline structure, rough side walls, etc.. Nevertheless, there is continued research into longer wavelength optical materials that will potentially have lower loss than presently available

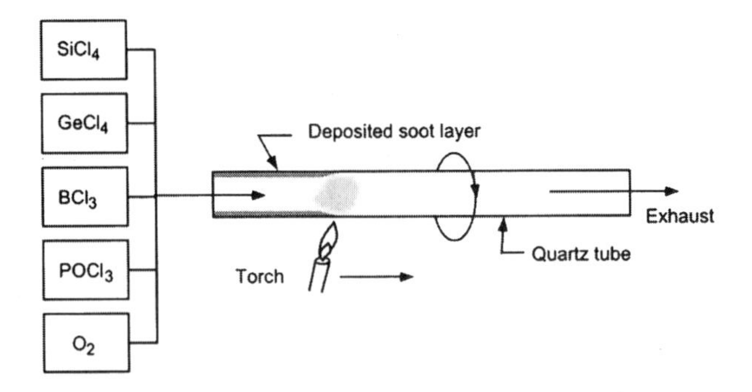

Figure 8.5. A preform is created by selectively depositing layers of doped glass powder on the inside of a quartz tube. The torch moves back and forth along the tube, initiating the hydrolyzation reaction that forms the soot of glass particles.

fused silica fibers. It should be noted, however, that the recent introduction of low cost, high performance optical amplifiers that are compatible with fused silica fiber systems have slowed the thrust into these new materials.

4. Optical Fiber Manufacture

To produce fibers with tight tolerances on dimensions and low loss, it is necessary to pull fibers from a glass "preform". The manufacture of the preform is the critical part of the process.

We will describe the basic process developed in the early 1970's based on Chemical Vapor Deposition (CVD). This technique relies on hydrolyzing a mixture of $SiCl_4$ and O_2 with a mixture of an additive such as $GeCl_4$, BCl_3, or $POCl_3$ to produce a "soot" of doped SiO_2. This soot looks like a fine snow, and is deposited on the outside or inside of a glass tube to create an index profile.

Fig. 8.5 depicts a simple preform lathe that uses inside deposition. A manifold connecting sources of ultrapure dopant gasses mixes the desired ratios of $SiCl_4$, O_2 and the dopants, and feeds the gas into one end of the rotating tube. The outside of the quartz tube is heated by a torch which moves back and forth along the entire length of the tube. Once the gas mixture comes in contact with the heated zone of the tube, a hydrolyzation reaction occurs, creating a fine snowfall of doped glass particles. These fall to the wall of the tube, and are fused into a layer on top of previous soot layers. The inside of the tube is built up layer by layer. Once the soot is deposited in the desired profile, the tube is heated and evacuated. This causes the tube to collapse on itself, forming a solid cylinder of glass with the desired index profile. The size of this solid preform can range from approximately 2-8 cm in diameter, depending on the process.

The index profile of the preform is controlled by adjusting the gas mixture flowing into the tube. Table 8.2 lists the gas reactions and their effect on the index of refraction of the soot.

Table 8.2. Glass mixtures for waveguide preform formation

Mixture	Glass	Refractive index
$SiCl_4, O_2$	SiO_2	n_0
$GeCl_4, O_2$	GeO_2	$n > n_0$
$POCl_3, O_2$	P_2O_5	$n > n_0$
BCl_3, O_2	B_2O_3	$n < n_0$

Gasses are used as a source material for two reasons. First, it is easier to chemically purify a gas than a solid or liquid. In this way, undesired impurities can be eliminated from the source material before it is incorporated into the preform. The second reason involves the use of chlorine to eliminate residual OH^{-1} from the hydrolyzing reactions. Chlorine is very effective in removing OH^{-1} from the soot. One characteristic of preform-making facilities is the need for large scrubbers to remove the chlorine from the exhaust gas before discharging it into the environment.

To form a fiber, the preform is lowered into a tube furnace which is operating at a temperature sufficient to bring the tip of the preform temperature to the melting point. The molten tip is pulled down and stretched into a thin glass fiber. This thin fiber leaves the furnace and solidifies, then is fed through a diameter measuring diagnostic, and across a mandrel that sets the pulling rate of material from the molten preform. The rotation speed of the mandrel is adjusted to maintain the diameter of the fiber at the desired value. Most commercial fiber is pulled at a rate of several meters per second, and has an outside diameter of 125 μm. A plastic coating is applied to the fiber add strength and protect the glass from mechanical scratches which could lead to fractures. The fiber is then wound on a spool. The ultimate length of a single fiber is limited on how long the preform can be made. Typical values are in the 10 km range for a single fiber. Longer fiber lengths are created by splicing shorter sections together.

The index profile in the preform is preserved in exact dimensional proportion in the pulled fiber. The glass flows in a pattern exactly as it was formed, making it possible to create the precise index profiles in the small fiber. For example, if one wanted to create a step-index fiber with a core diameter of 12.5μm, and a cladding layer extending to a diameter of 125μm, one would make a preform where the higher index core comprised the inner 10% of the diameter, with the lower index cladding material surrounding it. The spatial proportions are maintained when pulling the fiber down to its final dimension.

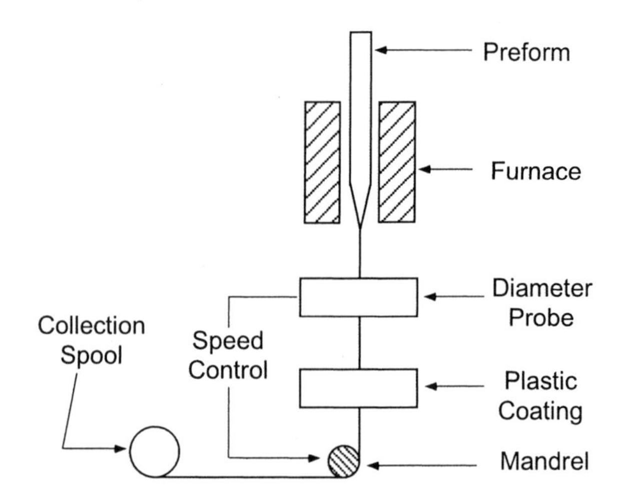

Figure 8.6. A fiber pulling station consists of a furnace to melt the tip of the preform, a diameter probe to provide speed control information to the pulling mandrel, and a plastic coater to add a protective outer jacket to the fiber.

5. Losses in Rectangular Waveguides

As we noted above, the losses in optical fibers tend to be in the 0.2 dB/km range, while for integrated waveguides in materials such as GaAs, the losses tend to much higher, such as 5dB/cm. Why is there such a dramatic difference? In this section we will describe how typical planar waveguide structures are fabricated, and will show that most of the loss arises from scattering at the sidewalls.

For highly integrated systems it is important to be able to guide light in waveguides with cross sectional dimensions on the order of 1 micron, and even sub-micron in some cases. This requires using waveguides with high index of refraction contrast between the core and the cladding. In an optical fiber the index difference of $\Delta n = 0.002$ was significant; in integrated structures the index difference can be 1000 times larger. This dramatically increases the sensitivity of the waveguide to scattering losses at the surface. Furthermore, if a single mode waveguide is needed, as it often is for modulators, switches, and couplers, a large index contrast between the core and cladding creates the need for small core dimensions. This can be seen from Eq. 3.40 by considering a slab waveguide. For a TE single mode, the waveguide thickness, h, must satisfy

$$\frac{\pi}{2} \leq hk(n_f^2 - n_s^2)^{1/2} \leq \frac{3\pi}{2} \tag{8.9}$$

Therefore for $hk(n_f^2 - n_s^2)^{1/2} \approx \pi$, the size of the waveguide will be given by:

$$h \approx \frac{\pi}{k} \frac{1}{(n_f^2 - n_s^2)^{1/2}} \tag{8.10}$$

High index of refraction contrast waveguides are usually buried waveguides that use a semiconductor core (such as GaAs, InP or Si with n 2.5-3.5) and an oxide cladding (AlO, InO and SiO with n 1.5). From Eq. 8.10 typical dimensions for a single mode Si waveguide clad with SiO_2 operating at $\lambda = 1.5\mu$m are

$$h \approx \frac{\pi}{k} \frac{1}{(n_f^2 - n_s^2)^{1/2}} = \frac{\pi}{2\pi/1.5} \frac{1}{(3.5^2 - 1.5^2)^{1/2}} = 0.2\mu\text{m} \tag{8.11}$$

The manufacture of such waveguides takes advantage of the processing technology developed for the semiconductor electronics industry. Most integrate photic devices are designed to operate on a semiconductor substrate, which typically have large refractive indices. The waveguide must be isolated from the high-index substrate, otherwise the optical energy would simple couple to the substrate and disperse. Isolation is accomplished by depositing a lower refractive index cladding onto the substate, typically using a Chemical Vapor Deposition scheme such as Plasma Enhanced Chemical Vapor Deposition (PECVD) or Organo-Metallic Vapor Phase Epitaxy (OMVPE). Fig. 8.7(a). shows the result. A waveguide layer of high refractive index is then deposited by sputtering. Sputtering is simply vaporizing a solid and directing the vapors toward the substrate, where they deposit and coalesce back into a solid. The net result is a high-index layer sitting atop a low index layer.

To make waveguides and devices from this layer, portions of the guiding film must be physically removed from the substrate. This is done using photolithographic processes. First a photoresist is spun onto the coated substate. The photoresist is a polymer material that can be cross-linked by exposure to blue light. The desired pattern of the waveguides and devices in the guiding layer is transferred by exposing the photoresist to blue light through a mask which has the a two-dimensional pattern of the device. This is shown schematically in Fig 8.7c. The photoresist is then developed. Depending on whether it is a positive resist or a negative resist, the developer will remove the resist from the regions which were exposed, or which were left in the dark, respectively. The developed photoresist acts as a barrier for the next step, which is the selective removal of the guiding layer. This step, called etching, can be done with wet chemicals, such as hydrofluoric acid or KOH solution, or via a reactive ion plasma (RIE– Reactive Ion Etching). After etching, only the waveguides and devices are left in the guiding layer. A final overgrowth of lower index material is deposited using PECVD.

Losses in these waveguides are usually on the order of 0.2-5dB/cm. These losses are highly dependent on the fabrication process of such waveguides. The

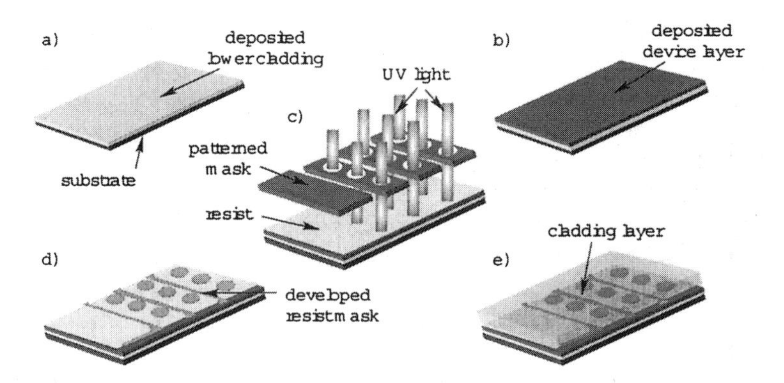

Figure 8.7. a. A low refractive index is deposited, b. a high index guiding layer is deposited, c. a pattern is transferred to the photoresist, d. the exposed layer is developed and etched, e. an overcoat is deposited over the patterned high-index layer

main source of the losses is the scattering off the sidewalls due to roughness. The top and bottom of the waveguide tend to be very smooth. The deposition process results in a even and uniform film. But the sidewalls of the waveguide are formed by chemically etching through a mask that was photographically transferred to the film. The dynamics of the etching always lead to small bumps and ridges in the sidewall of the waveguide. These small bumps lead to significant scattering. Physically the effect of small perturbations on the waveguide walls can be viewed as being identical to the small random fluctuations which lead to Rayleigh scattering in glass (refer to Eq. 8.8), except in the waveguide case the index difference is large, so the resulting attenuation component is dramatically larger. The power radiated due to scattering off a random fluctuation in the interface between the high and low index is proportional to $(n_f^2 - n_s^2)$.

To compound the problem, because the typical waveguide has a small cross sectional area to begin with, the sidewall roughness can represent a significant percentage of the waveguide cross sectional area. Losses can be reduced using geometries that minimize the mode overlaps with the rough surfaces. An example of such tailoring is shown in Fig. 8.8, where a numerically simulated mode is shown in two possible configurations of rectangular waveguide. The top and bottom surfaces of the waveguide were defined by the deposition system and are generally smooth, down to a few atomic layers. The vertical sides, however, of the waveguides were defined by the etching and photolithography process and are generally rough, with roughness on the order of a few nanometers. In Fig. 8.8a the horizontal geometry of the waveguide ensures that the mode overlap with the rough sidewalls is minimal. Losses in such a waveguide are expected to be low. In contrast, Fig. 8.8b shows a vertical geometry waveguide in which the mode overlap with the rough sidewalls is high. In such a waveguide the

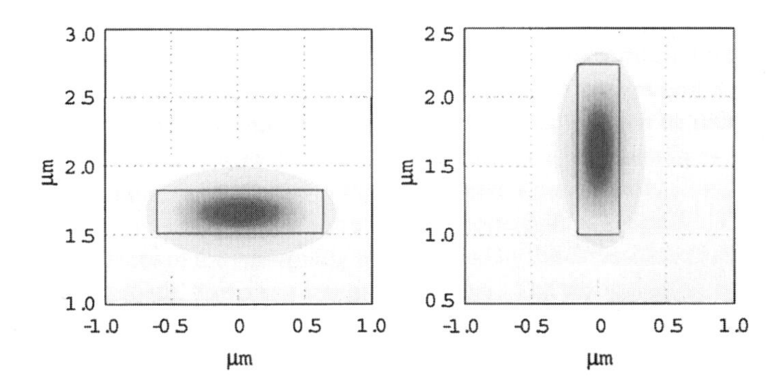

Figure 8.8. a. Low loss waveguide where the mode overlap with the rough sidewalls has been minimized. b. High loss waveguide where the mode overlap with the rough sidewalls is maximum.

losses are expected to be high. Considering the high-index of most integrated waveguide structures, until new processes are developed which provide atomically smooth sidewalls, designers will have to be very conscience to minimize mode overlap with the etched surfaces.

6. Mechanical Losses

Optical power can be lost due to leakage due to bending, and from defects at connections between waveguides. We will defer discussion of connection loss until we discuss coupled modes, however in this section we will discuss the effect of bends and packaging on the total attenuation. We will consider a single mode fiber in these examples.

6.1 Bending Loss

Consider Fig. 8.9, showing the mode field distribution of the LP_{01} mode in a fiber. The exponential tails of the field extend out away from the core, and theoretically never reaches zero, although practically speaking there is virtually no power in the tails beyond a few characteristic lengths ($1/\gamma$) from the core. If the fiber is bent, the spatial mode is not appreciably changed in shape compared to the straight fiber. However, the plane wavefronts associated with the mode are now pivoted about the center of curvature of the bend. To keep up with the mode, the phase front on the outside of the bend must travel a little faster than the phase front in the core. At some critical distance from the core of the fiber, the phase front will have to travel faster than the local speed of light, c/n_{clad}. Since this is not possible, the field beyond this critical radius breaks away and enters a radiating mode. The power that breaks away is a loss to the waveguide.

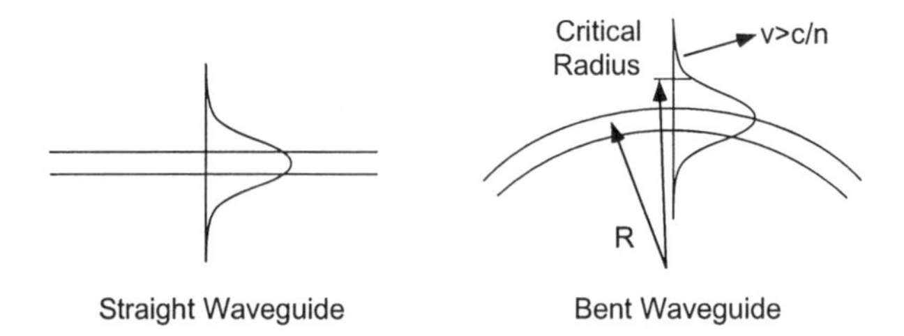

Figure 8.9. The plane wavefront in a bent waveguide is pivoted about the center of radius of curvature of the bend. At some critical radius, the phase velocity must exceed the velocity of light, and breaks away.

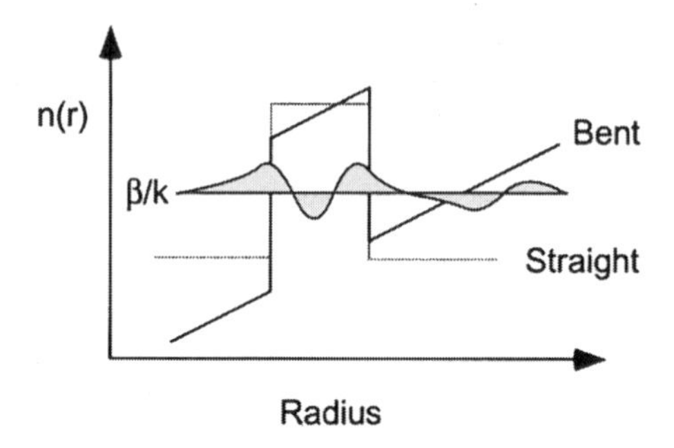

Figure 8.10. The effective index profile of a bent fiber (solid line) is distorted from that of the straight fiber (dashed line). The guided mode at value β can tunnel across the small barrier and couple to the radiation modes.

A more quantitative explanation of this effect can be derived modelling the fiber as having a distorted refractive index profile[6], which we simply quote here

$$n'(r) = n(r)(1 + \frac{r}{R}\cos\Phi) \tag{8.12}$$

where $n(r)$ is the actual index profile, R is the radius of curvature of the bend, and Φ is the azimuthal angle about the fiber axis. This profile is plotted in Fig. 8.10

A guided mode is indicated by its value of β/k_0. A weakly guided mode will be near cutoff, and hence will lie at lower values of β/k_0. The tail of the mode will extend into the region where $\beta/k_0 < n(r)$, and hence will radiate. The lower the value of β, the more radiation loss will occur for the bend. Modes with relatively large values of β will have a further distance to tunnel, and will not experience as much loss. The description here is exactly analogous to that describing leaky modes. We see that modes near cut-off experience far greater bending loss.

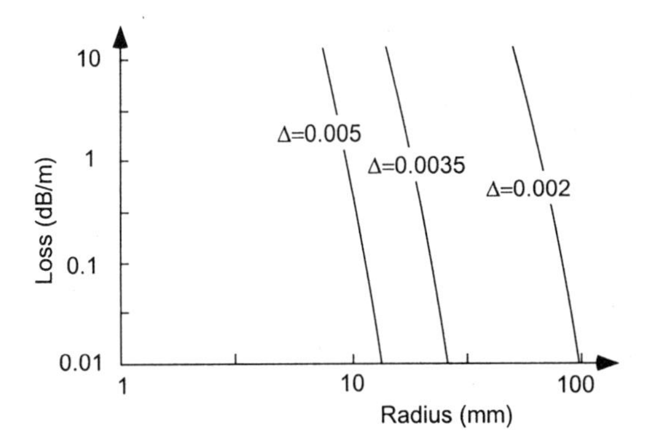

Figure 8.11. Bending loss calculated from Eq. 8.13 for three different fibers. The cut-off wavelength for the fiber is set at 1.2 μm, and the operating wavelength is 1.5 μm.

Quantitatively, the attenuation in a bend can be described as[7]

$$\alpha = \frac{1}{2}\left(\frac{\pi}{aV^3}\right)^{1/2}\left[\frac{\kappa a}{\gamma a K_1(\gamma a)}\right]^2 R^{-1/2} e^{-UR} \tag{8.13}$$

where

$$U = \frac{4\Delta(\gamma a)^3}{3aV^2 n_{clad}} \tag{8.14}$$

and a is the core radius, and Δ is the normalized core-cladding index difference. Some representative losses are plotted in Fig. 8.11 from this equation.

It is evident that the losses increase dramatically as the radius is reduced, and as the core-cladding index difference, Δ, is reduced. Note that this is one reason why integrated waveguides, which have to make small-radius turns on substrates, tend to use high-index contrast for waveguiding. Bending loss is reduced dramatically as Δn increases. Keep in mind that the bending losses shown in Fig. 8.11 are per unit length, and that a single bend of short radius will not have significant length. Bending losses become significant when the bend is extended over a long distance. This effect is seen in microbending.

6.2 Microbending Loss

In the process of putting an optical fiber in a cable, many small bends and curves are introduced. Cabling is necessary to protect the fiber and to provide sufficient mechanical strength to allow a fiber to be pulled through or strung along existing electrical wiring. Typically, an optical fiber is loosely wrapped around a strong cord made of nylon or steel, and the entire assembly is then encased in a pliant plastic jacket. The jacket protects the fiber from abrasion, while the central cord picks up any tensile stress that might occur during installation, leaving the optical fiber unstressed and hence unbroken.

Inevitably, there will be small bends in the optical fiber as it is wrapped about the central cord. This introduces a systematic bending loss into the fiber. Tests of fiber attenuation before and after cabling show that cabled fiber attenuation is larger. The degree of attenuation depends on the specific cabling geometry, because the loss depends critically on the bend radius. A typical cable is designed to minimize the total bending that the fiber must do around the tensile strength member. Nevertheless, microbending attenuation can become significant because of the long length of the fiber; even small effects eventually add up. Looking at the data in Fig. 8.11, one can see that increasing Δ reduces the effect of bending loss. Therefore it is desirable to use a single mode fiber with as large a numerical aperture as possible in order to minimize loss in long-distance optical communication links.

7. Nonlinear Effects in Dielectrics

Up to now, we have assumed that the material constants ϵ and μ were independent of the field strength. Linearity implies that the permittivity experienced by a field with strength $E = 10$ V/m would be the same as the permittivity experienced for a strong field with $E = 10^9$ V/m. In the case of most dielectrics, for field strengths in this range the assumption is reasonably accurate; the value of the permittivity changes only slightly over these huge ranges of applied field. But the essential fact is that it often does change slightly, and this slight change can lead to some spectacular effects. Among the notable nonlinear interactions are the generation of second harmonic radiation of optical frequencies and frequency mixing. Here we want to study the stimulated scattering that leads to excess loss, and imposes power limitations on the optical waveguide.

On a fundamental level, nonlinearities arise from an anharmonic motion of the electrons in response to an applied field. Consider the simple illustration below of how a nonlinearity can lead to the generation of a second harmonic field. A one-dimensional crystal is shown in Fig. 8.12. The crystal does not display inversion symmetry.

The electron, as it is pushed right and left by an applied field, will see different potential barriers restricting its motion. For example, the electron will be strongly inhibited from moving to the right, but will see less inhibitive force when going to the left. This leads to a nonlinear dipole moment in response to the applied field.

If an electromagnetic wave is incident on the crystal, the electron will respond by moving back and forth along the crystal in synchronicity with the field. When the electron moves to the left, it will see the potential field of atom a, and when it moves to the right it will see the potential of atom b. Since atoms a and b are different, the potentials they generate will be different. From the

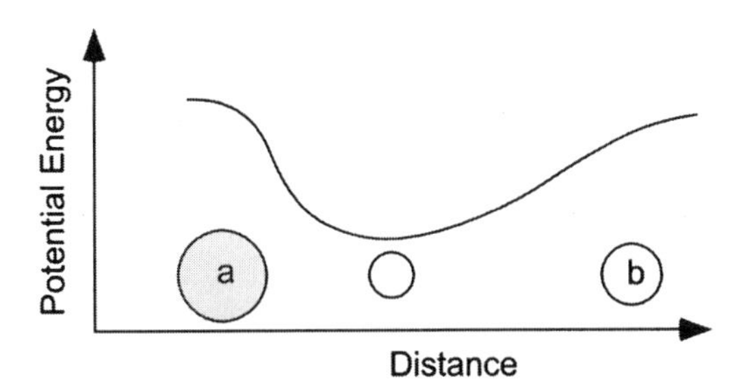

Figure 8.12. A molecule with no inversion symmetry.

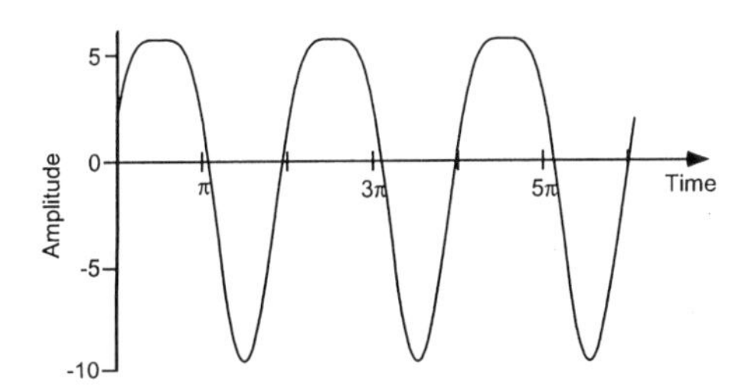

Figure 8.13. The motion of the electron under the influence of a sinusoidal driving field in the presence of a non-symmetric crystal.

electron's point-of-view, it sees an asymmetric potential well, as shown in the figure. In response to an applied field, the electron will move further to the right than the left. The motion of the electron, as plotted in Fig. 8.13, follows a distorted sinusoidal path. Using Fourier's theorem, the electron's motion in this example could be described as a superposition of several sinusoidal motions.

This simple illustration relied on a system with no inversion symmetry. Try to repeat the arguments made above with a system that is symmetric, and you will see why non-inversion symmetry is a necessary condition for second harmonic generation, although other (odd) harmonics can be generated. Whenever the moving electrons fail to maintain a strictly linear relation between applied E and position r, there will be a nonlinear term in the polarization. The field strength must usually be significant for these effects to occur. The binding potential of an electron to a nucleus or molecule is on the order of 10^{10} V/cm, so for applied fields that are orders of magnitude smaller than this value, the electrons will not be significantly affected by distortions of the charge distributions. Once the field strength approaches 1% or more of the binding potential, the nonlinearities begin to become significant. This is the reason why we do not experience nonlinear optical effects in our everyday terrestrial experience.

Nonlinear optics was not a widely investigated field prior to the development of the laser. With lasers, field strengths exceeding 10^8 V/cm are easily generated.

On a more formal level, the induced polarization from the applied electric field can be expressed in general form

$$\begin{aligned} P/\epsilon_0 &= \chi^{(1)} \cdot E \quad \text{Linear optics} \\ &+ \chi^{(2)} \cdot EE \quad \text{Sum Frequency Mixing, Pockels Effect} \\ &+ \chi^{(3)} \cdot EEE \quad \text{Third Harmonic, Self Phase Modulation, \ldots} \\ &+ \ldots \end{aligned} \tag{8.15}$$

where $\chi^{(i)}$ is the ith order susceptibility. To account for polarization effects, $\chi^{(i)}$ is a tensor of rank $i+1$. The linear susceptibility, $\chi^{(1)}$, is the polarization term we explored in Chapter 6 (see Eq. 6.14). The second order susceptibility $\chi^{(2)}$ is responsible for effects such as second-harmonic generation. As explained above, it exists only in materials with noninversion symmetry.

There are many optical nonlinearities that can be studied in optical waveguides. We will focus on three topics: Stimulated Raman Scattering, Stimulated Brillioun Scattering, and Self Phase Modulation. More detailed reviews can be found in specialized texts [see for example refs. [8] and [9].

8. Stimulated Raman Scattering

When an optical wave travels through a material system, the wave can be partially scattered by local imperfections. One example of such scattering is Rayleigh scattering, which was described above. Raman scattering is a second such mechanism. We will develop a classical picture of Raman scattering in this section, although we will invoke quantum concepts such as the *photon* and *phonon*. Only a qualitative description of the underlying physics of Raman scattering will be presented. Interested readers should consult the excellent texts listed in the references for more detail.[10-13]

A photon travelling through a material can excite a vibrational transition of the material, creating an optical phonon, *even if the frequency of the two quanta are dissimilar*. (A phonon is a quantum of vibrational energy in a lattice. It has energy $\hbar\omega$, where ω is the vibrational frequency of the lattice.) This is a nonresonant interaction. Fig. 8.14 shows an energy level diagram of the interaction. The incident photon with energy $\hbar\omega_1$ (ω_1 is the optical frequency) has a small but finite probability of exciting a single phonon of the molecular vibration, depositing $\hbar\omega_p$ energy in the molecule. To satisfy energy conservation, the photon will exit the system with a slightly reduced energy, $\hbar\omega_2$, where

$$\omega_2 = \omega_1 - \omega_p \tag{8.16}$$

The net effect of this interaction is that the molecule has been raised to a new vibrational state, and the photon energy (and frequency) has been reduced. This

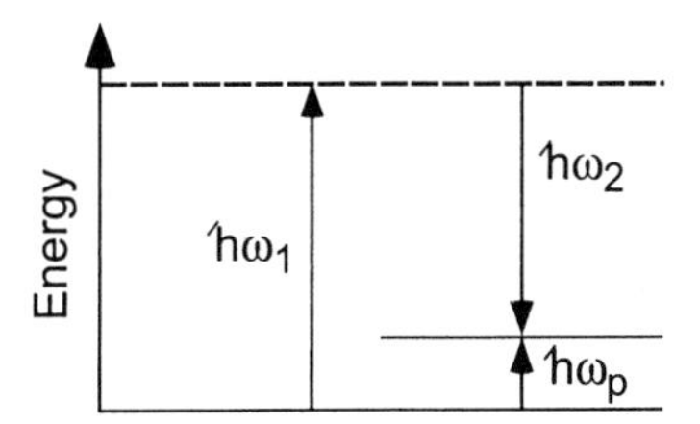

Figure 8.14. Energy level diagram for Raman scattering. An incident photon with energy $\hbar\omega_1$ excites a phonon with energy $\hbar\omega_p$, and is re-emitted with energy $\hbar\omega_2$.

process is called *Raman Scattering*. Energy is strictly conserved. Usually the photon and phonon frequencies are dramatically different, so the probability that the photon actually excites the phonon is very small. One can rely on everyday experience to notice that the wavelength of light does not noticeably shift when passing through a plate of glass. In other words, most of the light goes through a material without suffering Raman scattering. Only an extremely small fraction of any incident optical field actually undergoes Raman scattering in a material under low intensity conditions.

Example 8.1: Raman scattering from a variety of gasses

A laser beam with a well defined wavelength of $\lambda = 514.5$ nm is sent through a chamber of gas. Looking at the beam from the side, a monochrometer collects scattered light and disperses it according to its wavelength. The scattered light from the beam has two components: there is a lot of light at $\lambda = 514.5$ nm due to scattering from dust particles and Rayleigh scattering in the gas, and there is a second wavelength at $\lambda = 584.6$ nm. What is the gas in the cell?

Solution: The vibrational frequencies of several gasses are listed in Table 8.3. The vibrational frequency is listed in units of *wavenumber*, which is the number of wavelengths in a unit length, in this case 1 cm. Note that this definition is similar to that for the wavevector, k_0, except the wavevector is defined as the number of *radians* per unit length, while *wavenumbers* are listed in *waves* per unit length. Don't get confused by the units!

The incident wavelength (514.5 nm) corresponds to $1/\lambda = 19436$ cm^{-1}. The Raman scattered light at 584.6 nm corresponds to a wavenumber of $1/\lambda = 17105$ cm^{-1}. The difference in the incident and scattered light is 2331 cm^{-1}. Inspection of Table 8.3 indicates that the gas must be N_2. This illustrates the diagnostic power of Raman scattering—the composition of a material can be determined by remote (non-contact) sensing.

Table 8.3. Raman frequencies for several gasses

Gas	Vibrational Frequency (cm^{-1})
N_2	2331
O_2	1556
H_2	4161
CO	2145
CO_2 (ν_1)	1388
CO_2 (ν_2)	1286
N_2O (ν_1)	1285
N_2O (ν_2)	2224

8.1 Mechanism Behind Raman Scattering

Since the interaction responsible for Raman scattering is nonresonant, how does Raman scattering occur? In the classical picture the interaction occurs through a slight modulation of the index of refraction due to the molecular vibrations of the material. Consider the schematic representation of a molecule in Fig. 8.15, consisting of two atoms separated a distance x_0 by a spring. If an electric field is applied to this molecule as shown, there will be a slight change in the relative position, x_0, of the two charges. The induced polarization for the molecule is defined as

$$p = qx, \tag{8.17}$$

where q is the net charge that moves in response to the field, and x is the difference in distance between the charge centers under the influence of the field and at equilibrium. Often the polarizability of a material is written in terms of the applied electric field

$$p = \alpha E, \tag{8.18}$$

where α is called the complex polarizability of the material. This is a microscopic version of the bulk polarization expression,

$$P = \epsilon_0 \chi^{(1)} E \tag{8.19}$$

where the term $\chi^{(1)}$ is used to describe bulk polarizability, as opposed to the microscopic polarizability, α.

If the molecular length increases, the charge separation increases, so the polarizability, α, increases. In practice, molecules at finite temperature vibrate due to thermal energy. It is this small vibration, and its effect on the polarizability, that couples energy from the optical field to the vibrational field.

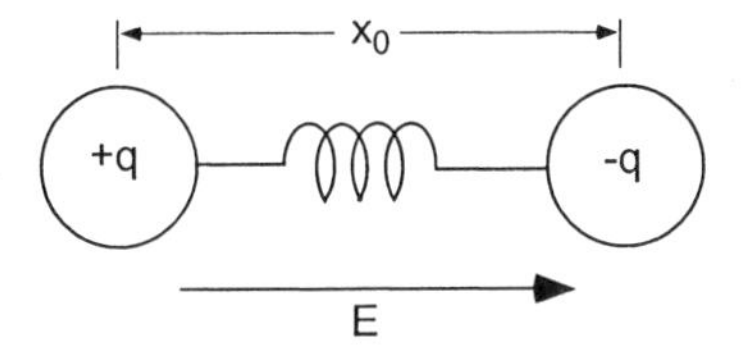

Figure 8.15. When a molecule is polarized by an external electric field, the two polar ends can be further separated, leading to a change in diploe moment.

The molecular polarizability of an atom that is vibrating can be described through a Taylor series expansion

$$\alpha(x) = \alpha_0 + \left.\frac{\partial \alpha}{\partial x}\right|_{x_0} \delta x \tag{8.20}$$

where δx is the displacement of the molecular length from its equilibrium value. If the molecule is vibrating at its resonant frequency, ω_p, then the displacement will be a periodic function

$$\delta x(t) = \delta x_0 \; e^{\pm j\omega_p t} \tag{8.21}$$

Combining all these terms, the polarization of the molecule becomes

$$\begin{aligned} p(t) &= \alpha(t)E(t) \\ &= \left(\alpha_0 + \left.\frac{\partial \alpha}{\partial x}\right|_{x_0} \delta x_0 \; e^{\pm j\omega_p t}\right) E_0 e^{j\omega_1 t} \\ &= \alpha_0 E_0 e^{j\omega_1 t} + \left.\frac{\partial \alpha}{\partial x}\right|_{x_0} \delta x_0 \; E_0 e^{j(\omega_1 \pm \omega_p)t} \end{aligned} \tag{8.22}$$

Notice that there are now two frequency components to the polarization: one term oscillates synchronously with the electric field, and is responsible for the dielectric constant of the material. The second term oscillates at a different frequency, given by the sum or difference of the applied frequency and the vibrational frequency, and acts as a source for the generation of radiation at the new frequency.

Recall that polarization acts as a driving force for new electromagnetic fields, according to Maxwell's equations

$$\nabla \times H = \frac{\epsilon_0 \partial E}{\partial t} + \frac{\partial P}{\partial t} \tag{8.23}$$

The polarization term at the frequency $(\omega_1 \pm \omega_p)$ acts as the driving source to generate E and H fields at these shifted frequencies. Classically, the scattered

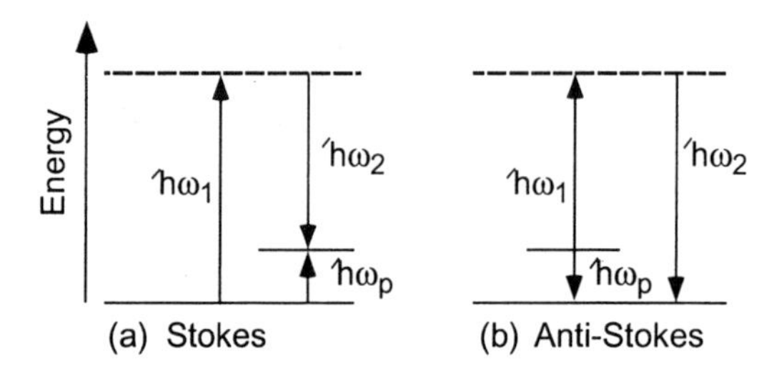

Figure 8.16. (a.) The incident photon excites a phonon from the ground state, and leaves with less energy. (b.) The incident photon scatters off of a molecule in the first vibrational excited state. The molecule drops back to the ground state, releasing one quanta of energy which appears in the exiting photon. The photon leaves with higher energy.

field can be shifted to higher or lower frequencies than the driving field. In practice the frequency depends on the vibrational state of the molecule. Fig. 8.16 shows an energy level diagram of both processes. A photon with energy $\hbar\omega_1$ is incident on a molecule with vibrational frequency ω_p. If the molecule is in the ground state initially (Fig. 8.16a), it can absorb energy from the applied field, and the scattered light (ω_2) will be at a lower energy and frequency. This photon is called a "Stokes" photon, indicating that it has a lower frequency than the input photon.

If, on the other hand, the molecule is already in an excited vibrational state, as in Fig.8.16b, then it can give up one phonon's worth of energy to the field, and drop to the next lower vibrational energy level. The optical photon exits with higher energy and frequency, and is called an "anti-Stokes" photon. Both processes can be observed in Raman scattering. Due to the Maxwell-Boltzmann statistics which describe thermal distributions, there are usually far more molecules in the ground state than in the excited state in thermal equilibrium, so Stoke's radiation usually dominates.

8.2 Amplification Using Stimulated Raman Scattering

Close examination of the second term of Eq. 8.22 gives some insight into the nature of the Raman process. The induced asynchronous dipole moment is given by

$$p(t)_{\omega_1 \pm \omega_p} = \frac{\partial \alpha}{\partial x}\, \delta x_0\, E_0\, e^{j(\omega_1 \pm \omega_p)t} \tag{8.24}$$

This is a nonlinear term (the output frequency is different from the input frequency), and it depends on the product of two coupled parameters: the applied electric field amplitude, E_0, and the amplitude of the molecular vibration, δx_0. The overall size of this term is usually small compared to the linear polarization. For this reason, we do not expect, nor in practice do we see, strong

Stokes-shifted signals coming from optical materials using low intensity light sources. Nevertheless, there is always some Raman scattered light produced when light travels through a material.

The molecules which contribute to this Raman process are, for the most part, independently vibrating. There is no coherence between the various molecules, so the phase of the Stokes light that is generated by each molecule is statistically distributed over 2π radians. The resulting light is therefore incoherent, and comes out in many possible directions. This is called *spontaneous Raman scattering.* Under much stronger driving force (i.e. larger electric field amplitudes), the interaction between the electromagnetic fields and the polarizability of the material can lead to the imposition of some order among the oscillators, and a coherent wave at the Stokes frequency can be generated.

The polarization at the Stokes frequency is proportional to the *product* of the applied electric field, E_0, and the amplitude of the molecular vibration, δx_0. Each time the electric field scatters a photon off of a molecule, the vibrational amplitude increases. From Eq. 8.22, we see that an increased amplitude directly increases the polarization at the Stokes frequency. The process can feed on itself— the creation of each Stokes photon makes δx_0 larger, making it easier to generate even more Stokes photons. For very low intensities, the thermal disorder of the system keeps the molecules out of phase, so there is no coherent build-up of the Stokes wave. At high intensities, the vibration of each molecule can fall in-phase with each other, leading to a large coherent array of vibrating oscillators. This regime is called *Stimulated Raman Scattering.* Once a stimulated field begins to form, it will exponentially grow until it has saturated the pump field.

Stimulated Raman Scattering (SRS) can be viewed as a problem, or as an effect to be exploited. SRS can be a serious problem for optical communication: if a sufficiently intense optical field is sent into a fiber, it can generate other wavelengths, and eventually deplete the energy at the original input wavelength. The creation of additional wavelengths in the communication link will, at the least, add to more dispersive pulse spreading, and at the worst, cause the information to be lost altogether. Thus, Stimulated Raman Scattering establishes an operational maximum limit on the amount of power that can be put into an optical fiber.

In the stimulated Raman regime, the Stokes wave will be amplified as it travels through the medium. As explained above, the polarization at the Stokes frequency is proportional to the applied field amplitude and the amplitude of the molecular vibration. Describing this mathematically in terms of intensities, the process is characterized by a simple differential equation

$$\frac{dI_2}{dz} = G_r I_2 I_1 \tag{8.25}$$

Table 8.4. Raman frequencies and gain coefficients[14] ($\lambda = 0.694\mu$m)

Material	Frequency Shift (cm^{-1})	Raman Gain (m/W)	Linewidth $\Delta\nu$ (cm^{-1})
$LiNbO_3$	258	28.7×10^{-13}	7
Benzene	992	2.8×10^{-13}	2.2
SiO_2	467 (peak)	0.9×10^{-13}	200

where I_2 is the intensity of the Stokes-shifted light($\omega_s = \omega_p - \omega_{vib}$), I_1 is the intensity of the pump light(ω_p), and G_r is the Raman gain term, which includes all the material terms such as $\partial\alpha/\partial x$ and frequency terms which arise in the conversion from polarization (Eq. 8.24) to intensity (Eq. 8.25). The gain coefficient, G_r, decreases proportional to $1/\lambda^2$. The differential equation can be solved in the case of weak Stokes intensity, $I_2 \ll I_1$, to be

$$I_2(z) = I_2(0)e^{G_r \cdot I_1 \cdot z} \tag{8.26}$$

We have assumed that the pump intensity is not significantly depleted through SRS. Table 8.4 lists values of G and ω_p for various materials of interest in optoelectronics. The gain values are specified for a pump wavelength of 0.694 μm, which corresponds to the ruby laser output wavelength. The key points to understand are that the gain is proportional to the intensity of the pump light, and that optical gain occurs at a frequency that is different from the pump frequency, shifted by ω_p.

Recall that 1 cm^{-1} is equal to 30 GHz.

Example 8.2 Raman gain in a single mode fiber

Consider the Raman gain that is generated by sending light down a single mode fiber made of fused silica. In this case, assume that the fiber has an effective core area of 10^{-6}cm^2 (this corresponds to a mode field diameter of 11μm, which is typical for a single mode fiber). A laser couples 100 mW of light at $\lambda = 1\mu$m onto the fiber. If the fiber is 1 km long, what is the magnitude of the Raman gain, and at what wavelength does the gain appear?

Solution: From Table 8.4 we see that SiO_2 has a vibrational frequency of 467cm^{-1}. (The spectrum is actually a broad distribution with a peak at 467 cm^{-1}). To convert this into frequency, the wavenumber must be multiplied by the speed of light, c.

$$\begin{aligned} \nu &= 467\text{cm}^{-1} \cdot 3 \times 10^{10}\text{cm/sec} \\ &= 14 \times 10^{12}\text{Hz} \end{aligned}$$

The vibrational frequency of SiO_2 is approximately 14 THz, so the Raman gain will exist for light at a frequency that is 14 THz *below* the pump frequency. The pump frequency is $\nu = c/\lambda = 300$ THz.

$$\nu_{gain} = 300 \times 10^{12} - 14 \times 10^{12} = 286 \times 10^{12}$$

This frequency corresponds to a wavelength of

$$\lambda = \frac{c}{\nu} = \frac{3 \times 10^{10}}{286 \times 10^{12}} = 1.049\mu\text{m} \tag{8.27}$$

The Raman gain will be maximum for light with wavelength 1.049μm. The magnitude of the gain depends on the intensity of the pump light in the core of the fiber. The intensity is $I = 0.1W/10^{-6}\text{cm}^2 = 10^5$ W/cm^2. The Raman gain is then

$$\begin{aligned} \frac{I(z)}{I(0)} &= e^{G_r \cdot I \cdot z} \\ &= exp(0.9 \times 10^{-13}\text{m/W} \cdot 10^5\text{W/cm}^2 \cdot 10^4\text{cm}^2/\text{m}^2 \cdot 1000\text{m}) \\ &= exp(0.09) = 1.094 \end{aligned}$$

A weak signal with wavelength $\lambda = 1.049\mu$m that enters the fiber with the pump beam will see approximately 9% gain in one kilometer.

The 1 km optical fiber used in Example 8.2 is effectively an amplifier for light at $\lambda = 1.049\mu$m, with a net gain of about 9%. In practice, the vibrational spectrum of SiO_2 peaks at 467 cm^{-1}, but is in fact rather broad, decreasing approximately linearly with frequency toward lower frequency. Thus, there is gain over a large frequency range in a SiO_2 Raman amplifier, although it is maximum near 467 cm^{-1}. If the pump wave does not attenuate seriously after travelling this distance, the amplification could easily be increased by extending the length of the fiber. Such amplifiers are widely used in long distance optical communication links, where they serve as a pre-amplifier at the receiving end of the fiber to boost a weak optical signal. They also can provide gain over a wide number of wavelengths, since only the frequency of the pump laser needs be adjusted to change the center wavelength of the gain.

Because the pump power decreases along the fiber due to linear absorption and scattering, the Raman gain is greater at the input end. To account for this, the operator dz in Eq. 8.25 should be replaced by $e^{-\alpha z}dz$. An *effective gain length*, L_{eff}, drops out of the integral solution to Eq. 8.25, defined as

$$L_{eff} = \frac{1 - e^{-\alpha \ell}}{\alpha} \tag{8.28}$$

where ℓ is the actual fiber length, and α is the linear attenuation coefficient for the fiber.

8.3 System Limitation of Stimulated Raman Scattering

A concern in optical fiber communication links is the limitation imposed by the presence of gain in the fiber created by the signal itself. This answers the question we posed at the beginning of the chapter, "Why not increase the input power of the signal to overcome fiber losses?" Consider the case where an optical fiber is used to transmit a signal between two points. The mere presence of the signal establishes gain at the Stokes shifted wavelength. Normally, there will be no input at the Stokes wavelength, so there will be no build-up or amplification of signal at this undesired wavelength. Unfortunately, spontaneous Raman scattering occurs all the time, and it is possible for a spontaneous Stokes photons to be launched in the guided mode of the fiber. These spontaneous photons will see the optical gain caused by the signal light, and will be amplified. Essentially, these spontaneous photons will rob power from the signal and convert it into light at a shifted wavelength.

How much optical power is generated in the Stokes wave? This depends on the loss of the fiber, the Raman gain, and the number of modes carried by the fiber. Spontaneous Raman scattering can occur at any wavelength within the bandwidth of the vibrational transition. Fused silica, for example, has a bandwidth of about 6 THz, and a central vibration frequency of 14 THz, implying that the scattered light will be frequency downshifted by approximately 14 ± 3 THz. As the bandwidth of the spontaneous scattering increases, the number of modes that the Stokes light can couple to increases. Each Stokes photon, no matter which mode it is in, will see gain. Thus, the output of the fiber will contain a large number of Stokes photons spread over the bandwidth of the medium. In practice, the spontaneous Stokes photons occur throughout the fiber length, and see different total gains depending on location. It can be rigorously shown that the net effect of this distributed noise source is equivalent to injecting one fictitious photon per mode at the beginning of the fiber[15]. The effective number of modes is given by

$$N = \frac{\sqrt{\pi}}{2} \frac{\Delta\nu_{FWHM}}{[I_p G_r/\alpha_p]^{1/2}} \tag{8.29}$$

where I_p is the intensity of the pump light, and α_p is the loss coefficient for the pump light.

An absolute upper limit for input power into a communication link can be defined in terms of the point at which the Stokes power, P_r, equals the signal power, P_{sig}. It can be shown that the limiting power is

$$P = \frac{16\pi w_0^2}{G_r L_{eff}} \tag{8.30}$$

where w_0 is the mode radius in the fiber.

Consider for example a single mode fiber operating at $\lambda_p = 1.55\mu$m. The mode radius for a typical fiber is 5 μm, yielding a mode area of approximately $80\mu m^2$. The loss coefficient for good fibers is approximately 0.2 dB/km, yielding an effective length $L_{eff} = 20$ km. Using the data from Table 8.4, the power limit for a signal at $\lambda = 1.55\mu$m is approximately 700 mW. This is quite large relative to the commonly used signal powers on the order of 1 mW.

9. Stimulated Brillouin Scattering

Brillouin scattering is similar to Raman scattering, except acoustic phonons are involved instead of optical phonons. Acoustic phonons consist of collective vibrations of the atoms in a solid, while optical phonons tend to involve vibrations only between a few individual atoms[10]. Acoustic vibrations occur at a much lower frequency than optical phonons, being on the order of $1\,\text{cm}^{-1}$ ($\approx$ 30 GHz). Brillouin scattering occurs when optical waves interact with the small periodic change in the index of refraction caused by these collective vibrations. The gain for Stimulated Brillouin Scattering in glass is about two orders of magnitude greater than for SRS, but it occurs over a much narrower frequency bandwidth. If the pump radiation has a linewidth larger than the Brillouin linewidth, the gain is proportionally reduced. Brillouin linewidths are on the order of 60 MHz, compared to Raman gain bandwidths of 10's of cm^{-1}.

Brillouin scattering is essentially caused by a reflection of the input light from a moving index variation caused by an acoustic wave in the material. The frequency of the scattered light is given by

$$\delta\nu = 2nV_s/\lambda \tag{8.31}$$

where V_s is the velocity of sound waves in the material. This formula effectively describes a Doppler shift of light bouncing off of a moving index variation in the solid. For the case where the pump linewidth is much narrower than the Brillouin linewidth, the gain coefficient is

$$G_B = \frac{2\pi n^7 p_{12}^2}{c\lambda^2 \rho V_s \Delta\nu_B}\ \text{cm/W} \tag{8.32}$$

where ρ is the density, p_{12} is the elasto-optic coefficient, and $\Delta\nu_B$ is the Brillouin linewidth. For plane waves in fused silica in the visible region of the spectrum, the frequency shift is approximately 35 GHz, and the gain coefficient is $G = 4.5 \times 10^{-9}$ cm/W near $\lambda = 1\mu$m. The Brillouin linewidth, $\Delta\nu_B$, is on the order of 135 MHz. If the pump linewidth, $\Delta\nu_p$, is larger than the gain linewidth, then

$$G_B \approx G_{B0}(\Delta\nu_B/\Delta\nu_p) \tag{8.33}$$

The major problem a system designer faces with Stimulated Brillouin Scattering (SBS) is the chance that input power will be reflected backward out of the waveguide. The SBS gain can cause a large back reflection, effectively reducing power transmission. SBS is usually observed with extremely narrowline pump input power, so a common solution is to increase the bandwidth of the input signal. Generally for temporally short pulses, the linewidth is much greater than $\Delta\nu_b$, so the effective Brillouin gain is reduced. Similar to the case for SRS, to avoid SBS, one must stay below a critical gain threshold. The critical power for backward stimulated Brillouin scattering is given by [15]

$$P = \frac{21 A_{eff}}{G_B L_{eff}} \tag{8.34}$$

SBS can be avoided if the input power has sufficient bandwidth to reduce the power spectral density below that established by Eq. 8.34. For optical communication systems with information bandwidths exceeding hundreds of Megabits/second, this bandwidth restriction is almost trivial to meet, so SBS is not generally a problem.

10. Self-Phase Modulation

The last significant nonlinearity we will consider in amorphous solids is due to $\chi^{(3)}$, which is responsible for phenomena such as third harmonic generation, four wave mixing, and self refraction. Unless special efforts are made to match the phase velocities of the harmonic frequencies (e.g. for second harmonic generation, phasematching requires that $2k_\omega n_\omega = k_{2\omega} n_{2\omega}$), harmonic generation is not efficient. Due to the dispersive nature of glass, most harmonic generation effects are negligible, leaving nonlinear refraction as the predominant nonlinearity. Consider the case of a wave, $E(r) = E_0 e^{j(\omega t - k_0 r)} + c.c.$, propagating in a fixed direction in glass. The polarization resulting from the $\chi^{(3)}$ term is described by

$$\begin{aligned}
\frac{P}{\epsilon_0} &= \chi^{(1)} E_0 e^{j(\omega t - k_0 r)} + c.c. + \chi^{(3)} [E_0 e^{j(\omega t - k_0 r)} + c.c.]^3 \\
&= \chi^{(1)} E_0 e^{j(\omega t - k_0 r)} + c.c. + \chi^{(3)} E_0^3 [2(e^{j(\omega t - k_0 r)} + c.c) \\
&\quad + (e^{3j(\omega t - k_0 r)} + c.c)] \\
&= \left[\chi^{(1)} + 2\chi^{(3)} E_0^2\right] E_0 e^{j(\omega t - k_0 r)} + c.c. + \text{3rd harm. terms}
\end{aligned} \tag{8.35}$$

The term in the brackets is the effective polarizability for the medium and as you can see, it depends on the intensity of the applied field. Converting this expression to intensity, and using the fact that $\chi^{(3)}$ is extremely small, the index of refraction, n, of a material can accurately be written as

$$n(I) = n_0 + n_2 \cdot I \tag{8.36}$$

where n_0 is the normal index of refraction, and n_2 is the nonlinear refractive component. For most materials n_2 is small: $n_2 = 3.2 \times 10^{-16}$ cm$^2/W$ for fused silica, and 3.8×10^{-16} cm$^2/W$ for sapphire. Clearly, $n(I)$ is nearly equal to n_0 under common terrestrial conditions (sunlight, etc.), and only deviates under extreme high intensities. However, these are exactly the conditions that can be easily achieved in an optical fiber: the small core size and long path length of optical fiber systems allow even modest fields to create significant nonlinear effects.

How does the $\chi^{(3)}$ term lead to self-phase modulation? First, let's define phase modulation. The instantaneous phase of a wave is defined by the argument of the exponent in the wave formulation. For example, a plane wave propagating through a dielectric material is described by

$$E(z) = E_o e^{j(\omega_0 t - k_0 n z)} = E_0 e^{j\phi} \tag{8.37}$$

Noting that $k_0 = \omega_0/c$, the instantaneous phase, ϕ, can be described as

$$\phi = \omega_0 t - \frac{\omega_0}{c} n z \tag{8.38}$$

If the wave is a single frequency sine wave, the phase can be expected to accumulate at a steady rate, defined by the angular frequency as ω radians per second. Phase modulation is the term describing any alteration of the phase from its linear predicted pattern. This can be done intentionally for communication purposes using a modulator, or the wave can modulate itself via a nonlinear interaction. The latter effect is called self-phase modulation.

Substituting the expression for the intensity dependent index of refraction, Eq. 8.36, into Eq. 8.38 yields

$$\phi = \omega_0 t - \frac{\omega_0 z}{c}[n_0 + n_2 I(t)] \tag{8.39}$$

The instantaneous frequency of the field is determined from

$$\omega(t) = \frac{d\phi}{dt} = \omega_0 \left[1 - \frac{z}{c} n_2 \frac{d}{dt} I(t)\right] \tag{8.40}$$

Inspection of this equation shows that when the intensity is increasing, the instantaneous frequency of the wave is reduced (assuming that n_2 is positive), and when the intensity decreases, the frequency of the wave is increased. The time-dependent index of refraction acts like a phase modulator.

Consider the optical pulse in Fig. 8.17. The intensity is a function of time, rapidly rising from zero intensity to a maximum value, and then returning to zero. Due to self-phase modulation, the index of refraction at the peak of the pulse will be slightly different than the value in the wings of the pulse. If n_2 is positive, the index at the peak will be slightly larger than in the wings. The

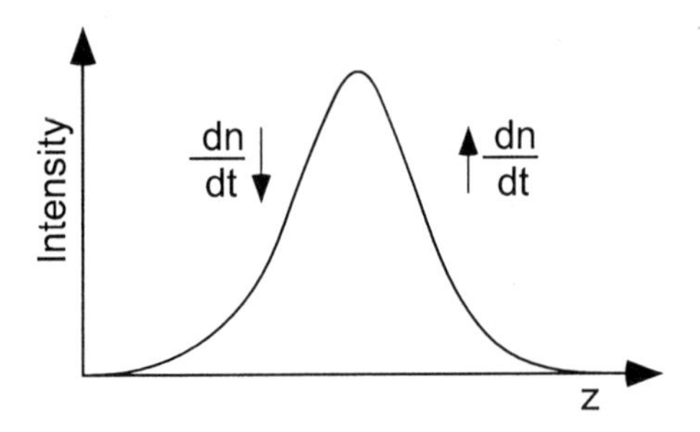

Figure 8.17. An optical pulse travelling along the z-axis will experience a time-dependent increase and decrease in the index of refraction. The leading edge will see an increase in index, while the trailing edge will see a decreasing index.

leading edge of the pulse will see a positive dn/dt, while the trailing edge will see a negative dn/dt. The phase of the optical wave depends on the refractive index , $\Phi = kn\ell$, so the time-varying index of refraction will lead to a time-dependent phase shift, $d\Phi/dt$. This phase modulation leads to the creation of additional frequency components.

A simple way to view this process is to look at the wavelength of the fields on both sides of the pulse. Since the leading edge of the pulse has lower intensity than the peak, the index of refraction will be lower, and therefore the phase velocity, $v = c/n$, of the field will be slightly faster than at the peak. Phase fronts on the leading edge will move away from the center of the pulse, effectively stretching out the waves, and lowering the frequency of the light. The leading edge of the pulse is *frequency down-shifted.*

Similarly the trailing edge of the pulse has a lower index than the peak, so the trailing edge waves move slightly faster than those at the peak. As they propagate, they catch up with the peak, or effectively compress their wavelength. Compressed waves lead to shorter wavelengths, which is equivalent to a higher frequency optical wave. The trailing edge of the pulse is *frequency up-shifted*.

The pulse frequency bandwidth increases due to self-phase modulation, and the pulse develops a "chirp" where the frequency of the pulse monotonically increases across the pulse. The magnitude of the frequency chirp experienced by the pulse is

$$\delta\omega = \Delta k L = \frac{2\pi L}{\lambda}\frac{d\delta n}{dt} \tag{8.41}$$

where $\delta n = n_2 I$. The total accumulated phase for this pulse is

$$\Delta\phi = \frac{2\pi L}{\lambda}\delta n \tag{8.42}$$

where δn is the maximum increase in index

$$\delta n = n_2 I_{max} \tag{8.43}$$

This excess bandwidth generally has a detrimental effect on pulse propagation, as it leads to increased dispersion effects. The pulses are no longer "transform limited", meaning that the actual bandwidth of the pulse is larger than that predicted by a Fourier transform of the temporal envelope.

The key problem with SPM is the increased dispersion that results when the bandwidth is increased. We can derive a *critical length*, L_{crit}, in which the high frequency component of the pulse is retarded by one pulse length time from the low frequency component. In other words, the effective temporal pulse width doubles. As a rough estimate, let's assume that the rate of change of intensity for a pulse can be given as $dI/dt = 2I/\tau$, where τ is the pulsewidth and I is the peak intensity. The increase in the pulse bandwidth (from Eq. 8.41) traveling a distance L_{crit} is

$$\Delta\omega = \frac{2\pi L_{crit}}{\lambda}\frac{n_2 I}{\tau} \tag{8.44}$$

Define the group velocities of the two different wavelength pulses as v_1 and v_2, respectively. In a time T, each pulse travels a distance $z_1 = v_1 T$ and $z_2 = v_2 T$. The time T required for the pulses to separate a distance equal to their pulse widths (given by $\tau c/n$) is

$$T = \frac{\tau c/n}{v_2 - v_1} \tag{8.45}$$

Noting that the critical length is $L_{crit} = Tc/n$, this equation can be recast as

$$L_{crit} = \frac{\tau c^2}{(v_2 - v_1)n^2} \tag{8.46}$$

The difference term in the denominator can be replaced by $v_2 - v_1 = dv_g/d\omega \cdot \Delta\omega$.

$$\begin{aligned} v_g &= c[n_0 + \omega dn/d\omega]^{-1} \\ \frac{dv_g}{d\omega} &= \frac{-1}{2\pi}\frac{1}{n_0 + \omega dn/d\omega}\lambda^2\frac{d^2n}{d\lambda^2} \approx \frac{-1}{2\pi}\frac{1}{n_0}\lambda^2\frac{d^2n}{d\lambda^2} \end{aligned} \tag{8.47}$$

Substituting this and Eq. 8.44 into the expression for L_{crit} yields

$$L_{crit} = \left[\lambda^2\frac{d^2n}{d\lambda^2}\right]^{-1/2} c\tau \tag{8.48}$$

Example 8.3: Critical length for a pulse

A SiO_2 fiber with a 10 μm diameter core is excited with a 10 psec pulse at wavelength 1 μm. The pulse has a peak power of 10 W. What is the critical length of this pulse?

Solution: The peak change in index is obtained from Eq. 8.42, using $n_2 = 3.2 \times 10^{-16}$ cm^2/W, and calculating the effective intensity in the core of the fiber

$$\delta n \quad = 3.2 \times 10^{-16}\text{cm}^2/\text{W} \cdot 10\text{W}/(\pi(5 \times 10^{-4})^2) \quad = 4 \times 10^{-9}$$

We must determine the dispersion term. By evaluation of the Sellmeier equation, or from Fig. 8.18 we can find that

$$D(1\mu\text{m}) = -\frac{\lambda}{c}\frac{d^2n}{d\lambda^2}|_{1\mu\text{m}} = -38\text{ps/km/nm}$$

To convert this to a numeric expression for $\lambda^2 d^2n/d\lambda^2$, we need to multiply by λc and convert units in the proper fashion. We leave it as an exercise to show that the dimensionless quantity $\lambda^2 d^2n/d\lambda^2$ is

$$\lambda^2\frac{d^2n}{d\lambda^2}|_{1\mu m} = 0.0114$$

Plugging these values into Eq. 8.48 yields

$$\begin{aligned} L_{crit} &= (0.011 \cdot 4 \times 10^{-9})^{-1/2} \quad 3 \times 10^8 m/s \cdot 10^{-11} s \\ &= 452\text{m} \end{aligned}$$

The pulse will double in width due strictly to self-phase modulation in a distance of 0.45 km.

11. Optical Solitons

While Mother Nature sometimes makes our lives difficult with problems such as self-phase modulation or stimulated Raman scattering, we can find ways to exploit these "problems" to our advantage. The optical amplifier based on the stimulated Raman effect is one example of such exploitation. Self-phase modulation leads to another possible scheme which is rich with potential application, namely *optical solitons*[16],[17]. A soliton, by definition, is a solution to a wave equation that propagates without distortion. As we know, when we launch a pulse in a real optical material, dispersion will lead to a temporal broadening of the pulse. This broadening is a form of distortion.

To generate an optical soliton it is necessary to cancel out the effect of pulse broadening due to dispersion. This can be done with very careful balancing of self-phase modulation and negative dispersion. Recall in our discussion

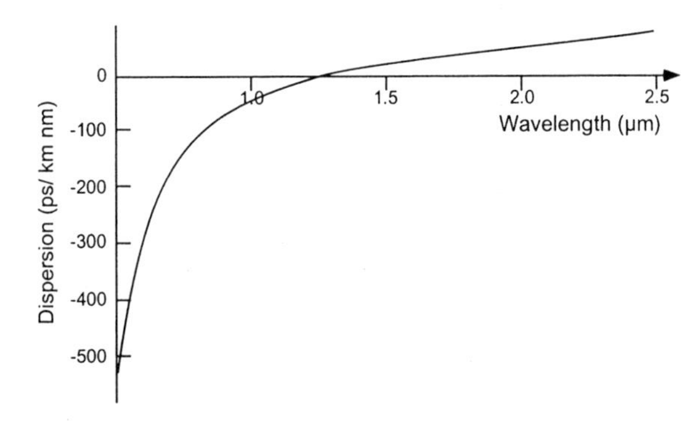

Figure 8.18. The dispersion in SiO_2. The region above 1.3 μm is called the negative dispersion region.

of the optical properties of matter, that the material dispersion term, $D = -\lambda/c \; d^2n/d\lambda^2$, went from a negative value for short wavelengths, through zero near λ_0, and then became positive for wavelengths longer than λ_0. The negative sign in the expression for dispersion adds a misfortunate confusion to the language. In the dispersion curve calculated for SiO_2 shown in Fig. 8.18, the "positive dispersion" region occurs for wavelengths less than 1.3 μm. The "negative dispersion" region occurs for $\lambda > 1.3\mu$m. Another way of defining negative dispersion is to note that material displays negative dispersion when $dv_g/d\lambda < 0$. In normal dispersion, lower frequencies travel slightly faster than higher frequencies. For example, red light travels faster than blue light in glass. The fact that they travel at different velocities is dispersion, and the fact that the low frequencies travel faster than the high frequencies is termed "positive dispersion." For wavelengths longer than λ_0, lower frequencies travel *slower* than the high frequency components.

For a transform-limited pulse with bandwidth $\Delta\nu$, negative dispersion will have the same type of effect as positive dispersion, namely it will cause the pulse to temporally spread. However, if the pulse has a frequency chirp caused by SPM, negative dispersion can actually compress the pulse. Consider the effect of negative dispersion on a pulse that has been phase shifted by self-phase modulation. Assume the central wavelength of the pulse is chosen to operate in the negative dispersion regime of the material. The leading edge of the pulse, which has lower frequency components, will travel slightly slower than the rest of the pulse. Similarly, the trailing edge will advance with respect to the pulse envelope. The pulse will tend to collapse upon itself as shown in Fig. 8.20. Thus, provided that the frequency chirp is large enough, dispersion — the former pulse broadener-- now leads to pulse narrowing.

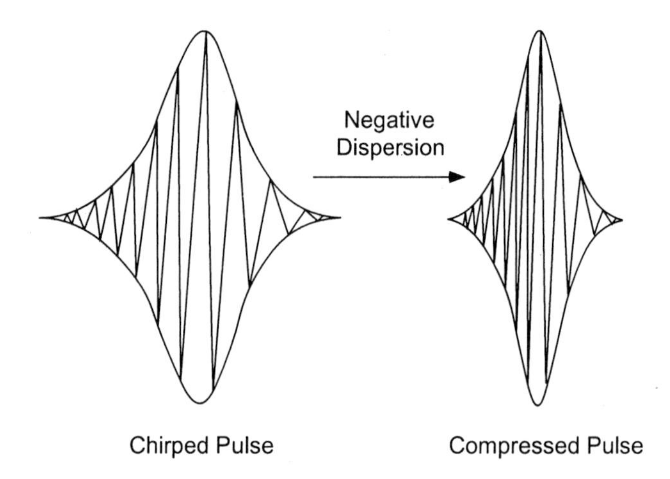

Figure 8.19. A chirped pulse will be temporally compressed by negative dispersion.

The simple description above can explain pulse compression, it is not sufficient to account for solitons. The correct description requires solution of the wave equation. If the pulse is described as

$$E(z) = u(z,t)e^{j(\omega t - kz)} \tag{8.49}$$

it can be shown[17] that the amplitude, $u(z,t)$, satisfies the nonlinear wave equation when SPM is included

$$j\left[\frac{\partial u}{\partial z} + \frac{\partial k}{\partial \omega}\frac{\partial u}{\partial t}\right] = \frac{-1}{2}\frac{\partial^2 k}{\partial \omega^2}\frac{\partial^2 u}{\partial t^2} + \frac{1}{2}k_0\frac{n_2}{n_0}|u|^2 u \tag{8.50}$$

The nonlinear interaction is contained in the last term of the equation. The first step in solving a nonlinear differential equation is to reduce it to dimensionless form. This can be done with the transformations

$$\begin{aligned} s &= \frac{1}{\tau}\left[t - \frac{\partial k}{\partial \omega}z\right] \\ \eta &= \left|\frac{\partial^2 k}{\partial \omega^2}\right|\frac{1}{\tau^2}z \\ v &= \tau\left[\frac{(k_0 n_2)/(2n_0)}{\left|\frac{\partial^2 k}{\partial \omega^2}\right|}\right]^{1/2} u \end{aligned} \tag{8.51}$$

where τ is the pulse width. Substituting this into Eq. 8.50 yields

$$j\frac{\partial v}{\partial \eta} = \frac{1}{2}\frac{\partial^2 v}{\partial s^2} + |v|^2 v \tag{8.52}$$

In general, this equation must be solved numerically. However, there are a few solitary solutions based on pulses with specific amplitudes and hyperbolic secant shape. For example, the fundamental soliton has a envelope described by a "secant-squared" shape,

$$u(t) = \frac{A}{\cosh^2(t/\tau)} \tag{8.53}$$

where A is a critical amplitude. Since the dispersion and self-phase modulation must be carefully balanced, the peak intensity of the optical pulse must be critically set to be

$$I_0 = \frac{\lambda^3 |D|}{1.288\pi^2 c\tau^2 n_2} \tag{8.54}$$

where τ is the full width at half maximum for the pulse, and $|D|$ is the material dispersion. As the pulse gets shorter, the required peak intensity increases quadratically. The absolute value of the dispersion, $|D|$, is specified only because of the inconsistent use of the sign of the dispersion in various references. The absolute value has no physical implications.

Fig.8.20 shows graphically the evolution of the pulse intensity for several solutions to Eq. 8.52. The fundamental soliton is a pulse that does not change shape as it propagates. It represents a pulse with just the right amplitude so that the pulse-spreading dispersion effects are exactly cancelled by the pulse narrowing effects of the nonlinearity. For higher order solitons, the input pulse amplitude must be related to the fundamental soliton amplitude by an integer number, n. The peak intensity of the higher order solitons then have n^2 the peak intensity of the fundamental soliton, given by Eq. 8.54. The higher order solitons exhibit complex behavior, temporally reducing, sometimes breaking into several peaks, and then expanding back to their original form after travelling a distance, z_0, called the *soliton period*

$$z_0 = \frac{0.322\pi^2 c\tau^2}{|D|\lambda^2} \tag{8.55}$$

The soliton period is independent of the order of the soliton.

Another fascinating property of solitons is their attraction and repulsion to one another. Because of the intensity-induced increase in refractive index, the soliton can be viewed as a wave trapped in its own "potential well" along the z-axis of propagation. When two solitons get close enough to each other so that their fields begin to overlap, they can attract or repel (depending on relative phase) due to the potential wells. This interaction can introduce errors in communications links if the pulses represent binary information that is placed

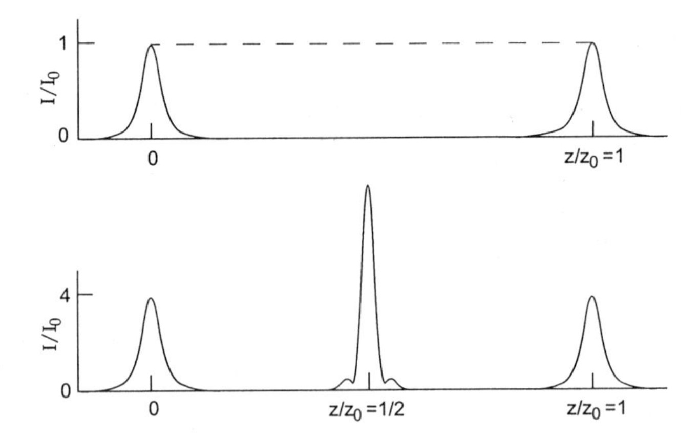

Figure 8.20. The theoretical behavior of the $n = 1$ and $n = 2$ soliton as they propagate down a fiber. I_0 is the peak intensity required for the fundamental soliton, and z_0 is the soliton period. The fundamental soliton never changes shape, while the higher order soliton exhibits a compression and then restoration to its original size after one period.

in certain temporal windows. Generally, interaction is negligible if the solitons are separated by approximately 10τ.

Example 8.4 Peak power for a soliton

Calculate the peak power needed to send a 10 psec and a 20 psec optical soliton through a fused silica fiber with a mode field diameter of 5 μm. The wavelength is 1.5 μm, and n_2 for fused silica is 3.2×10^{-16} cm²/W.

Solution: From Fig. 8.18, we can estimate that the negative dispersion has a magnitude of 20 ps/nm km. Using Eq. 8.54, we can calculate the necessary intensity for each pulse. We must be careful to keep all units consistent, so we will convert all units to cm or sec. The dispersion can be expressed as

$$D = 20 \text{ ps/km nm} = 2 \times 10^{-9}\text{sec/cm}^2$$

Plugging values into Eq. 8.54 for the 10 psec pulse yields

$$\begin{aligned} I_0 &= \frac{(1.5\times10^{-4})^3 \cdot 2\times10^{-9}}{1.288\pi^2(3\times10^{10})(10^{-11})^2 \cdot 3.2\times10^{-16}} \\ &= 553\text{kW/cm}^2 \end{aligned}$$

The peak power necessary to create this intensity in the optical fiber is

$$\begin{aligned} P &= I \cdot \pi r^2 = 553 \times 10^3 \pi (5 \times 10^{-4})^2 \\ &= 0.434\text{W} \end{aligned}$$

Inspection of Eq. 8.54 tells us that the intensity, and thus power, scales as the inverse square of the pulse length. Therefore a 20 psec soliton will require an intensity in the fiber of only 138 kW/cm^2, which corresponds to a peak power of 109 mW.

An alert reader will notice that the intensities in the optical fiber are sufficient for Stimulated Raman Scattering to become significant. This is indeed the case. Experiments have observed what is called the "Soliton self frequency shift", where the pulse intensity creates Raman gain, and the pulse bandwidth is large enough that the low frequency components of the pulse can experience gain from the high frequency components[17]. As the soliton travels, it slower transfers energy from the high frequency components to lower frequency components, leading to a steady shift toward longer wavelengths.

The soliton has tremendous potential application to communications and optical switching. The ability to send a pulse a long distance without distortion leads to the possibility of optical links that can span the oceans without regeneration. You might wonder how soliton propagation can exist in lossy fibers, since the intensity is critical to the dispersion cancelling effect. For example, even the best of fibers displays loss of 0.2 dB per kilometer, so after 30 km the power will be down by 6 dB, i.e. having only 25% of the original power. It has been experimentally observed, and theoretically confirmed, that the soliton adiabatically adjusts in temporal width as the amplitude decreases. As the power decreases due to attenuation, the pulse duration increases in such a way as to satisfy Eq. 8.54. The "soliton-like" behavior of the pulse is maintained, even as the power slowly decreases. The soliton is easily amplified using a distributed amplifier such as the Raman amplifier or Er-doped amplifier. Amplified links have experimentally demonstrated soliton propagation over distances exceeding 10,000 km.

12. Summary

In this chapter, we briefly touched on three important nonlinear effects which play significant roles in optical fiber links. The discussion was motivated by the question "Why not compensate for attenuation by coupling more power initially into the fiber?" We have found that at sufficient intensities, nonlinear effects can appear, and can totally dominate the optical system. In the case of Raman scattering, the nonlinearity will lead to the generation of new wavelengths which will dramatically increase the spectral bandwidth of the signal, making dispersion a severe limitation. We found that Brillouin scattering could effectively reflect the input power back out of the fiber, reducing the forward wave. And we saw how self-phase modulation increases the bandwidth of a pulse, again making pulse distortion due to dispersion a serious limitation. The bottom line of these effects are that there are physical limitations on the amount of power that can be coupled into a fiber if linear operation is desired.

We also tried to show that there is always a way to exploit such effects. The problem of stimulated Raman scattering can be turned into an advantage if it is used to amplify weak signals. Self-phase modulation can be used to make optical solitons, which propagate without distortion. Such nonlinear systems are certainly more complicated than simple linear pulse propagation along a fiber, but they offer many new advantages that may make the increased complexity well worthwhile. It is important for students of optoelectronics to be aware of these developments, as they are likely to become major tools of the trade for tomorrow.

References

[1] N. Ashcroft and D. Mermin, *Solid State Physics*, Holt, Rinehart, Winston, USA, (1976)

[2] *Semiconductors and Semimetals, Vol. 3*, ed. R. K. Willardson and A. C. Beer, p. 518

[3] F. P. Kapron, D. B. Keck, and R. D. Mauer, *Appl. Phys. Lett.*, vol. 17, pp. 423-425 (1970)

[4] S. G. Lipson and H. Lipson, *Optical Physics, 2nd ed.*, pp. 313, Cambridge University Press, 1981 (USA)

[5] K. A. Stacey, *Light Scattering in Physical Chemistry*, Academic Press, New York, (1956)

[6] M. Heiblum and J. Harris, *IEEE Jour. Quant. Elec. QE-11*, 75-83 (1975)

[7] W. A. Gambling, H. Matsumura, and C. M. Ragdale, *Opt. Quantum Electron. 11*,43-59 (1979)

[8] R. Stolen, in *Optical Fiber Telecommunications*, S. Miller and A. Chynoweth, eds., Academic Press, San Diego, pp. 125-150 (1979),

[9] G. P. Agrawal, *Nonlinear Fiber Optics,* Academic Press, USA, (1989)

[10] G. Baldwin, *Introduction to Nonlinear Optics*, Plenum Publishing Corp., New York, (1974)

[11] P. Milonni and J. Eberly, *Lasers*, John Wiley and Sons, New York, (1988)

[12] Y. R. Shen, *Principle of Nonlinear Optics*, John Wiley and Sons, New York, (1984)

[13] A. Yariv, *Quantum Electronics, 3rd Ed.*, John Wiley and Sons, New York, (1989)

[14] W. Kaiser, and M. Maier, *Stimulated Rayleigh, Brillouin and Raman Spectroscopy,* in *Laser Handbook*, F.T. Arecchi and E. O. Schultz-Dubois, eds., pp. 1077, North Holland, Amsterdam, (1972),

[15] R. L. Smith, Appl. Opt. 11, 2489 (1972)

[16] A. Hasegawa and F. Tappert, *Applied Physics Letters* 23, 142 (1973)

[17] L. F. Mollenauer and R. Stolen, "Solitons in optical fibers," Laser Focus, April 1982, pp. 193-198, and L. F., Mollenauer, R. H. Stolen, and J. P. Gordon, *Phys Rev. Lett.*, 45, 1095 (1980)

Practice Problems

1. A single mode fiber with a mode field radius of $5\mu m$ and a core of pure GeO_2 is to be used to make a Raman amplifier. GeO_2 has a characteristic phonon frequency of 420 cm^{-1}, and a Raman gain of $5 \times 10^{-13}/$ (m/W). A signal with wavelength 1.55 μm and input power of 10 nanowatts is to be amplified.

   (a) What wavelength should the pump light be?

   (b) What intensity of pump light is required if the signal is to be boosted to an exit power of 1 mW in a length of 500 meters?

   (c) What should the *power* of the pump light be to satisfy part *b*? Assume the MFR is the same for the pump and signal beams.

   (d) How many photons at 1.55 μm are injected per second into the amplifier? How many exit? Compare this to the number of pump photons per second. Do we have to worry about pump depletion in this amplifier?

   (e) If the bandwidth of the Raman gain is 6 THz, how many noise photons due to Spontaneous Raman Scattering would you expect to see at the end of the amplifier?

2. Design an optical amplifier based on the Raman effect that will boost a 1.3 μm signal by 20 dB. Assume the single mode fiber is identical to that used in Prob. 1, but let the length be 1000 meters. Specify the pump wavelength and power necessary to create such an amplifier. What are the power limitations for the signal? In other words, if the amplifier can boost a 1 nanowatt signal by 20 dB, can it boost a 1W input signal by 20 dB? When does gain saturation start to become important?

3. For the fiber described in Prob. 1, what is the maximum input signal power at 1.55 μm that can be coupled into the fiber before Stimulated Raman Scattering begins to become a serious problem? Assume that the fiber has 0.2 dB/km of attenuation at 1.55μm.

4. A 5 psec pulse at 1.55μm is coupled onto a SiO_2 fiber with a 5μm mode field radius. The pulse has a Gaussian temporal profile. If the fiber is 10 meters long, what peak intensity will lead to the accumulation of π radians of additional phase between the leading edge and peak of the pulse by the time the pulse leaves the fiber?

5. Nonlinear interactions between two fields are usually dramatically enhanced when the two fields travel at the same velocity in the medium. If an intense pulse from a laser at 1.06 μm is launched into fused silica, what

other wavelength(s) will travel at the same group velocity in the material? If the group velocities are identical, will the phase velocities be identical?

6. The complex dielectric constant of a certain material is given by $\epsilon/\epsilon_0 = 2.5 - j0.010$ at $\lambda = 1.5\mu$m. Find the attenuation coefficient for this wavelength, and determine the phase velocity.

7. A transform-limited pulse of temporal duration $\tau = 500$ psec is launched into an medium with material dispersion of D = -50 psec/nm·km. How far must the pulse travel before the pulse envelope doubles in temporal width? Note that you will have to convert the dispersion, D, into a numeric value for $\lambda^2 d^2n/d\lambda^2$. Is the pulse still transform limited at the end of this distance? Justify your answer.

8. A certain material is made into an optical fiber, and this material has an attenuation coefficient of 0.7 dB/km. If 1 mW of light is coupled into the fiber, what is the power of the light after traveling 10 km?

9. What peak power is required to form a soliton with 5 psec duration on a fiber with a MFR of 5μm? Assume the fiber is made of fused silica, and the operating wavelength is 1.5 μm. What is the total energy in the solitary pulse?

10. Using Eq. 8.8, calculate the expected attenuation coefficient for green light, $\lambda = 0.5\mu$m, travelling through the atmosphere. Assume that the density of air displays a Poisson distribution such that if a volume of air contains $N = nV$ molecules, there will be a statistical variance of $\Delta N = \sqrt{nV}$, where n is the number density of molecules per unit volume. Assume the index of refraction of air is 1.0003 at standard pressure and temperature, and that is depends linearly on density.

11. It is desired to make a Raman amplifier for very short optical pulses using a SiO_2 fiber as the gain medium. To maximize the interaction between the pump pulse and the Stokes pulse, the two pulse must travel at the same velocity. Using the data on SiO_2 from Table 3.1, determine what the optimum pump and Raman wavelength will be. Note that the two wavelengths must have the same group velocity, and they must be separated by 467 cm^{-1}.

12. A 5 psec pulse at 1.55 μm is coupled onto a SiO_2 fiber with a 5 μm mode field radius. The peak power is 5W. The pulse has a Gaussian temporal profile. How far can the pulse travel before SPM has doubled the pulsewidth? What intensity is required if this distance is to be doubled?

13. Estimate the effective focal length of a 2mm glass slide when it is illuminated by a field

$$I(r) = I_0 e^{-r^2/w^2}$$

where $I_0 = 10^8$ W/cm^2, and $w = 1$ mm. The lens is caused by the nonlinear refractive index induced by the strong optical field. This effect is called self focussing, and is responsible for creating lots of damage in early solid state lasers.

14. A 10 psec pulse with $\lambda = 1\mu$m is transmitted through 30 cm of single mode fiber. The bandwidth of the pulse doubles. What is the intensity of the pulse?

15. Design an optical fiber amplifier based on stimulated Raman scattering. The desired gain of the amplifier is 20 dB, and the maximum power that should be extracted is 10 mW. The signal wavelength is 1.53 μm, and the fiber displays a 0.3 dB loss/km at the signal and pump wavelengths. Determine the optical pump power required, and the pump wavelength. Make sure that the amplifier delivers not only the desired gain, but can deliver it at the desired output power.

Chapter 9

NUMERICAL METHODS FOR ANALYZING OPTICAL WAVEGUIDES

1. Introduction

Analytical solutions of the wave equation exist for only a few waveguide structures. Direct numeric solution of the wave equation is possible for many structures, although this usually involves iteration to find the approximate eigenvalue. In this chapter we will discuss two methods used to study modes of structures which are not amenable to analytic or approximate solution. These methods are the Beam Propagation Method (BPM) and the Finite-Difference Time Domain method (FDTD). Numerical simulations are needed to evaluate special structures such as waveguides with bends or reflective mirrors, split Y-couplers, and coupled adjacent waveguides [1, 2, 3], or structures which have reflections, such as a grating. A Y-coupler, shown in Fig. 9.1, is a simple device that connects one waveguide to two waveguides. Since the mode dynamically changes as it enters the structure, it is very difficult to calculate the answers to questions such as "How much loss will the mode encounter?", "What is the optimum angle to split the waveguide?", and so forth. The numerical simulation methods allow us to determine these answers.

Both BPM and FDTD are *numerical simulations* of the field in a guide, in contrast to the numerical solution of the exact wave equation that we did in Chapter 6. Simulations are often the only way to determine the mode profile in

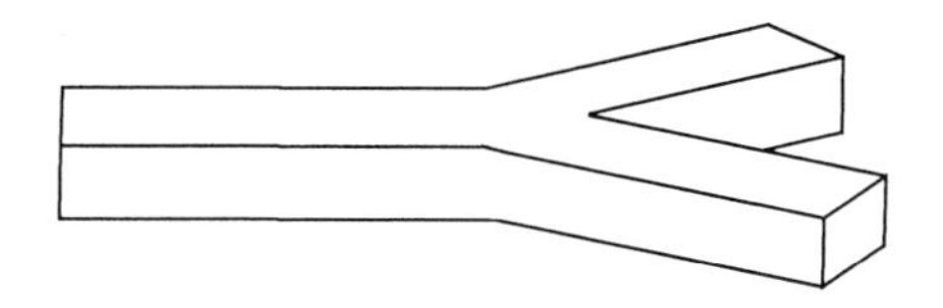

Figure 9.1. A Y-coupler is a simple device that couples one waveguide into two waveguides.

an unusual waveguide, and to map out the behavior of a mode as the index profile changes along z. This latter effect is very common in practical devices, such as waveguide tapers, or waveguides perturbed by a second nearby waveguide.

2. Beam Propagation Method

The Beam Propagation Method works by decomposing a spatial mode into a superposition of plane waves, each travelling in a slightly different direction. After advancing each wave a certain distance through the dielectric structure of interest, the plane waves are added back together to reconstruct the spatial mode. This process requires the use of Fourier transforms to convert from the spatial mode into the superposition of plane waves, and back again. Therefore, we will spend some time discussing the Fast Fourier Transform in the following sections.

We will describe only the scaler BPM technique. Interested readers can use this chapter as a springboard into the current literature which describes vector beam propagation methods, and other advanced techniques. Before we begin, let us apologize for the heavy use of acronyms in this chapter. Besides BPM and FDTD, we will discuss FFTs, and several other nondescript symbols of modern computing convenience. We will try to define acronyms several times so that the reader can find the translation without too much searching.

2.1 A Note About Numeric Computations

We will develop numerical procedures for describing beam propagation in this chapter. Problems at the end of the chapter illustrate the principles of the techniques. Like many of the examples or problems developed in this text, the exercises developed in this chapter were done with either *Mathematica*, or *Matlab* on a personal computer. If you become a serious numerical simulator of waveguides, you will probably want to develop or buy specialized software to increase the calculation speed. However, for initial exploration and learning how BPM works, we recommend that you simulate simple structures using a numerical package such as *Matlab*. You should first try the techniques described in this chapter with simple, one-dimensional structures that will demonstrate the algorithms without consuming a great deal of computation time.

3. Superposition of Waves

Consider the planar slab waveguide, where the index profile only varies in the x direction (Fig. 9.2). Due to symmetry the spatial field is functionally independent of the y-direction. A guided field in such a source-free dielectric structure must be a solution to the (by now familiar) wave equation

$$\nabla^2\Psi + k_0^2 n^2(x)\Psi = 0 \tag{9.1}$$

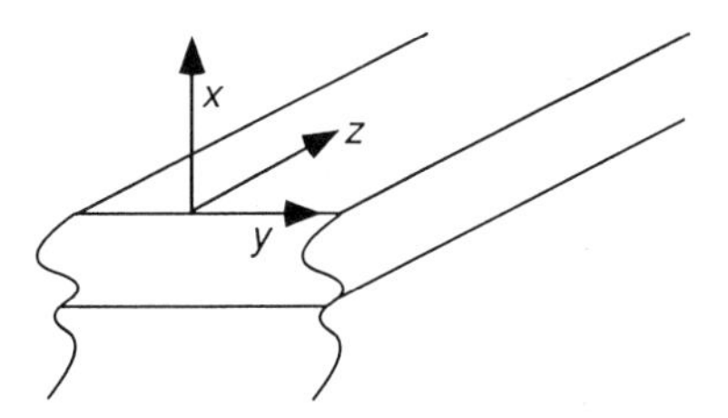

Figure 9.2. An infinite slab dielectric waveguide. The wave is presumed to travel in the z-direction.

where $\Psi(x, z, t)$ is a vector function describing the amplitude, polarization, and direction of field propagation. If the waveguide consists of isotropic regions, i.e. the index does not have a gradient profile, the spatial solution to the scalar wave equation in each region of space is simply a plane wave

$$\Psi_i(x, z, t) = A_i e^{-j(k_x x + k_z z)} e^{j\omega t} + c.c. \tag{9.2}$$

where A_i is the amplitude in the region i, and the k-vector (described in terms of its components) depends on the frequency of the wave and the local index of refraction, and can be real or imaginary. Note that the k vector components are the generalized forms of β, κ, and γ. Boundary conditions connect the solutions at the interfaces separating the different regions. Plane waves are the natural solution to the wave equation.

Since the wave equation (Eq. 9.1) is linear, any linear superposition of solutions will also constitute a valid solution. This important fact forms the foundation of the technique used to numerically analyze the fields in a waveguide. We will use a superposition of plane waves, each with identical angular frequency ω but different values of k, to describe the general mode of a waveguide. The plane waves form a *basis set* for the mode description.

4. The Fourier Transform in Guided Wave Optics

Describing a *spatial function*, $\Psi(x)$, in terms of a superposition of plane waves, $\exp(jkx)$,

$$\Psi(x) = \sum A_i e^{jk_i x} \tag{9.3}$$

should remind you of the Fourier Transform. To illustrate, consider a one-dimensional electric field distribution with a Gaussian distribution.

$$\Psi(x) = E_0 e^{-x^2/x_0^2} \tag{9.4}$$

Fig. 9.3 shows a Gaussian profile, where the characteristic width is chosen to be $x_0 = 8\mu$m.

The Gaussian profile describes the lowest order mode in a parabolic index profile waveguide (see Chap.2). Also, readers familiar with the modes of laser

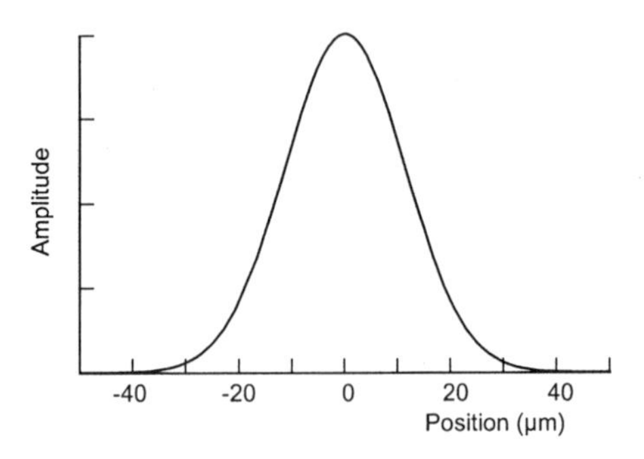

Figure 9.3. The amplitude distribution of a Gaussian mode.

beams in free space will recognize this profile as the fundamental TEM_{00} mode of a Hermite-Gaussian beam.

To describe this transverse *spatial mode* in terms of a superposition of *plane waves*, e^{jkx}, we employ the Fourier transform pair,

$$\begin{aligned} \Psi(x) &= \frac{1}{2\pi}\int_{-\infty}^{\infty} A(k_x)e^{jk_x x}dk_x \\ A(k_x) &= \int_{-\infty}^{\infty} \Psi(x)e^{-jk_x x}dx \end{aligned} \tag{9.5}$$

where k_x is the x-component of the wavevector, k. The transform of $\Psi(x)$ yields $A(k_x)$, which is a complex number that contains information about the amplitude and phase of each plane wave component. Eq. 9.5 can be readily evaluated to give the amplitudes

$$A_{k_x} = \frac{1}{\sqrt{\pi x_0^2}}\, e^{-\pi^2 x_0^2 k_x^2} \tag{9.6}$$

Recall that the *magnitude* of k is identical for all components in a mode; only the k_x and k_z components vary. Simple trigonometry provides the value of the z-component of the wavevector: $k_z = \sqrt{k^2 - k_x^2}$. The largest k-vector has $k_x = 0$, corresponding to a plane wave travelling along the z-axis. From Eq. 9.6, as the k_x component increases, the amplitude of the plane wave decreases. The amplitude and direction of the individual plane waves that comprise a spatial amplitude distribution are shown schematically in Fig. 9.4. The length of each arrow represents the plane wave amplitude, A_k, while the direction indicates the orientation of the k-vector.

The Fourier transform pair in Eq. 9.5 allow us to readily convert a wave described in the spatial domain ($\Psi(x)$ to a wave described in phase space domain ($\Psi(k)$. In the Beam Propagation Method (BPM) propagation effects are

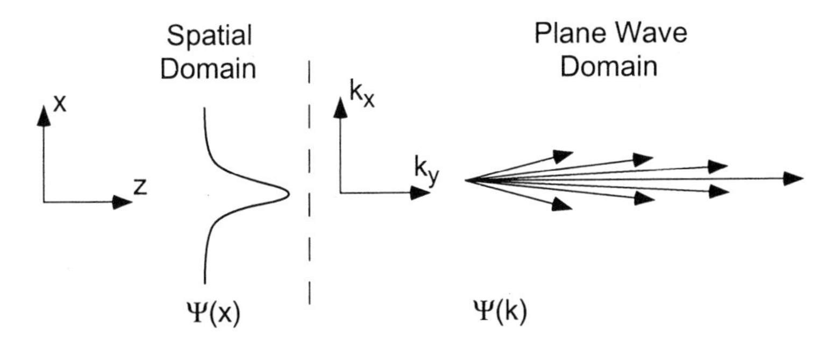

Figure 9.4. A spatial amplitude distribution, $\Psi(x)$, is equivalent to a superposition of plane waves, $\Psi(k)$, each with a slightly different amplitude (indicated by length of arrow) and direction.

calculated using the phase space representation, and phase shifts caused by the waveguide structure are introduced using the spatial representation of a mode. We will use both of these transforms to alternately convert a spatial field into a superposition of plane waves, and back again.

4.1 The Fast Fourier Transform (FFT)

To take advantage of numerical computers for calculating Fourier transforms, we will use the discrete Fourier Transforms based on what are generically called Fast Fourier Transforms (FFT). FFT algorithms are widely available in literature[4], and are common features in engineering and mathematical numerical software packages for workstations and personal computers. Application of the FFT to optical propagation problems is discussed in ref. [5]. The Fast Fourier Transform (FFT) is closer in operation to a Fourier *series* than to a Fourier *transform*. Recall that a Fourier series is used to describe a periodic function in terms of a discrete set of sinusoidal basis states. The FFT describes a distribution in terms of a large but finite number of discrete sinusoidal waves with appropriate amplitude. The effect of discrete sampling can lead to the creation of aliases of the waveform. This fact will introduce a complication in the BPM calculation.

To find the FFT of a spatial profile, the profile must first be represented as a numeric array. The sampling resolution must be fine enough to resolve all spatial features of the amplitude profile, yet at the same time be sparse enough to allow reasonable processing speed on a computer. This trade-off is obviously something each designer must address based on their personal computing capabilities and patience. In the calculations that follow, an array with 100 points proved adequate to see the desired behavior.

Let's begin a demonstration of the BPM using the Gaussian mode profile shown in Fig. 9.3. Since evanescent amplitudes follow an exponential decay,

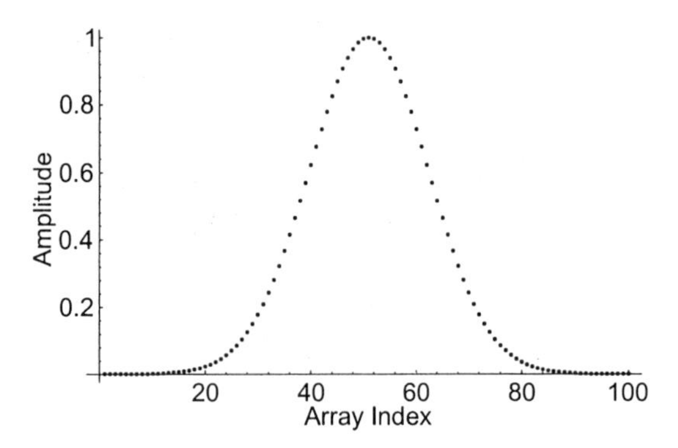

Figure 9.5. The sampled profile of a Gaussian mode, taken with 100 samples.

they never truly go to zero. Theoretically there will be some error introduced when we restrict the spatial domain to something less than infinity. If we extend the sampling domain out to three or four characteristic decay lengths (the decay length is defined as $1/\gamma$ for a slab waveguide), we will usually get satisfactory results. In this example, we sampled the profile in Fig. 9.3 at 100 equally spaced discrete points ranging from $x = -25\mu$m to $x = 24.5\mu$m. Since the profile is assumed to be periodic in an FFT (this means that the algorithm assumes that if it looked at points 101 to 200, it would find another gaussian wave of the same shape and amplitude centered near point 150), the data point at $x = -25\mu$m is the same as at $x = 25\mu$m, so it is important to *not* include this point twice in the array. That is why the domain is selected as shown. Fig. 9.5 graphically shows the resulting array. The abscissa is the array index, not the position.

Having established an array, we can compute the Discrete Fourier Transform of the spatial profile in order to determine the superposition of plane waves that comprise the mode. As before, we will use *Mathematica* for calculations; other packages such as *Matlab*, or *Maple* have equal abilities. To create the data for the Gaussian profile with $x_0 = 8\mu$m, to find the FFT, and to observe the intermediate results, we used the following *Mathematica* commands

```
mode= Table[Exp[-x^2/64], {x, -25, 24.5, 0.5}]//N;
ListPlot[mode]
fftmode=Chop[Fourier[mode]];
ListPlot[Abs[fftmode], PlotRange-> All] (* magnitude of FFT *)
ListPlot[Arg[fftmode], PlotRange -> All] (* phase of FFT *)
```

Any good *Mathematica* reference[6] describes these commands. Briefly, `Table` creates an array of 100 points which sample the Gaussian amplitude profile over the range from -25μm to +24.5μm. The data in this table were plotted in Fig. 9.5. `Chop` converts any number less than 10^{-10} to zero, reducing round-off noise. If we do not do this, the phase of the output, which is

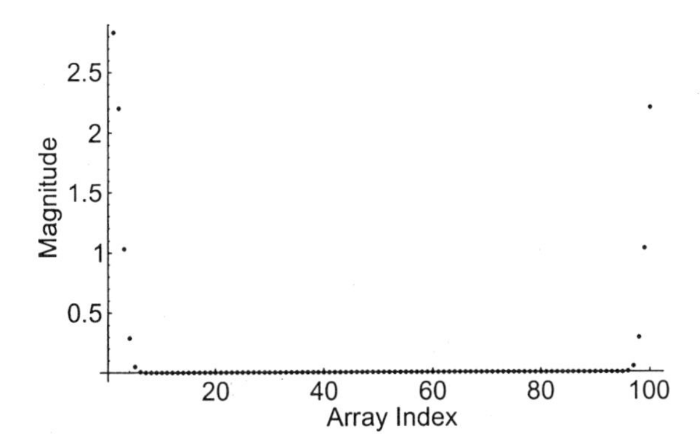

Figure 9.6. The FFT of the Gaussian profile is plotted as a function of the array index. Note the symmetry of the FFT; the magnitude of the value at $s = 2$ is identical to the value at $s = 100$, etc.

determined by the ratio of the imaginary and real components, can show extreme variations when, in fact, the magnitude of the actual amplitudes are negligible. `ListPlot` plots the values of an array. Since the Fourier transform of an arbitrary input will in general be complex, we need to specify the magnitude (`Abs`) or the phase (`Arg`) when plotting the output of a FFT computation.

The FFT of the N-point array is itself an N-point array, found by approximating the integral in Eq. 9.4 with a discrete summation. The N complex numbers which describe the spatial amplitude are converted into N complex numbers which correspond to the amplitude and phase of each planewave component with $k_x = 2\pi(s-1)/H$, where H is the spatial size of the sample, and s runs from 0 to N.

The magnitude of these amplitudes for the Gaussian profile are plotted in Fig. 9.6, where the abscissa is the s index. There are in fact 100 points in the FFT, exactly the same number as are in the amplitude array. The appearance of the FFT is a little strange at first glance. Instead of producing a smooth peak in k-space, we find a distribution with non-negligible values near $s \approx 0$ and $s \approx 100$, but very little magnitude at mid-range values.

The strange structure of the FFT arises because the Fourier transform is calculated on a *discrete* array of samples from the actual waveform. The value of the FFT at $s = 1$ corresponds to the average value of the spatial profile, the $k_x = 0$ term of the expansion. The next few terms describes the $k_x = 2\pi(s-1)/H$ components of the transverse k-vector, where H is the domain of the spatial wave. Each additional point corresponds to the next higher transverse component. In this example we chose $H = 50\mu$m. From the FFT, it is clear

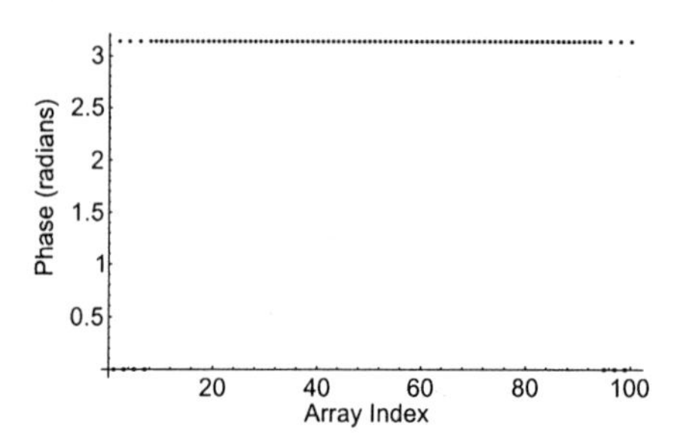

Figure 9.7. The phase of the DFT of the Gaussian wave shown in Fig. 9.5 is plotted as a function of the array number. The phase is either 0 or π, indicating that the wave is everywhere real.

that there are not many high order transverse components needed to describe this mode.

Now, what about those terms near $s = 100$? Due to the periodic sampling of the mode profile, the terms with $s = 99$ corresponds to the $s = -1$ term, or $k_x = -2\pi/H$. Similarly, $s = 98$ corresponds to $s = -2$, and so forth. Physically, these correspond to plane waves travelling with a slight downward inclination, while the plane waves with values such as $s = 2$ correspond to plane waves travelling with a slightly positive inclination.

Fig. 9.7 shows the plot of the *phase* of the FFT of the Gaussian beam. Notice that in the region where there is significant amplitude for E_s, the phase alternates between 0 and π. This is equivalent to multiplying every other term in the series expansion of the mode profile by -1. The phase tells us what the wavefront curvature of the beam is. In this case the field is everywhere real, indicating that it represents a plane wave.

4.2 Wavefront Curvature and Complex Numbers

A wavefront is a locus of points where the phase is constant. Complex numbers convey phase information in a wave. Since the equation which describes the Gaussian mode, Eq. 9.4, is purely real, the phase is constant as a function of r, so the mode has a planar wavefront. If the phasefront of the mode had some curvature, the phase would change with distance from the axis, and the proper description of this would involve using complex numbers. In general, the arrays used to describe the spatial waves and the Fourier amplitudes will be complex. To illustrate this, consider the mathematical description of a curved wavefront shown in Example 9.1.

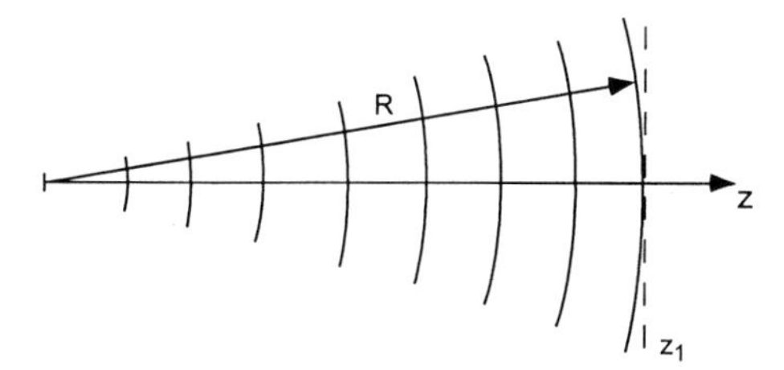

Figure 9.8. A wave originating from z=0 has a spherical wavefront as it propagates away from the origin. At a distance $z_1 = R$ along the z-axis, the wavefronts have a radius of curvature, R.

Example 9.1 Phase of a curved wave front

Consider the picture below of a spherical wave travelling in the z-direction. How would you describe this wave at the $z = z_1$ plane in terms of complex numbers?

Solution: The amplitude of a spherical wave is described by

$$E(r) = \frac{E_0}{R} e^{-j(kr-\omega t)}$$

where R is the absolute distance from the origin, and E_0 is an arbitrary amplitude. Ignoring the time dependence, and using the small angle approximation, we can expand this expression near the z-axis.

$$\begin{aligned} E(x,z) &= \frac{E_0}{R} e^{-jk(\sqrt{z^2+x^2})} \\ &= \frac{E_0}{R} e^{-jkz(\sqrt{1+x^2/z^2})} \\ &\approx \frac{E_0}{R} e^{-jkz(1+x^2/2z)} = \frac{E_0}{R} e^{-jkz} e^{-jkx^2/2z} \end{aligned}$$

Along a plane $z = z_1$, there will be a propagation phase term, e^{-jkz_1}, and a term which changes quadratically as the position increases from the z-axis, $e^{-jkx^2/2z_1}$. This represents the curvature of a spherical wave.

5. Beam Diffraction

So why are we belaboring Fourier transforms? By describing a real beam as a superposition of plane waves, we can develop an accurate method for simulating beam propagation which includes effects such as diffraction. In this section, we will show how the previous analysis of plane wave superposition can numerically determine the beam diffraction of a propagating field. This step, incidentally, is the first step in understanding the Beam Propagation Method.

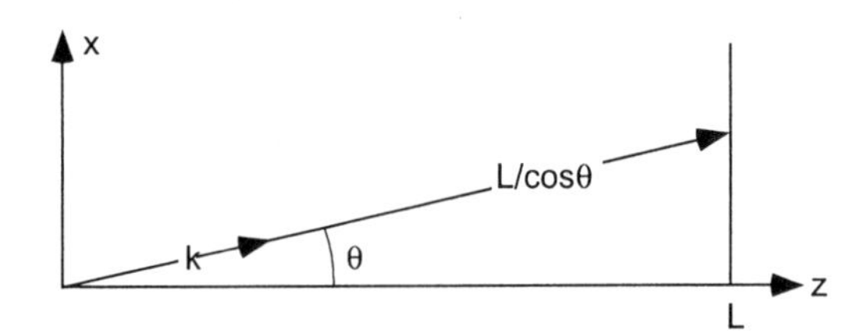

Figure 9.9. Geometrical picture of path a plane must follow to move a distance L along the z-axis.

Let's propagate a wave a distance L in the z-direction using the plane wave method. We first decompose the spatial profile into a superposition of plane waves, and then advance each plane wave component forward to the plane $z = L$. Once there, we will superimpose these plane waves back together to form the new spatial mode. Since each plane wave travels in a different direction, each will accumulate a different amount of phase due to the path length difference incurred travelling to the plane at $z = L$. How much phase is accumulated? Fig. 9.9 shows a geometric argument for the phase shift.

Every plane wave component of the expansion has a wavevector with magnitude k, that travels in a unique direction. A component travelling at an angle θ ($\theta = \sin^{-1}(k_x/|k|)$)with respect to the z-axis will travel a slightly longer distance, $L' = L/\cos\theta$, to get to plane at $Z = L$ than would a wave travelling parallel to the z-axis. Using the small angle approximation for θ, and the fact that in the FFT, $k_x = (s-1)2\pi/H$, and $|k| = 2\pi/\lambda$, the phase accumulated by each ray is given by

$$\phi_s = \frac{kL}{\cos\theta} = kL\left[1 + \frac{k_x^2}{2n^2k_0^2}\right] = kL\left[1 + \frac{\lambda^2}{2H^2}(s-1)^2\right] \tag{9.7}$$

where H is the size of the spatial domain of the amplitude profile. The term, s, refers to the index of the $(s-1)$th spatial frequency component of the plane wave superposition. For example, the $s = 1$ term corresponds to the $k_x = 0$ term of the Fourier expansion. Each Fourier component will accumulate a different amount of phase after travelling along the z-axis becasue they each traverse a slightly different path length.

We are now ready to propagate the optical mode a distance through space. We do this in two steps. First, we must determine the plane wave superposition that comprises the initial spatial field. This was described above using the FFT of the spatial function, $E(x)$. Next, we let each component of the wave propagate up to the plane at $z = L$. Since nothing alters the magnitude of the individual plane waves as they propagate, the amplitude of each component remains the same. However, since they accumulate different amounts of phase based on the difference in path length, we must add the proper amount of phase to each component. The example below illustrates the procedure.

Diffraction of a spatial mode in free space

Consider the Gaussian mode shown in Fig.9.5. We will assume that it represents a TEM_{00} laser beam with wavelength 0.8 μm, and furthermore we will assume that the profile represents the beam at a focus, so the wavefront is planar. We will use the BPM to calculate how the beam spreads and develops wavefront curvature as it travels through free space. Since diffraction theory has well established analytic expressions for describing Gaussian beam diffraction, we can use these to confirm the operation of the BPM technique.

We start with an expression for the beam at $z = 0$. The beam is a simple Gaussian profile with characteristic length of 8μm.

$$E(x, 0) = Ae^{-x^2/8^2} \tag{9.8}$$

where all dimensions are in μm, and A is an arbitrary amplitude which we will set to unity. Notice that the wave is everywhere *real*, showing that the field is a plane wave at $z = 0$.

We must determine the FFT of this mode by first creating an array of equally spaced samples of the amplitude. By inspection of Fig. 9.10, we can see that the mode has negligible amplitude beyond ± 20 μm from the core axis. Therefore we could probably set a spatial domain, H, equal to 40 μm, and not introduce significant clipping. However, since the beam will be expanding in the spatial dimension, we will choose a larger domain. We chose $H = 120$ μm after some iteration. The *Mathematica* file used to generate this array and calculate the DFT is listed below.

```
xo=8; h=120; del=h/100; wave=Exp[-x^2/xo^2];
discretewave=Table[wave, x,-h/2, h/2-del,del]//N;
fftwave= Chop[Fourier[discretewave]]
```

Fig. 9.10 shows a plot of the sampled mode profile. The last line of the code calculates the FFT of the spatial mode, and creates a complex array called `fftwave`. The magnitude (`Abs[dftwave]`) and phase (`Arg[dftwave]`)of the components of the FFT array are shown in Fig. 9.11 and Fig. 9.12.

Now we can simulate the effect of propagating a distance L. Eq. 9.7 describes the phase shift each plane wave will accumulate. In this example, where the wavelength of the light is 0.8 μm, let's first travel 500 μm in vacuum (the index of refraction, n, equals unity) and $k = 2\pi/\lambda = 2\pi/0.8\mu$m.

The phase shift accumulated for the $s = 1$ component ($k_x = 0$) is simply $k_0 L$. Due to the symmetry of the mode and the nature of the FFT, the phase shift $s = 2$ term ($k_x = 2\pi/L$) is the same as the $s = 100$ term. Similarly the components labelled $1 < s < 50$ are the mirror image of the upper terms,

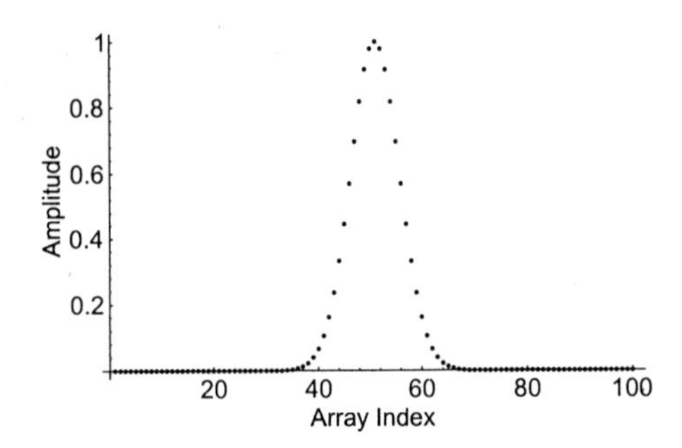

Figure 9.10. The waveform described in Eq. 9.8 is sampled at 100 discrete points, and the data is plotted to confirm the accuracy of the sampling. The horizontal axis corresponds to positions ranging from -60 μm to 60 μm.

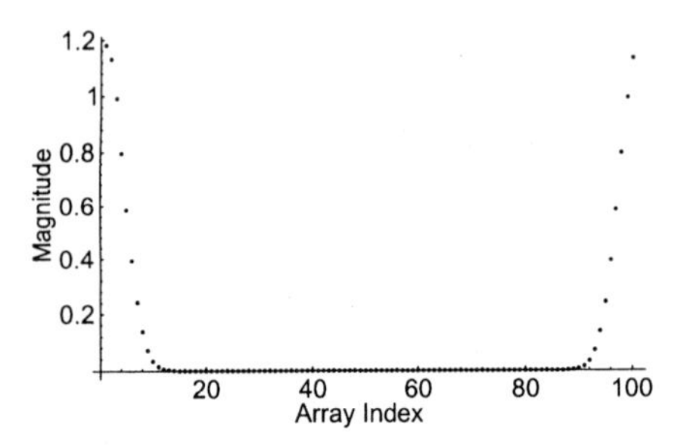

Figure 9.11. The magnitude of the FFT of array `dftwave` is shown as a function of the array index. In this case, the array index corresponds to transverse momentum of $k_x = 2\pi(s-1)/120$ μm^{-1}.

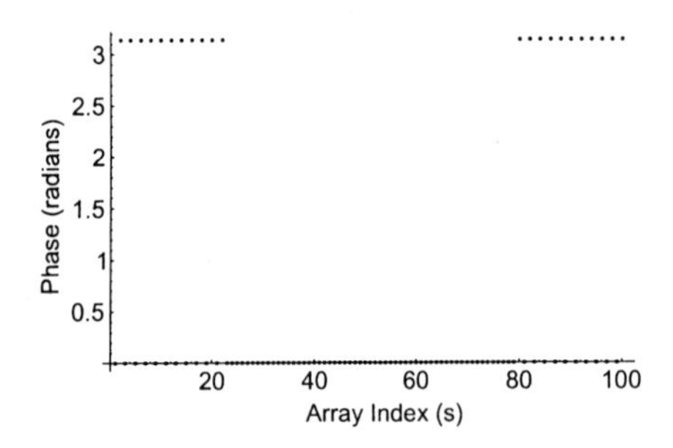

Figure 9.12. The phases of the Fourier components of the DFT of the spatial mode are shown as a function of array index. There is little to glean from this other than to note that all phases are either 0 or 180°, implying that the wave has no curvature, and is therefore a plane wave. This is a good test of our calculations.

$51 \leq s < 100$, because the mode is perfectly symmetric. Recall the Fourier components labelled by $s \geq 50$ correspond to transverse components travelling in the negative x-direction. Once again being careful about the array indices, we calculate the additional phase that each wave will accumulate, as a function of its index. The *Mathematica* code is

```
k=2 Pi / 0.8;
l=500;
phase1[s_]:= Table[k l ( 1+0.5 (0.8 /h)^2 (s-1)^2), {s,1,50}];
phase2[s_]:= Table[k l ( 1+0.5 (0.8 /h)^2 (101-s)^2), {s,51,100}];
phaseshift=Join[phase1[s], phase2[s]];
```

This set of commands creates an array of 100 points, each with a phase shift corresponding to that required by Eq. 9.8. The phase corrections are calculated in two 50 element arrays that are then concatenated to form a 100 element array called `phaseshift`.

Once the phase shifts are computed for each spatial frequency term, they must be added to each appropriate plane wave. The *Mathematica* statement below describes the transfer into a new wave called `newfft`, where the magnitude of the plane wave has not been affected, so the magnitude of the old wave components and new wave components are identical, but the phases are updated.

```
newfft = Abs[fftwave] Exp[I (Arg[fftwave] + phaseshift)]//N;
```

Note that `fftwave` and `newfft` are arrays, and that the operation listed above is implicitly an array operation. The `//N` at the end of the statement forces *Mathematica* to convert all symbolic values, such as π, into numeric values. Note also that the only difference between `fftwave` and `newdft` is that each component has accumulated additional phase due to propagation. To determine the spatial mode at plane $z = L$, we simply inverse Fourier transform the phase space superposition and plot the result.

```
newwave= Chop[InverseFourier[newdft]]
ListPlot[Abs[newwave], PlotRange->All]
ListPlot[Arg[newmode], PlotRange->All]
```

The output is shown below in Fig.9.13. As the mode propagates along the z-axis, it spreads out in the transverse dimension, as we would expect for diffraction. We can compute the mode shape further along the z-axis, and compare the relative magnitudes. Fig. 9.15 shows the calculated amplitude at $z = 0$, $z = 500$, and $z = 1000\mu$m. The amplitude decreases as the width increases, conserving total power.

The amplitude ripple apparent in Fig. 9.15 for the profile at $z = 1000\mu$m is an artifact of the FFT, and is not a true representation of the profile. Due to the periodic nature of the FFT, high spatial frequency components that travel out of the field on the right side of the spatial domain reappear on the left hand side, and vice-versa. These spatial frequency components interfere coherently with the same frequency components in the original wave and form small standing waves. One way to avoid this is to increase the domain size, H, but this just

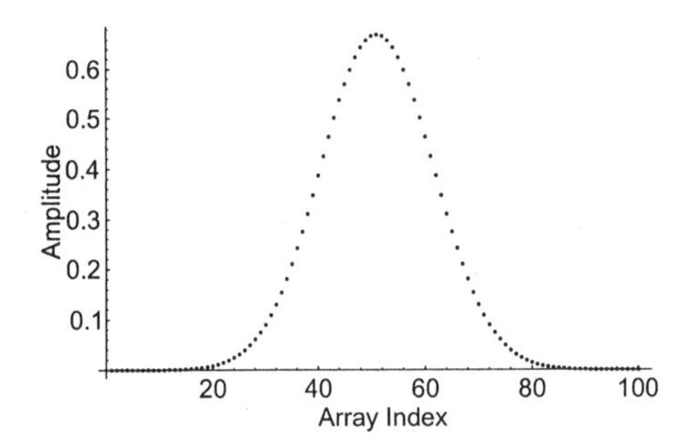

Figure 9.13. The amplitude of the Gaussian field after travelling 500 μm. The peak amplitude is decreased, but the width of the mode has increased. The horizontal axis corresponds to ± 60 μm.

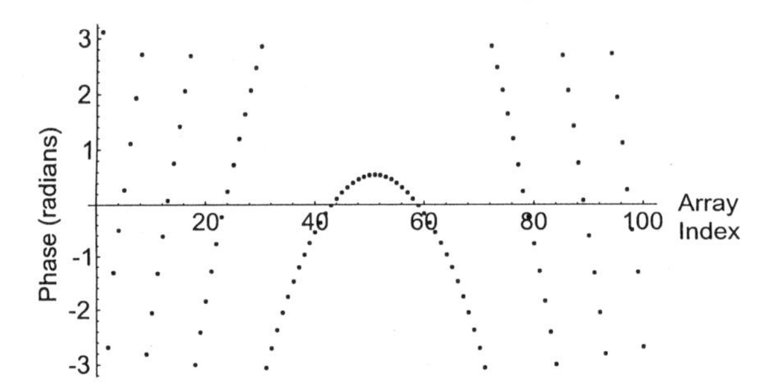

Figure 9.14. The phase of the mode after travelling 500 μm has picked up considerable curvature. The plot shows the phase shift *modulo* 2π. The phase shift increases quadratically with distance from the axis, however the amplitude rapidly decreases, so the phase information is only significant near the axis.

delays the onset of the problem. A second method is to apodize the domain, effectively adding an attenuation near the edge of the spatial domain. In the beam propagation method, we will introduce the latter method to dissipate these waves before they reappear on the other side.

6. The Beam Propagation Method

We can apply the principles of beam propagation to a guided wave problem. The beam propagation method is motivated by two physical properties of electromagnetic waves. First, as we have just seen, a wave travelling through any region of space will diffract. Second, the phase shift accumulated by the wave as it propagates in the forward direction depends on the local index of refraction. In an inhomogeneous medium, a wave will accumulate phase depending on the distance travelled *and* on the local index of refraction.

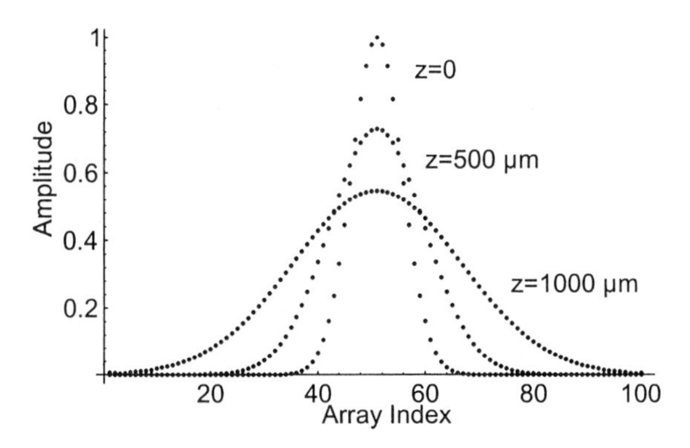

Figure 9.15. The amplitude profile of the mode at $z = 0$, $z = 500\mu$m, and $z = 1000\mu$m. Notice that the overall amplitude maintains the same area, but decreases in peak amplitude and increases in width.

The effects of propagation and local index act continuously on the phase as the wave travels, but we numerically simulate this process in a series of small steps. The local index is modelled as a sequence of "lenses", separated by short regions of homogeneous space with index $\overline{n}$, which is the average of the refractive index that the beam travels through between adjacent lenses.

The Beam Propagation Method uses a "split-step" process. In the first step, the transverse electric field at position z, $\Psi(x, y, z)$ is decomposed into a superposition of plane waves, $\Psi_i(k) = A_i e^{jk_i \cdot r}$, via the Fast Fourier Transform, and propagated a distance Δz as if it were travelling through an index $\overline{n}$. We have already discussed how to perform this first step. Following the propagation step, an inverse FFT converts the superposition of plane waves back into a spatial field.

The second step adds the phase correction needed to account for the spatial structure of the index profile. As the wave propagates from z to $z + \Delta z$, different parts of the phase front will experience different amounts of phase shift depending on the local index of refraction, $n(x, y, z)$. We adjust the step size so that the accumulated phase corrections are small following each step. Typical step sizes are on the order of a μm. The spatial phase correction is added to the spatial wave. The resulting field is a reasonable representation of the actual field distribution at location $z + \Delta z$. Fig. 9.16 shows this process schematically, where index inhomogeneity is lumped into a "lens" at the end of each discrete step. These lenses are not like conventional lenses, but are generalized to incorporate all the local index properties between adjacent planes of the medium. The new field serves as the source field for the next propagation step. The BPM repeats this two-step process until the wave has travelled the desired distance.

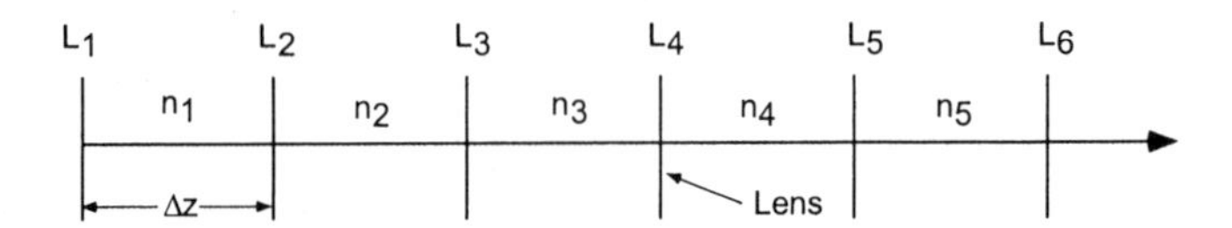

Figure 9.16. The optical path is broken into a sequence of finite steps through an average index, $\overline{n}$, and generalized lenses that incorporate phase shifts due to index inhomogeneities.

Introducing the lens step is surprisingly easy. For a field travelling along the z-axis a distance Δz, the phase can be approximately described as

$$\Phi(x, y, z) = e^{jk_0 n(x,y,z)\Delta z} \tag{9.9}$$

The total phase accumulated in propagating from z_1 to z_2 depends on the index of the media along the path. Since the free space propagation step already includes a phase shift $\exp(-jk_o\overline{n}\Delta z)$, the amount of phase shift due to the inhomogeneity is simply added to this

$$\Delta\Phi(x, y, z) = e^{jk_0(n(x,y,z)-\overline{n})\Delta z} \tag{9.10}$$

The influence of the local index distribution on the propagation of the wave is included by multiplying the spatial wave, $\Psi(x, y, \Delta z)$ by the phase correction $\Delta\Phi(x, y, z)$ after each free space propagation step. This process is then repeated using $\Phi(x, y, z + \Delta x)$ as the source field for the next propagation step.

Let's look at some examples of implementation. As before, we have chosen a one-dimensional example to allow reasonable speed on a small computer. We will describe the steps involved in setting up a simple BPM program using *Matlab*, a commercially available package that is excellent at performing numeric matrix calculations. (We depart from using *Mathematica* here, as *Matlab* was found to be much faster in dealing with FFTs.) This discussion is based on a more complete description found in reference [6].

The first step is to propagate the spatial mode a distance Δz. This requires using a FFT to determine the plane wave expansion, $\Phi(k_x)$, then advancing each plane wave a distance Δz, and then reconverting the phase space superposition into a spatial field using an inverse FFT. Formally, using continuous variables to describe the step, the field at position $z + \Delta z$ is

$$\Psi(x, y, \Delta z) = \frac{1}{2\pi}\int_{-\infty}^{\infty} \Phi(k_x)e^{-jk_x x + k_z \Delta z} dk_x \tag{9.11}$$

where k_z is the z-component of the k-vector *for each ray*. Each ray has a unique value of k_z, given by

$$k_z = \sqrt{\overline{n}^2 k_0^2 - k_x^2} \tag{9.12}$$

Evaluating the integral in Eq. 9.11 using an FFT is straightforward, however, the large magnitude of k_z makes the phase vary rapidly with Δz. Since the fast phase variation is of no interest to us, we usually separate it from the slow variation. This can be accomplished by writing k_z as

$$k_z = \overline{nk_0} - \frac{k_x^2}{\overline{nk_0} + \sqrt{n^2k_0^2 - k_x^2}} \tag{9.13}$$

Note that this is a more exact version of Eq. 9.8. In this form, the fast term $(\overline{n}k_0)$ is distinct from the slow terms. The wave, after propagating a distance Δz in the homogeneous region, can then be expressed as

$$\Psi(x, y, \Delta z) = \frac{e^{-j\overline{n}k_0\Delta z}}{2\pi} \int_{\beta>0} \Phi(k_x) \exp\left[\frac{-jk_x^2}{\overline{nk_0} + \sqrt{n^2k_0^2 - k_x^2}}\Delta z\right] dk_x \tag{9.14}$$

We changed the limits of integration to restrict the argument of the exponent to purely imaginary values, ensuring that no evanescent waves become included in the description of the wave propagation. Physically speaking, a complete description of a plane wave expansion requires evanescent waves, as they represent the loss mechanism to radiation modes. However, the lens step of the BPM technique assumes that the rays are travelling essentially parallel to the z-axis. For this reason, the expansion is limited to rays that make small angles to the z-axis. Evanescent waves are explicitly excluded from the expansion by restricting the possible values of k_z to real values only. The beam propagation method does not provide for energy loss due to radiation, so artificial means must be added to dissipate such modes. Restricting the plane wave expansion to beams with large area and to waveguides which are weakly guiding helps insure that the k vectors will be nearly parallel to the z-axis.

Finally, we add the contribution from the lens by multiplying the propagated spatial field, $\Psi(x, y, \Delta z)$ by $\Delta\Phi(x, y, z)$. The process repeats by advancing the field forward by one more Δz.

7. A MATLAB program for one-dimensional BPM

To help drive home these points, there is nothing like actually doing some examples. You are strongly encouraged to write a simple BPM program using a suitable computer programming language. In this section, we describe such a program based on the numerical program, *Matlab*[8]. There is nothing exclusive to *Matlab* which is required for BPM programming, so do not be discouraged if it is not available on your system. For our purposes, *Matlab* was found to calculate DFT's and multiply arrays much faster than most other numeric software packages.

We start with a very simple example that starts with a trial mode profile, and launches it down a waveguide. By iteration, eventually the true eigenmode of the waveguide emerges, while the rest of the energy is dissipated away. The *Matlab* program is described here in functional blocks. We wrote the program in block form, calling subroutines to do different functions. The program `Waveguide` is a short program of setup commands and subroutine calls:

```
[Waveguide]
format compact
format long e
hold off
units_stuff
wg_params
wg_setup
wg_iterate
```

The first three commands modify the output and can be adjusted to the users preference. The core of the program occurs in the next four statements. The routine `units_stuff` defines dimensional MKS units in their mathematical form:

```
[units_stuff]
cm=1e-2;
mm=1e-3;
um=1e-6;
nm=1e-9;
```

The routine `wg_params` defines the waveguide structure. In this example, we described a simple triangular-profile waveguide. We chose an array size of 512 data points to describe the index profile, the spatial mode, and the phase correction. A smaller array would be proportionally faster in calculation speed, but we found that this size provided tolerable throughput speed. We arbitrarily chose a index profile 150 μm wide, with a guiding region 10μm wide. The term `sig` is the $1/e$ length for the initial Gaussian mode profile. The step size in the calculation is 4 μm. Most other parameters are self-evident. The index difference is small, and $\overline{n}$ (`nave`) is set to the average value of the two indices.

```
[wg_params]
ns=1.499;
nf=1.5;
nave=(ns+nf)/2;
cladwidth=200*um;
wgwidth=10*um;
sig=3*um
dz=4*um;
atten=1500;
aper=40;
loopnum=250;
maxiterations=4000;
for j=1:512;
```

```
temp=(j-256)*cladwidth/512;
triangle(j)=ns+(nf-ns)*(abs(temp*2)<=wgwidth)*(1-abs(temp*2/wgwidth));
end
n=triangle;
```

The functions `aper` and `atten` apodize the beam. As we saw in the first part of this chapter, when an amplitude component moves off one side in the spatial FFT, it reappears on the other side of the spatial domain. This is in contrast to a real waveguide, where we expect energy to continue travelling away from the core once it has been shed. There are several methods of apodizing. Signal processing algorithms often use triangular filters or Hamming filters to proportionally attenuate the extreme spatial components. For BPM we require a flat transmission for the central portion of the waveguide, but wish to add attenuation in the cladding region to simulate radiation mode losses, and to prevent energy from wrapping around and re-entering the waveguide structure from the other side of the data array.

The exact form of the apodizer is up to the user. In this example, we chose to add a small attenuation to the cladding far from the region of the guided mode. The parameter `aper` defines the percentage of the aperture where the core and cladding are lossless. Here, we defined the clear aperture to be 40% of the total aperture. The aperture must be larger than the final spatial extent of the guided mode to prevent adding unrealistic loss to the simulation. The magnitude of the attenuation is set through trial and error: setting it too small obviously does not accomplish the desired results, but making it too large leads to a reflection at the transition from the lossless to lossy region. Such a reflection will return unwanted energy back to the guiding region, degrading the quality of the simulation. In the region of loss, the amplitude is attenuated by $exp(-\text{atten}\ \Delta z)$ each step. Since in this simulation, $\Delta z = 4\mu$m per step, and `atten` $= 1500$ m^{-1}, the effective attenuation leads to a decrease of roughly 0.6% of the amplitude outside the aperture per step. This will add up to significant loss after hundreds of steps, but does not act as a major perturbation to a field upon incidence to the loss region. You will have to explore different values to find a suitable value for new parameters. We found this value after trying several runs — in our experience the simulations were not strongly affected by the choice of the attenuation. You may wish to explore other types of apodizing function.

Choosing the step length, Δz, depends on the guiding structure and wavelength. To stay within the region of validity for the BPM, Δz should satisfy [9]

$$\Delta z \ll 6k_0(k_e + k_w)^{-2} \qquad (9.15)$$

where k_e is the largest transverse component of the wavevector describing the guided electric field, and k_w is the largest spatial frequency required to describe the index profile, if it were described as a Fourier superposition. To first order, the maximum spatial frequencies needed to describe both the electric field and the index profile can be approximated as

$$k_E \approx \frac{2\pi}{x_0} \quad k_n \approx \frac{2\pi}{w} \tag{9.16}$$

where x_0 is the characteristic half width of the mode, and w is the half width of the waveguide. If we apply this criterion to the waveguide in this example, where x_0 is approximately 10 μm for the eigenmode, and w is approximately 5 μm, we find that $\Delta z < 12\mu$m. Our choice of $\Delta z = 4\mu$m satisfies this requirement. The advantage of a larger step size is that the calculations run faster. If Δz is chosen to be too large, the simulation might provide an unreliable result. A good test is to repeat a calculation with a smaller Δz to see if the final solution changes.

The next subroutine is wg_setup, which defines most of the parameters used in the other subroutines, and defines the phaseshifts according to Eq. 9.14.

```
[wg_setup]
aper=round(512*aper/100);
iterations=0;
lambda=1*um; %define a wavelength
k0=2*pi/lambda;
od=atten*[ones(1,256-fix(aper/2)),zeros(1,aper),ones(1,256-fix((aper+1)/2
a=cladwidth/2/pi;
i=sqrt(-1);
k=[0:255 -256:-1]/a; %define the transverse wavevectors
x=cladwidth*(-0.5+(0:511)/512);
phase1=exp(i*dz*(k.^2)./(nave*k0+sqrt(max(0,nave^2*k0^2-k.^2))));
phase2=exp(-(od+i*(n-nave)*k0)*dz);
phase2=fftshift(phase2);
axis([-cladwidth/2 cladwidth/2 0 2]);
plot(x,(n-ns)/(nf-ns)/10+1,'-g',x,od/atten,'-b');
disp('Press a key...'); pause;
v=exp(-(x/sig).^2); %this is the initial amplitude profile
initialpower=sum(v.*conj(v));
ov=v;
plot(x,ov);
hold on;
```

The transverse wavevector, k_x is an array of 512 points, ranging from $k_x = -256 \times 2\pi/$ cladwidth, to $k_x = +255 \times 2\pi/$ cladwidth. We chose a Gaussian amplitude profile for convenience, and because it can be adjusted to provide a close approximation to most waveguide modes through the parameter sig. The array v is our initial guess at what the mode profile will be.

The phase terms are calculated as described in the text above. Note that the attenuation is included in the "lens" term of the phase corrections. We also calculate the total power residing in the mode by computing the integral of the square of the amplitude. This is done in the variable initialpower, and serves as a normalization point for future calculations.

Finally, we get to the main computational part of the BPM program in the final subroutine,`wg_iterate`. This takes the trial amplitude distribution, labelled as `v`, and performs an `fftshift` on the data in the array. This simply switches the first half of the array with the second half, putting the ends of the data array in the middle. This counters the effect we observed earlier where the FFT of an array ends up with the Fourier components at the two extremes of the array. The heart of the BPM program resides in the single loop, where the trial amplitude distribution is repeatedly propagated (`fft(v).*phase1`), multiplied with a phase correction (`*phase2`) and then inverse FFTed. The results are plotted on the computer screen after a reasonable number of iterations, so that the user can watch the evolution of the amplitude profile.

```
[wg_iterate]
v=fftshift(v);
while iterations<maxiterations;
for loop=1:loopnum,
 v=ifft(fft(v).*phase1).*phase2;
iterations=iterations+1;
end
clg
plot(x,abs(fftshift(v)));
end
v=fftshift(v);
```

We will use two examples to demonstrate the utility of the BPM. First we will use BPM to find the amplitude distribution in a graded index slab waveguide with a triangular profile. Then we will simulate a mode propagating through a coupled waveguide.

Let's first find the shape of the eigenmode for a triangular waveguide. We will use the profile `triangle` described in `wg_params`, and set the trial mode characteristic width to 3μm. We intentionally made the mode narrower than reasonably expected so the dynamics of the BPM process will be illustrated. Fig. 9.17 shows a sample of the output as a function of the distance the simulated mode has travelled down the waveguide. The initial field distribution is a narrow Gaussian spike located at $z = 0$. This initial spatial mode can be described as a superposition of guided and radiation modes of this waveguide. As the initial amplitude distribution propagates forward, the non-guided components begin to travel away from the guiding layer. The eigenmode becomes distinguished after travelling about one millimeter. The broad pedestal that the eigenmode sits on represents unguided energy that is radiating away from the waveguide. The unguided energy extends beyond the clear aperture, and suffers attenuation with each step of the calculation. Eventually the non-guided energy is totally dissipated.

We can see an artifact of the calculation in the ripples that form on the pedestal for the plots between $z = 1$ and $z = 2$ mm. These ripples arise from

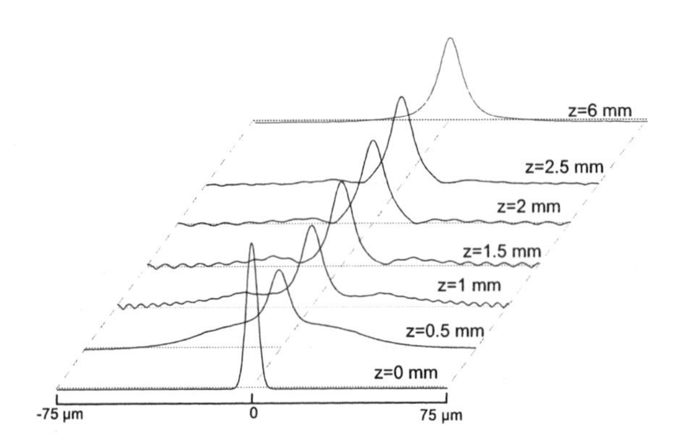

Figure 9.17. Results of a BPM run on a triangular index slab waveguide. A trial mode with a Gaussian profile is launched at $z = 0$. As it travels down the guide, the unguided energy radiates away, while a guided mode emerges after suitable distance.

interference between the outward bound waves and those that have wrapped around from the other side that were not totally attenuated before reaching the boundary of the domain. After sufficient propagation, these interference features are damped out.

The BPM successfully determined the shape of the mode for the triangular waveguide. The waveguide acted as a spatial filter to the input field distribution, eliminating all energy except that in the waveguides fundamental mode. We could substitute any reasonable index profile into the program, and use the same technique to find the eigenmode.

8. Waveguide Coupler

The beam propagation method is often used to evaluate the performance of either a coupled waveguide, a Y-junction, or some other complex structure. We know that the field of a confined mode extends out beyond the core region. These evanescent tails can transfer energy from one waveguide to another if the dielectric structure is suitable. We will explore the *theory* of energy transfer and mode coupling in the next chapter. Here, we want to use BPM to "experiment" with a coupled waveguide structure. As an example, we will examine the propagation of a mode on a waveguide which is located adjacent to an identical

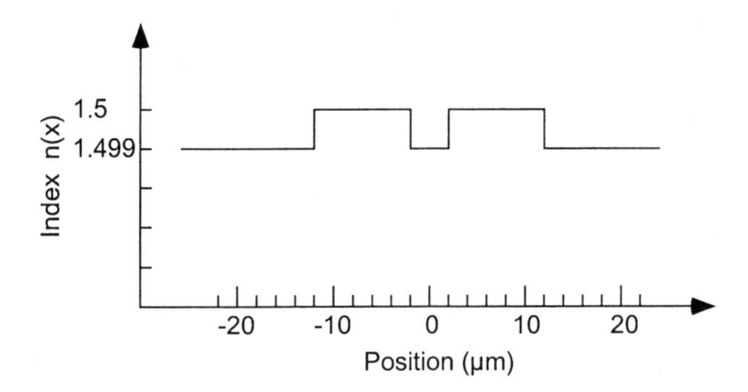

Figure 9.18. Index profile of two slab waveguides separated by only 4 μm.

guide. The index profile for this structure is plotted in Fig. 9.18. The coupler consists of two identical step-index waveguides with core thickness of 10 μm situated approximately 4 μm from each other. The evanescent field of either guide extends into the other guide.

To begin the analysis, we launch a mode which is close to being an eigenmode of one of the individual waveguides. We found by simple trial and error that a Gaussian profile with a $1/e$ length (`sig`) = 7 μm gave an excellent approximation to the actual mode of the waveguide for 1 μm wavelength light. We wrote the program to launch the initial mode into the right hand waveguide at $z = 0$. Fig. 9.19 shows the evolution of the mode profile as it propagates down the waveguides. Each snapshot of the mode is taken after the field has propagated 1 mm down the waveguide. We see an evolution of the mode energy as it propagates along the coupled waveguide structure. After travelling approximately 4 mm down the waveguide, the energy has completely transferred over to the left hand guide. As the beam continues to propagate, the energy transfers back to the original guide. This process will continue indefinitely so long as the waveguides do not change their relative position or dimension. This is an extremely useful effect which can be exploited to make many practical devices such as couplers, taps, interferometers, and wavelength selective filters. For example, if the waveguides were brought together for only 1 mm, approximately 10% of the power from the first waveguide could be tapped, while the remaining 90% of the energy would continue along the main channel. The BPM method can be used to explore the effect of waveguide separation or mismatch on the coupling rate and efficiency. We will explore the theoretical basis for this behavior in the next chapter. We leave it as an exercise to show that the transfer rate of energy decreases as the waveguide separation is increased.

The relevant *Matlab* code for the index profile and initial amplitude distribution are listed below.

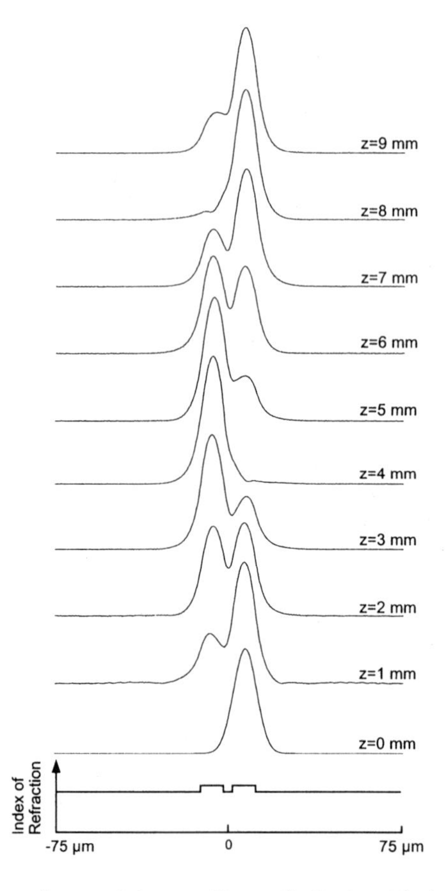

Figure 9.19. The power in a coupled waveguide transfers back and forth as it propagates along the guide. The index profile is shown below the mode profiles.

```
wgsep=14*micron;
for j=1:512;
coupledindex(j)=ns+(nf-ns)*(abs(abs(j-256)-wgsep*
256/cladwidth)<wgwidth*256/cladwidth);
end;
```

`v=2*exp(-((x-wgsep/2)/sig).^2);` where `wgsep` is the distance between the centers of the coupled waveguides (in μms), and the initial amplitude profile is offset to overlap the right hand waveguide.

9. The Finite-Difference Time-Domain ethod

The BPM technique provides an excellent way to evaluate the spatial structure of a mode as it propagates through a waveguide. But if the waveguide has sharp changes in index that could result in a reflection, the BPM fails to account for the reflected wave, or for possible interference effects between the forward and backward wave. Since many optical devices rely on interference based on reflections to function. Other techniques are needed to simulate such structures. The Finite-Difference Time Domain (FDTD) technique is a powerful method for such simulations. FDTD basically calculates a numeric solution to Maxwell's equations, therefore so long as errors due to round-off do not arise, the simulations are exact. We will outline the basic steps of the FDTD method,and develop a simple code to demonstrate how it operates on time domain problems. The purpose is not to develop a program that competes with commercial code, but simply to provide the user with an operating insight into how the program works. We will follow the development by Sullivan in reference [10]

The time dependent Maxwell curl equations in free space are

$$\begin{aligned} \nabla \times \mathbf{E} &= -\frac{\partial \mathbf{B}}{\partial t} &= -\mu_0 \frac{\partial \mathbf{H}}{\partial t} \\ \nabla \times \mathbf{H} &= \frac{\partial \mathbf{D}}{\partial t} &= \epsilon_0 \frac{\partial \mathbf{E}}{\partial t} \end{aligned} \tag{9.17}$$

We will consider the simplest, one dimensional case, where the wave is propagating in the z direction, and the fields are E_x and H_y. In this one-dimensional system the curl equations dramatically simplify to

$$\begin{aligned} \frac{\partial E_x}{\partial t} &= -\frac{1}{\epsilon_0} \frac{\partial H_y}{\partial t} \\ \frac{\partial H_y}{\partial t} &= -\frac{1}{\mu_0} \frac{\partial E_x}{\partial t} \end{aligned} \tag{9.18}$$

These are exact equations. We now want to put them in discrete form so that a computer can evaluate them. From basic calculus we know the derivatives come from the limiting values of finite differences. Using this concept we can express the curl equations as

$$\begin{aligned} \frac{E_x(t+\Delta t, z) - E_x(t-\Delta t, z)}{\Delta t} &= -\frac{1}{\epsilon_0} \frac{H_y(t, z+\Delta z) - H_y(t, z-\Delta z)}{\Delta z} \\ \frac{H_y(t+\Delta t, z) - H_y(t-\Delta t, z)}{\Delta t} &= -\frac{1}{\mu_0} \frac{E_x(t, z+\Delta z) - E_x(t, z-\Delta z)}{\Delta z} \end{aligned} \tag{9.19}$$

To implement this on a computer, the fields $E_x(z)$ and $H_y(z)$ will be stored as an array of points which represent the spatial distribution of the fields at a given point in time. To describe the temporal evolution of the fields,the data in each array must be updated according to Eq. 9.19. Using the superscript n to denote the time step (real time is $t = n \cdot \Delta t$), and an array index k to represent position ($z = k \cdot \Delta z$), Eq. 9.19 can be rearranged into an iterative algorithm

$$
\begin{aligned}
E_x^{n+1}(k) &= E_x^n(k) - \frac{\Delta t}{\epsilon_0 \Delta z}[H_y^n(k) - H_y^n(k-1)] \\
H_y^{n+1}(k) &= H_y^n(k) - \frac{\Delta t}{\mu_0 \Delta z}[E_x^n(k+1) - E_x^n(k)]
\end{aligned} \tag{9.20}
$$

The two arrays are interleaved in space and time. Notice that the next value for E_x is found from the present value and the most recent values of H_y. Similarly, H_y is determined by its present value and the most recent values of E_x.

The equations for E and H are very similar, but the magnitudes of E and H will differ by several orders of magnitudes, because the impedance of the medium, η, is typically 100—337Ω. If we make a change of variables using the impedance to normalize the fields

$$
\hat{E} = \sqrt{\frac{\epsilon_0}{\mu_o}} E \tag{9.21}
$$

we can form a normalized set of equations

$$
\begin{aligned}
\hat{E}_x^{n+1}(k) &= \hat{E}_x^n(k) - \frac{1}{\sqrt{\epsilon_0 \mu_o}} \frac{\Delta t}{\Delta z}[H_y^n(k) - H_y^n(k-1)] \\
H_y^{n+1}(k) &= H_y^n(k) - \frac{1}{\sqrt{\epsilon_0 \mu_o}} \frac{\Delta t}{\Delta z}[\hat{E}_x^n(k+1) - \hat{E}_x^n(k)]
\end{aligned} \tag{9.22}
$$

The relation between Δz and Δt is fixed by the speed of light in free space, c. In a one-dimensional case the time step is usually set to

$$
\Delta t = \frac{\Delta x}{2c} \tag{9.23}
$$

The factor of two arises because it requires two cycles of iteration in the algorithm above before a field point can actually be updated. The normalizing term then simplifies to

$$
\frac{1}{\sqrt{\epsilon_0 \mu_o}} \frac{\Delta t}{\Delta z} = \frac{1}{2} \tag{9.24}
$$

Rewriting Eqs.9.20 in *Mathematica* code yields

```
Do[
  Do[
```

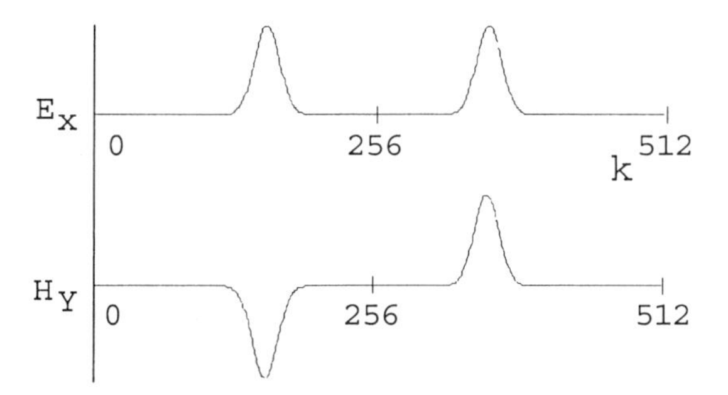

Figure 9.20. The gaussian pulse propagates in both directions. The electric field positive in both directions, but the magnetic field is negative in the negative propagating direction.

```
  ex[[k]] = ex[[k]] - 0.5*(hy[[k]] - hy[[k - 1]]), {k, 2, kmax - 2}
  ];
  Do[
  hy[[k]] = hy[[k]] - 0.5(ex[[k + 1]] - ex[[k]]),{k, 1, kmax - 1}
  ],
50]
```

There are three Do loops. The first loop advances the "time" by one step each time it runs, so n this case we are advancing the field 50 steps. The number of steps can be adjusted as desired. The second Do loop updates the value of the electric field, which is in array ex. The third loop updates the magnetic field, hy, using the new values of ex.

The array has to be initialized with an electric field distribution before it can propagate a wave. As a first example, consider a gaussian shaped pulse. This can be created as shown in the next lines

```
kmax = 512;
ex = Table[ Exp[-(k - kmax/2.)^2/200], {k, 1, kmax}];
hy = Table[0., {k, 1, kmax}];
```

The gaussian electric field after 200 "steps" is shown in Fig. 9.20. Notice the pulses propagates in both directions. E_x is positive in both directions, but H_y is negative for the negative propagating pulse. Had we set up an initial distribution for the magnetic field that was self-consistent with a pulse travelling in one direction, the pulse would not have divided. Because we set $H_y = 0$ initially, the first magnetic field was derived from the initial electric field distribution, so two directions emerged.

The FDTD techniques gets much more interesting when we add some structure to the propagation path. We can introduce material parameters via the permittivity, ϵ. Since the time step is defined in terms of ϵ_0, we can introduce the effect of different refractive indices by redefining the timing step as

$$\frac{1}{\sqrt{\epsilon\mu_o}}\frac{\Delta t}{\Delta z} = \frac{1}{\sqrt{\epsilon_r}\,2} \tag{9.25}$$

where $\epsilon_r = \epsilon/\epsilon_0$. Before we add that feature to the code, we need to take care of another detail. Normally to calculate the E_x field, we need to know the surrounding H_y fields. This is the fundamental assumption of the FDTD process. But, at the end of the array we will not have the value on one side, and a reflection results. To eliminate the reflection we need a boundary condition at the ends. Because the current value of $E(k)$ is determined by a previous version of $H(k)$, which itself was determined by an even earlier version of $E(k)$, an acceptable boundary condition to prevent reflection is to set $E^n(1) = E^{n-2}(2)$ for the $k = 1$ end of the array, and a similar condition for the k_{max} end. This provides the "look-ahead" that the fields are expecting as they propagate. The code below contains the addition of the boundary condition and the inclusion of an array that describes a step increase of the index of refraction at $k = 250$ to a value of $n = \sqrt{\epsilon_r} = 3$.

```
kmax = 512;
cb = Table[If[k >250,0.33,1], k,1,kmax];
Do[
 Do[
 ex[[k]] = ex[[k]]-0.5*cb[[k]]*(hy[[k]]-hy[[k-1]]),{k,2,kmax-2}
 ];
 ex[[1]] = exm2;
 exm2 = exm1;
 exm1 = ex[[2]];
 Do[
 hy[[k]] = hy[[k]]-0.5(ex[[k+1]]-ex[[k]]),{k,1, kmax-1}
 ],
{200}]
ListPlot[ex, PlotRange -> {-1,1}, PlotJoined -> True]
```

To illustate the impact of the index change, we launched a two-cycle sinusoidal electric field at the interface. Fig. 9.21 shows the wave this example we included a portion of a sine wave (defined in the original array for ex), and had it propagate into a dielectric interface. You can see the reflection that results, the relative changes in amplitude as the wave enters a higher index medium (note especially the magnetic field increases in the dielectric), and the shortening of the wavelength in the dielectric.

The FDTD technique provides a powerful and accurate method for determining the fields of a waveguide structure, and is especially useful for situations where there are reflections. This one-dimensional example is trivial, but the technique can be extended to two-dimensional structures without much complication.

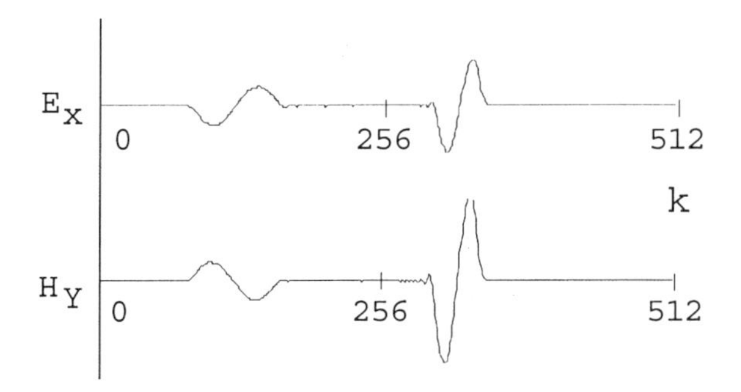

Figure 9.21. A sinusoidal wave is propagated across a dielectric boundary at $k = 250$.

10. Summary

In this section we introduced the use of numeric techniques to evaluate difficult index profiles, and to simulate the behavior of modes in a coupled waveguide. We began with a description of the Fast Fourier Transform, which is the standard tool available for performing Fourier Transforms on a computer. Using the FFT, we showed that a spatial wave can be described as a superposition of plane waves. Once we have a plane wave expansion, we can predict how a wave will propagate. We applied this knowledge to free-space propagation, and demonstrated how a wave with finite transverse dimension will diffract as it propagates.

We then added one more piece of information to our propagation model. We allowed the local index of refraction to modify the cumulative phase of a propagating wave. By adding a phase correction to the spatial waveform, the combined effect of diffraction and guiding was described.

We listed a simple one-dimensional BPM program written for *Matlab*, and demonstrated the program by finding the mode profile of a triangular waveguide, and by demonstrating the mode coupling that occurs between two identical waveguides. These are very trivial examples which demonstrate the operation of the BPM technique. More sophisticated models, such as 2-dimensional analysis, and including vector analysis, have been developed and are described in current literature.

References

[1] J. Van Roey, J. Van der Donk, and P. E. Lagasse, *J. Opt. Soc. Amer.*, 71, pp.803 (1981),

[2] D. Yevick and L. Thylen, *J. Opt. Soc. Amer.*, 72,pp.1084 (1982)

[3] R. Baets and P. E. Lagasse, *Appl. Opt.*, 21, pp.1972 (1982),

[4] William Press, Brian P. Flannery, Saul A. Teukolsky, and William T. Vetterling, *Numerical Methods*,Chapter 12, Cambridge University Press, USA (1986)

[5] B. Buckman, *Guided Wave Optics*, Saunders College Pub., USA, (1992)

[6] Stephen Wolfram, *The Mathematica Book, 4th ed*, Wolfram Media/Cambridge University Press, 1999, USA

[7] D. Marcuse, Theory of Dielectric Waveguides, 2nd Ed., Academic Press, USA (1991)

[8] The Mathworks, *The Student Edition of Matlab*, Prentice-Hall, USA (1992)

[9] Lars Thylen, *Optical and Quantum Electronics*, 15, pp.433 (1983),

[10] D. M. Sulivan, *Electromagnetic Simulation Using the FDTD method*, IEEE Press Series on RF and Microwave Technology, IEEE Press, New York

Practice problems

1. Using a personal computer, explore the appropriate commands for evaluating the FFT of a smooth spatial profile. Consider the gaussian function

$$\Psi(x) = e^{-\frac{x^2}{25\mu m^2}}$$

   (a) Plot this function over the range from $x = -20\mu$m to $x = +20\mu$m.

   (b) Create an array containing 50 points that are uniformly sampled over the range from -25μm to 24.5μm.

   (c) Calculate the FFT of this array. Observe the amplitude and phase of the FFT.

   (d) Explore the effect of altering the domain of the function. For example, decrease the domain of the array from -10μm to $+10\mu$m, and repeat part c. Does the change in the FFT make sense?

   (e) Explore the effect of shifting the array index on the resulting phase of the FFT. Modify the array in part b to include 51 points, including the point at $+25\mu$m. What happens to the phase of the DFT? Can you explain this?

2. Repeat the BPM calculations for the triangular waveguide using a mode profile that is exponential in shape (as might arise from a weakly bound mode in a symmetric waveguide. Specifically, let

$$E(x) = Ae^{-|x|/x_0}$$

where $x_0 = 8\mu$m.

3. Explore the effect of beam size on diffraction. Modify the gaussian profile in Fig.9.10 to effectively double its spatial domain. Follow through with the calculation of the beam diffraction based on a FFT analysis as described in Sect. 9.4.4. Does the larger beam diffract differently than the smaller beam? Is the difference consistent with the diffraction of a Gaussian mode?

4. Explore the effect of wavelength on the diffraction of a beam. Modify the mode in Fig.9.10 so that it has a wavelength of 1.5 μm, but still has the original characteristic width (8 μm) at $z = 0$. How does increasing wavelength affect diffraction? Is your simulated result consistent with diffraction theory for a gaussian mode?

5. The cause of the ripples in Fig. 9.15 was attributed to the finite size of the domain. Try increasing the initial value of H, and see if this in fact the case.

6. What happens if the distance separating the waveguides described in Fig. 9.18 is changed? Repeat the simulation using a separation of 16 μm and 12 μm.

7. The total energy exchange demonstrated in the coupled waveguides in Fig. 9.17 is a result of their being identical in structure. What happens if the waveguides are not identical? Reduce the dimension of the left hand waveguide by 2 μm, and then run a similar BPM analysis, observing the coupling of energy between the waveguides. Comment on the period of the exchange, and on the completeness of the exchange.

8. Use the BPM to find the mode shape of a step index profile with a total width of 6μm, and a symmetric index profile defined as $n_f = 1.5$ and $n_s = 1.498$. Compare your result with the exact mode shape determined from direct solution of the wave equation.

9. Extend the FDTD technique to evaluate the propagation of a sine wave through a dielectric slab. Try to observe the resonance condition, where there is perfect transmission and no reflection.

Chapter 10

COUPLED MODE THEORY AND APPLICATION

1. Introduction

Mutual coupling between optical modes is essential in the design of integrated optic devices. In this chapter we will describe how optical energy couples between modes within and between optical waveguides. Up to now, we have treated the waveguide as an ideal optical wire, which conveys light from one point to another in the form of a "mode". We have implicitly assumed that these modes, once formed, are unchanging except perhaps through attenuation due to absorption. In reality, simple mechanisms can lead to significant energy exchange among the various modes of a structure. Coupled mode theory describes this energy exchange, and serves as the primary tool for designing optical couplers, switches, and filters.

We will explore a coupling technique that describes the scalar electromagnetic field of a perturbed waveguide in terms of a superposition of modes of the ideal waveguide. More advanced coupled theories are being developed everyday, based on vector equations and other considerations. The theory we present here will serve as a basic step in understanding these advanced theories. Interested readers should explore the references cited.

2. Derivation of the Coupling Equation Using Ideal Modes

Consider two proximate single mode optical waveguides, as shown in Fig. 10.1. In each waveguide there are two waves: one propagates in the forward direction, and one in the backward direction. Energy transfer, i.e. coupling, can occur if the evanescent field from one waveguide extends into the core of the neighboring waveguide

The degree to which two modes exchange energy depends on the design of the coupler, and the mode structure of the two waveguides. For example, it

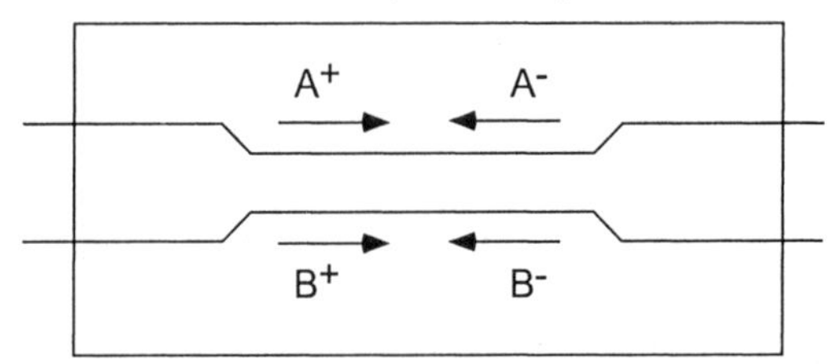

Figure 10.1. Two waveguides in proximity can couple energy. This device is sometimes called a 2x2 coupler. There are four fields that must be considered in the analysis.

seems plausible in the structure shown in Fig. 10.1 that the forward wave in one waveguide (say A^+) will primarily couple to the forward wave in the adjacent waveguide (B^+). The coupling from the forward wave (A^+) to the backward travelling wave of the other waveguide (B^-) will be insignificant unless the coupling incorporates some means of reflection.

Using the principles of superposition and completeness, we can describe any waveguide amplitude distribution in terms of a superposition of ideal waveguide modes. By "ideal," we mean a waveguide with no perturbations. The amplitudes of the modes in the superposition will change only when a perturbation is present. As we have often done with new topics, we will restrict our discussion to planar or rectangular waveguides in the discussion below. Staying in the rectilinear coordinate system makes it much easier to calculate node coupling, and is conceptually easier to understand and follow. We will follow the development of the coupled mode theory as described by Yariv[1].

Consider the planar step-index waveguide structure shown in Fig. 10.2 with film index n_f of thickness h, and substrate and cover indices, n_s and n_c, respectively. The waveguide can support a finite number of guided modes and an infinite number of radiation modes. For this example, assume that only TE modes are carried. The electric field of the eigenmodes of this structure satisfy the wave equation,

$$\nabla^2 E_y(x,z,t) = \epsilon\mu \frac{\partial^2 E_y(x,z,t)}{\partial t^2} \tag{10.1}$$

If ϵ and μ are a time-invariant quantities, each mode solution to Eq. 10.1 will have the familiar form

$$E_{y_i}(x,z,t) = \frac{1}{2} A_i \mathcal{E}_{y_i}(x) e^{-j(\beta_i z - \omega t)} + \text{c.c.} \tag{10.2}$$

where A_i is the amplitude for mode i, and $\mathcal{E}_{y_i}(x)$ is the *normalized* amplitude distribution for mode i, given by Eq. 3.20 for the asymmetric slab waveguide.

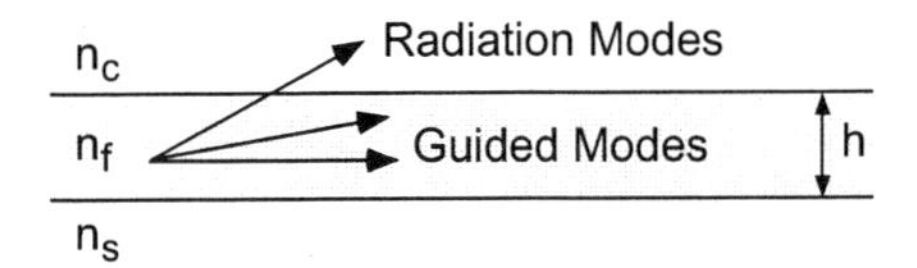

Figure 10.2. A waveguide can support both guided and unguided modes. These are nominally independent of one another in the absence of any coupling.

If there are no perturbations in the waveguide, (e.g., if the dimensions of the waveguide do not change with position, and there are no changes in the dielectric constant of the three films) the modes will be completely independent from one another. However, any deviation from this well defined waveguide structure perturbs the modes and couples energy between them. In the analysis which follows, we describe the perturbation in terms of a distributed *polarization source*, P_{pert}, which accounts for the deviation of the dielectric polarization from that which accompanies an unperturbed mode. Perturbations can arise through two mechanisms: either the dielectric constant of the structure is modified ($\Delta\epsilon$) from what the mode expects to see, or an electric field from a second source appears in the waveguide, and excites a mode of the structure.

Why is polarization the preferred method for solving these problems? Recall the constitutive relation for electric flux

$$D = \epsilon E = \epsilon_0 E + P \tag{10.3}$$

The effect of the media is to increase the local displacement flux by a value of P. Any *deviation* from the normal dielectric constant of a guided wave structure leads to a perturbation in the polarization, P_{pert}, defined through the expression

$$D = \epsilon E + P_{pert} \tag{10.4}$$

Substituting Eq. 10.4 into the wave equation yields

$$\nabla^2 E_y = \mu\epsilon \frac{\partial^2 E_y}{\partial t^2} + \mu \frac{\partial^2 P_{pert}}{\partial t^2} \tag{10.5}$$

where the polarization perturbation clearly stands out as a driving force in the equation.

To solve this equation, we use standard perturbation theory techniques. First, the term P_{pert} is set equal to 0, and the eigenmodes of the unperturbed structure are found. The eigenmodes are symbolized by the notation, $\mathcal{E}$. Then we seek a solution to the perturbed wave equation in terms of a superposition of the orthogonal eigenmodes. Since the eigenmodes of the waveguide form a complete set, any continuous electric field distribution, $E(x)$, can be described in terms of the modes of the system

$$E(x) = \tfrac{1}{2}\sum_i A_i^+(z)\mathcal{E}_i(x)e^{-j(\beta_i z-\omega t)} + \tfrac{1}{2}\sum_i A_i^-(z)\mathcal{E}_i(x)e^{j(\beta_i z+\omega t)} + \tfrac{1}{2}\int_{k_0 n_s}^{k_0 n_f} A(\beta,z)\mathcal{E}_\beta(x)e^{-j(\beta z-\omega t)}d\beta + c.c \qquad (10.6)$$

where $\mathcal{E}_i(x)$ are the spatial amplitude distributions for each mode, and $A^\pm$ are the amplitudes of the forward or backward travelling wave, respectively. The terms β_i represent the mode propagation coefficients for individual modes. The factors $1/2$ in front of the amplitude terms arise from describing the real electric fields in terms of their complex exponential form and complex conjugate. The complex conjugate terms are implicit in these calculations. Note that the superposition includes a sum over modes travelling to the right (A^+), as well as the left (A^-). Each pair of oppositely directed modes has the same spatial distribution, $\mathcal{E}_i(x)$, and the same magnitude propagation coefficient, β, but they differ in the direction of propagation. The superposition also includes the so-called "radiation modes." These modes are essential to describe any arbitrary electric field as a superposition of eigenmodes of the system. All examples considered in this chapter concern coupling between discrete guided modes, so we will neglect radiation fields. However, radiation modes are important when dealing with grating or prism couplers.

Our goal is to develop an expression for the coupling between the amplitudes of the individual modes of the waveguide. If there is no coupling between the modes, the individual amplitudes, A_i, will be constant in time and position. If coupling exists, the amplitudes will vary with position. To derive an equation of motion for the amplitudes, we plug the general solution, Eq.10.6, into the perturbed wave equation, Eq. 10.6. The eigenmodes, $\mathcal{E}_i(x)$, satisfy the unperturbed equation, so many terms drop out. The details of the algebra are left as an exercise. The perturbed wave equation reduces to

$$\frac{1}{2}\sum_i\left[\frac{\partial^2 A_i^+}{\partial z^2} - 2j\beta_i\frac{\partial A_i^+}{\partial z}\right]\mathcal{E}_i(x)e^{-j(\beta_i z-\omega t)} + c.c. + \tfrac{1}{2}\sum_i\left[\frac{\partial^2 A_i^-}{\partial z^2} + 2j\beta_i\frac{\partial A_i^-}{\partial z}\right]\mathcal{E}_i(x)e^{j(\beta_i z+\omega t)} + c.c. = \mu\frac{\partial^2}{\partial t^2}P_{pert} \qquad (10.7)$$

The effect of the perturbation is to change the amplitudes of the modes in the superposition. If we assume that the perturbation causes "slow variations" in the amplitude, then the second derivative terms are negligible ($\partial^2 A_i/\partial z^2 \ll \beta\partial A_i/\partial z$). The final equation of motion for the amplitude is given by

$$\frac{1}{2}\sum_i\left[-2j\beta_i\frac{\partial A_i^+}{\partial z}\right]\mathcal{E}_i(x)e^{-j(\beta_i z-\omega t)}$$

$$+\frac{1}{2}\sum_i \left[+2j\beta_i \frac{\partial A_i^-}{\partial z}\right] \mathcal{E}_i(x) e^{j(\beta_i z+\omega t)} + c.c.$$

$$= \mu \frac{\partial^2}{\partial t^2} P_{pert} \tag{10.8}$$

This is a complicated equation: it involves the sum over a large number of modes. We can simplify the equation using mode orthogonality. The normalized eigensolutions to the wave equation satisfy

$$\frac{1}{2}\int \mathcal{E}_i(x) \times \mathcal{H}_j(x) dA = \delta_{ij} \tag{10.9}$$

where $\mathcal{H}$ is the normalized magnetic field eigenmode. The term δ_{ij} is the Kroenecker delta function which is unity if $i = j$. If the fields can be described by a scalar equation (i.e. there is no coupling between the x, y, and z fields in this structure), Eq. 10.9 can be cast into simpler form requiring only a spatial description of the electric field. Assuming that we are using a TE mode, Maxwell's equations can be used to express $\mathcal{H}_x$ in terms of $\mathcal{E}_y$, for normalized fields

$$\int_{-\infty}^{\infty} \mathcal{E}_i(x)\mathcal{E}_j(x)dx = \frac{2\omega\mu}{\beta_i}\delta_{ij} \tag{10.10}$$

with units of W per unit length (time average power). A similar condition can be derived for TM modes.

Using mode orthogonality, we can simplify the series terms in Eq. 10.8 by multiplying both sides by $\mathcal{E}_j(x)$, and integrating over x.

$$\frac{\partial A_j^-}{\partial z} e^{j(\beta z+\omega t)} - \frac{\partial A_j^+}{\partial z} e^{-j(\beta z-\omega t)} + c.c. = \frac{-j}{2\omega}\frac{\partial^2}{\partial t^2}\int_{-\infty}^{\infty} P_{pert}(x)\cdot \mathcal{E}_j(x)dx \tag{10.11}$$

This is the primary equation for determining mode coupling. Notice that the degree to which the amplitude of mode j changes is directly proportional to the *overlap* of the perturbation (P_{pert}) and the modal distribution of mode j ($\mathcal{E}_j$). The more complete the overlap, the stronger is the coupling. Also note that if the perturbation is null, the mode amplitudes will remain constant.

In spite of the simplifications, Eq. 10.11 is still a difficult differential equation to solve. The right hand side of the equation is the driving force for changing the amplitude of the forward and backward waves. The differential equation is solved by integration over z, keeping in mind a few tricks. If the driving term and guided mode have different temporal frequencies, the interaction will average out to zero over a time long compared to their difference in frequencies. Therefore, only terms of similar frequency need be retained in the equation. Second, the driving term and guided mode should have nearly the same spatial phase dependence so that the interaction does not average out to zero over distance. Terms which do not satisfy this condition have negligible impact on

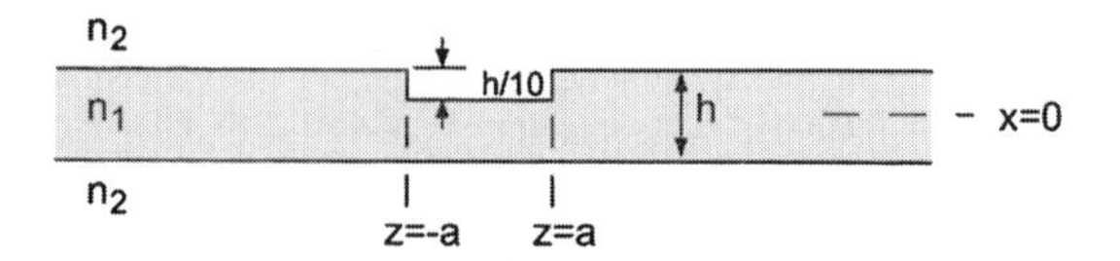

Figure 10.3. A dielectric waveguide with a notch defect in it acts as a perturbation to the guided modes. The notch extends a distance 2a, and is h/10 deep.

the solution. Based on these considerations, we can usually disregard one of the terms on the left hand side of Eq. 10.11.

Finally, there are two regimes of coupled mode theory: weakly coupled, and strongly coupled. In a weakly coupled system, we can to first order ignore changes in the amplitude of the driving field. The amplitude of the driving field will only change in second order. In strongly coupled situations, the mode amplitudes will oscillate, sometimes exchanging all of their energy between each other. These cases require exact solution of the coupling equations. We will examine examples of both.

3. Nondegenerate Coupling Between Modes in a Waveguide

When two orthonormal modes have identical values of β, we must use degenerate perturbation methods for solving the equation of motion. Such examples arise in circular fibers where one spatial mode can have two orthogonal polarizations, or in codirectional couplers was shown in Fig. 10.1. "Non-degenerate" coupled mode theory applies when two modes with different βs are coupled. Non-degenerate modes generally display weak coupling. We will explore non-degenerate coupling in this section.

3.1 Coupling Due to a Dielectric Perturbation.

We begin with the simple case of a perturbed single mode waveguide, and see how a dielectric defect can couple a mode's forward wave to its backward wave. Consider the symmetric slab waveguide shown in Fig. 10.3. The waveguide consists of a guiding film, n_1, of thickness h, surrounded by a cladding index, n_2. A small dielectric notch of depth $h/10$ and length $2a$ in the core region perturbs the waveguide structure. Assume that the waveguide is a single mode structure, and that for $z > a$ there is no backward travelling wave, $A^-(z > a) = 0$.

The mode expects to see an index of n_1 when it gets to the notch region, but instead finds the index is n_2. This is a perturbation. To begin our analysis, we must determine the unperturbed modes of the waveguide. In the case of the symmetric waveguide, we have already calculated the exact form of the amplitude distribution (Eq. 3.26). We can use Eq. 3.29 to describe the eigenmode of the waveguide

$$\begin{aligned}\mathcal{E}_y &= Ce^{-\gamma(x-h/2)} \qquad \text{for} x \geq h/2 \\ \mathcal{E}_y &= C\cos\kappa x/\cos\kappa h/2 \quad \text{for} - h/2 \leq x \leq h/2 \\ \mathcal{E}_y &= Ce^{\gamma(x+h/2)} \qquad \text{for } x \leq -h/2 \end{aligned} \tag{10.12}$$

where the coefficient, C, is chosen to normalize the power in the waveguide to a value of 1 W per unit length (in agreement with Eq. 10.9). The specific values of κ, γ, and β are determined from the eigenvalue equation (Eq. 3.30 or 3.31) for this structure once h, n_1, and n_2 are specified. We assume this has been done, and all eigenvalues are known.

To determine the coupling, we must describe the perturbation term. In this example, the polarization perturbation, $P_{pert}(x)$, is the product of the change in dielectric constant, $\epsilon_0(n_2^2 - n_1^2)$, and the electric field of the forward wave, $E(x)$

$$\begin{aligned}P_{pert}(x) &= \Delta\epsilon E_1 \\ &= \epsilon_0(n_2^2 - n_1^2)\left[\frac{1}{2}A^+\mathcal{E}(x)e^{-j(\beta z-\omega t)} + c.c.\right] \\ &\quad \text{for } 0.4h < x < 0.5h, -a < z < a \\ &= 0 \quad \text{elsewhere.}\end{aligned} \tag{10.13}$$

Notice that the perturbation term exists *only in the notch region.* Everywhere else in the waveguide the mode sees an index distribution that it would normally see without the perturbation. The amplitude equation of motion becomes

$$\begin{aligned}&\frac{\partial A^-}{\partial z}e^{j(\beta z+\omega t)} - \frac{\partial A^+}{\partial z}e^{-j(\beta z-\omega t)} + c.c. \\ &= \int_{0.4h}^{0.5h} P_{pert}\mathcal{E}_y(x)\,dx \qquad (10.14) \\ &= \frac{-j}{2\omega}\frac{\partial^2}{\partial t^2}\int_{0.4h}^{0.5h} \epsilon_0(n_2^2 - n_1^2)|C|^2\left(\frac{\cos\kappa x}{\cos\kappa h/2}\right)^2\left[\frac{1}{2}A^+e^{-j(\beta z-\omega t)}dx + c.c.\right] \\ &= \frac{j\omega\epsilon_0(n_2^2 - n_1^2)}{2}\left[\frac{1}{2}A^+e^{-j(\beta z-\omega t)}dx + c.c.\right]\int_{0.4h}^{0.5h}|C|^2\left(\frac{\cos\kappa x}{\cos\kappa h/2}\right)^2 dx \\ &= j\mathcal{K}(A^+e^{-j(\beta z-\omega t)} + c.c.)\end{aligned}$$

where $\mathcal{K}$ is the coupling constant defined as

$$\begin{aligned}\mathcal{K} &= \frac{\omega\epsilon_0(n_2^2 - n_1^2)}{4}\int_{0.4h}^{0.5h} |C|^2 \left(\frac{\cos\kappa x}{\cos\kappa h/2}\right)^2 dx \quad \text{for } -a < z < a \\ &= 0 \qquad \text{elsewhere.}\end{aligned} \tag{10.15}$$

Notice that the coupling constant is determined by the overlap of the coupled modes in the perturbation region. In this one dimensional problem, we only integrate over the x coordinate to determine the coupling coefficient. Multiplying both sides of Eq. 10.14 by $e^{j(\beta z - \omega t)}$ yields an equation of motion for the amplitudes

$$\frac{\partial A^-}{\partial z}e^{2j\beta z} - \frac{\partial A^+}{\partial z} + c.c. = j\mathcal{K}A^+ \tag{10.16}$$

To solve Eq. 10.16, we must integrate both sides. The first term average to a small value when integrated over a distance large compared to $(2\beta)^{-1}$. Thus to first order we can drop the A^- term, leaving the simplified equation of motion

$$-\frac{dA^+}{dz} = j\mathcal{K}A^+ \tag{10.17}$$

which, when integrated over z, and applying the initial condition that $A_0 = A(-a)$, has solution

$$A^+(z) = A^+(-a)e^{-j\mathcal{K}z} \tag{10.18}$$

Thus we see that the *amplitude* of the field does not change (to first order), but the perturbation (a small region of lower index material) alters the phase of the wave. The forward wave in the region of the perturbation will be (note $\mathcal{K}$ is negative)

$$E(z) = \frac{A^+(-a)}{2}e^{-j(\beta+\mathcal{K})z} \quad \text{for } -a < z < a \tag{10.19}$$

You should see an analogy between this solution and that obtained when we used perturbation theory to clean-up the solutions to rectangular waveguide structures in Chapter 5, Section 5.3.

What about the reflected component, A^- ? To find a first order solution for A^-, we must assume that the forward wave, A^+, does not change, so $\partial A^+/\partial z = 0$. Except for a slight phase change, this is true. The coupling equation becomes

$$\frac{\partial A^-}{\partial z} = j\mathcal{K}A^+e^{-2j\beta z} \qquad -a < z < a \tag{10.20}$$

This can be directly integrated, noting that $A^-(a) = 0$, to yield

$$A^-(-a) = \frac{-j\mathcal{K}A^+}{\beta}\sin[2\beta a] \tag{10.21}$$

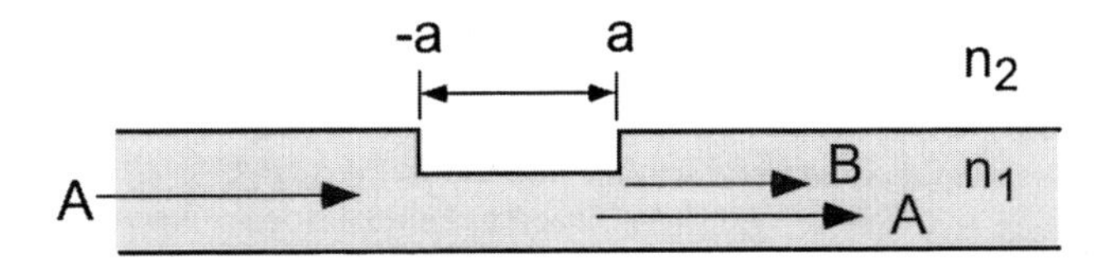

Figure 10.4. A dielectric perturbation can couple one mode to another within the waveguide.

To first order, the coupling of the forward wave to the backward wave is maximized when $2\beta a = (q + 1/2)\pi$, where q is an integer. In terms of a, maximum reflection occurs when

$$a = \frac{(q + 1/2)\pi}{2\beta} \tag{10.22}$$

The behavior of this structure is similar to that of a half-wave dielectric stack used to make mirrors. Similarly, choosing a quarter-wave length for a will minimize the reflected term. The maximum amplitude of the reflection depends on the ratio $\mathcal{K}/\beta$, which is usually small. Generally speaking, maximum coupling will occur between a forward and backward wave when the *spatial frequency* of the perturbation is approximately 2β.

3.2 Intermode Coupling

Consider the same structure as in Fig. 10.3, but let it now be a multimode waveguide. Assume that mode A consists of a forward wave, and it strikes the defect at $z = -a$. How much power does this defect couple from the first mode into a second mode?

As before, we first solve the unperturbed wave equation to determine the eigenmodes of the waveguide. These modes will be of the same form as in Eq. 2.29, but each mode will have a distinct value of β. The specific form for the two spatial modes is:

$$\begin{aligned} E_a(x,t) &= \left[(A/2)\mathcal{E}_a(x)e^{-j(\beta_a z-\omega t)} + c.c.\right] \\ E_b(x,t) &= \left[(B/2)\mathcal{E}_b(x)e^{-j(\beta_b z-\omega t)} + c.c.\right] \end{aligned} \tag{10.23}$$

We assume that for $z < -a$, all the optical energy is in mode A. The coupled mode equation for the amplitude of mode B is then

$$-\frac{\partial B^+}{\partial z}e^{-j(\beta_b z-\omega t)} + \frac{\partial B^-}{\partial z}e^{j(\beta_b z+\omega t)} + c.c. = -\frac{j}{2\omega}\frac{\partial^2}{\partial t^2}\int_{-\infty}^{\infty} P_{pert}\mathcal{E}_b(x)\,dx \tag{10.24}$$

In this case, the perturbation term arises from the presence of electric field from mode A in the dielectric notch, $P_{pert} = \epsilon_0(n_2^2 - n_1^2)E_a(x,t)$. The constants and integral are combined in the coupling constant $\mathcal{K}$,

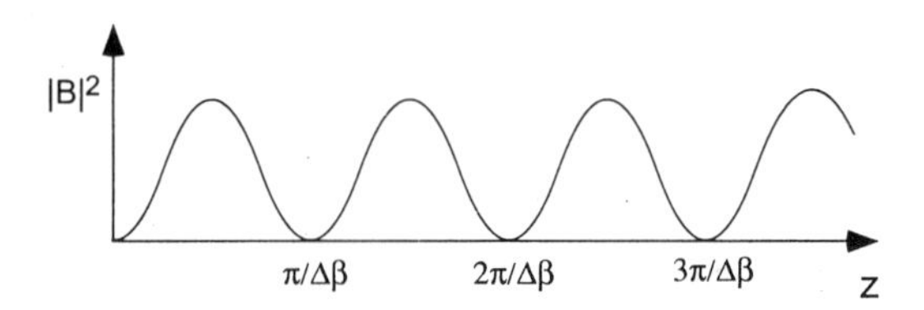

Figure 10.5. Power in mode B as a function of the perturbation length.

$$\begin{aligned} \mathcal{K} &= \frac{\omega\epsilon_0(n_2^2 - n_1^2)}{4}\int_{0.4h}^{0.5h} \mathcal{E}_a(x)\mathcal{E}_b(x)dx \qquad \text{for } -a < z < a \\ &= 0 \qquad \text{elsewhere.} \end{aligned} \tag{10.25}$$

Substituting the coupling coefficient into the equation, and multiplying both sides with $e^{j(\beta_b z - \omega t)}$ yields

$$-\frac{\partial B^+}{\partial z} + \frac{\partial B^-}{\partial z}e^{2j\beta_b z} = j\mathcal{K}A^+e^{-j(\beta_a-\beta_b)z} \tag{10.26}$$

We solve this equation by direct integration over z, from $-a$ to a. If the perturbation length, a, is long enough to satisfy $\beta_b a > \pi$, then the term $(\partial B^-/\partial z)e^{2j\beta_b z}$ will average to zero due to the rapid oscillation of the phase term. The term is relatively small anyway. The term $e^{j(\beta_a-\beta_b)z}$ on the right hand side oscillates at a slower rate because it depends on the *difference* between the two values of β, not the sum. Solving the integral, assuming that $B(-a) = 0$, yields

$$\begin{aligned} B^+(a) &= -j\mathcal{K}A^+\int_{-a}^{a} e^{-j(\beta_a-\beta_b)z}dz \\ &= \frac{\mathcal{K}A^+}{(\beta_a-\beta_b)}\left[e^{-j(\beta_a-\beta_b)a} - e^{j(\beta_a-\beta_b)a}\right] \\ &= \frac{-2j\mathcal{K}A^+}{(\beta_a-\beta_b)}\sin[(\beta_a-\beta_b)a)] \end{aligned} \tag{10.27}$$

The coupled power periodically goes from zero to maximum over the length of the interaction region. Fig. 10.5 illustrates the power (proportional to $|B|^2$) in the second mode at position $z = a$ as a function of the perturbation region length.

The above formula illustrate several important characteristics of mode coupling. First, if the two modes have dramatically different values of β, the overall power coupling will be weak due to the inverse square dependence on the difference in the propagation constants. Only modes that are closely related in β

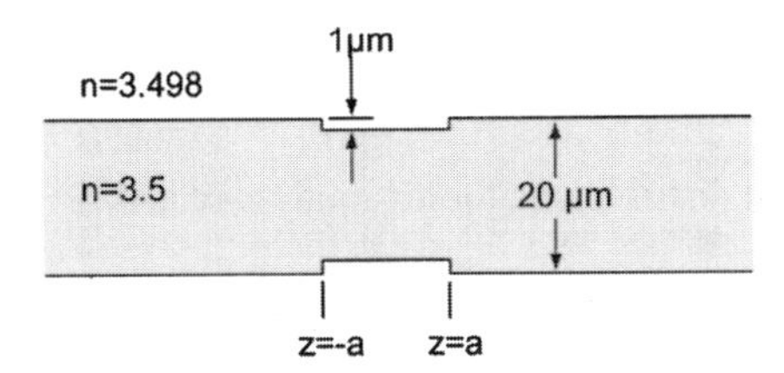

Figure 10.6. A symmetric waveguide is perturbed by a dielectric notch on the top and bottom of the guiding film. The film is 20 μm thick, and the notches are each 1 μm deep.

will couple significantly; modes with large differences in β do not couple effectively, unless the spatial structure of the perturbation somehow makes up the difference, as in a diffraction grating. Second, the coupling depends directly on the value of the coupling constant, $\mathcal{K}$. Third, as before, the coupling is periodic. Energy flows back and forth between the two modes in the perturbation region, and the ultimate coupling depends on the termination point of the perturbation. Coupling is maximum when

$$\beta_a = \beta_b + q\pi/2a \quad (q \text{ odd}) \tag{10.28}$$

When the spatial period of the perturbation makes up the difference between the two propagation constants, there will be enhanced coupling. We will see an example of this with the diffraction grating coupler.

Note what happens when the two modes become degenerate: as the value of β_a approaches β_b, the coupled mode amplitude approaches infinity, which is unphysical. Clearly, perturbation theory as we have applied it to this problem fails for degenerate modes. To solve problems involving degeneracy, we must find exact solutions to the perturbed equations.

Example 10.1 Coupling due to a symmetric notch in a slab waveguide

Consider the symmetric slab waveguide shown in Fig.10.6. A notch is symmetrically located about $z = 0$. In this waveguide, we will determine the normalized modes of the structure, and calculate the coupling between the fundamental TE mode and the other TE modes of the structure. Assume that $\lambda = 1.3\mu$m, $n_1 = 3.5$, and $n_2 = 3.498$.

Solution: The first step is to determine the normalized modes of the ideal waveguide. We find the eigenvalues κ and γ using Eq. 3.28

$$\begin{aligned} \tan \kappa h/2 &= \frac{\gamma}{\kappa} \text{ for even modes} \\ &= \frac{-\kappa}{\gamma} \text{ for odd modes} \end{aligned}$$

There are two even modes and two odd modes supported by this structure. Numerically evaluated values for κ and β for each of mode are listed below in Table 10.1.

The electric field distributions for the modes are given by Eq. 3.27

$$\begin{aligned}
\mathcal{E}_y &= Ce^{-\gamma(x-h/2)} \qquad \text{for } x > h/2 \\
\mathcal{E}_y &= C\frac{\cos\kappa x}{\cos\kappa h/2} \text{ or } C\frac{\sin\kappa x}{\sin\kappa h/2} \qquad \text{for } -h/2 < x \leq h/2 \\
\mathcal{E}_y &= \pm Ce^{\gamma(x+h/2)} \qquad \text{for } x \leq -h/2
\end{aligned}$$

We know every variable in these equations except the normalization coefficient, C. To normalize, we must adjust the amplitude, C, to satisfy

$$\int_S \mathcal{E}_n(x)\mathcal{E}_m(x)dx = \frac{2\omega\mu}{\beta_n}\delta_{nm}$$

The necessary values of C are listed in Table 10.1. (Note: to use consistent units in the calculation, the value for μ_0 had units of Henry/cm, not Henry/meter.) These mode amplitudes will produce a normalized power of 1 W per cm of width in the slab waveguide. Fig. 10.7 shows the four normalized modes taken from this data shows the calculated mode distributions.

Having found the modes of the ideal waveguide, we can now calculate the coupling constant for each pair of modes. The polarization perturbation in this example is generated by the electric field of the TE$_0$ mode in the dielectric notches. Explicitly, the perturbation is (for $9\mu\text{m} < |x| < 10\mu\text{m}$, and $-a < z < a$)

$$\begin{aligned}
P_{pert} &= \epsilon_0(n_2^2 - n_1^2)E_0(x,z,t) \\
&= \epsilon_0(n_2^2 - n_1^2)C\frac{\cos(\kappa_0 x)}{\cos(\kappa_0 h/2)}\left[\frac{A_0^+}{2}e^{-j(\beta_0 z - \omega t)} + c.c.\right]
\end{aligned}$$

Table 10.1. Propagation coefficients for the four allowed modes.

Mode Designation	κ	β	Normalization Amplitude
TE$_0$	1335.12 cm^{-1}	169157 cm^{-1}	99.7 V/cm
TE$_1$	2658.1 cm^{-1}	169142 cm^{-1}	197.2 V/cm
TE$_2$	3949.8 cm^{-1}	169117 cm^{-1}	287.7 V/cm
TE$_3$	5158.45 cm^{-1}	169084 cm^{-1}	353.3 V/cm

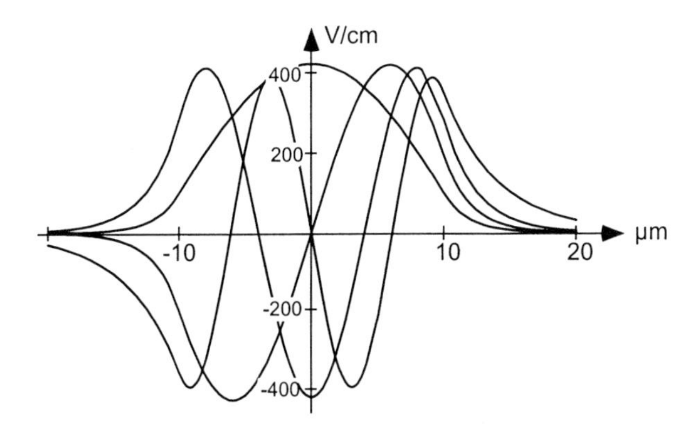

Figure 10.7. The four mode amplitudes are plotted on the same scale. The horizontal axis is in units of cm.

$$= \quad -2.647 \times 10^{-13} \cos(1335.12x) \left[\frac{A_0^+}{2} e^{-j(\beta_0 z - \omega t)} + c.c. \right]$$

Knowing P_{pert}, we can next calculate the coupling coefficients between the modes. Plugging the result into Eq. 10.11, the formula for the general coupling coefficient between the TE_0 and TE_i mode is

$$\begin{aligned} \mathcal{K}_{0i} &= \frac{\omega}{2} \left[\int_{0.45h}^{0.5h} -2.647 \times 10^{-13} \cos(1335.12x)\ \mathcal{E}_i dx \right. \\ &\quad \left. + \int_{-0.5h}^{-0.45h} -2.65 \times 10^{-13} \cos(1335.12x) \mathcal{E}_i dx \right] \end{aligned}$$

There are two integrals because there are two dielectric regions in the perturbation. By symmetry, we can see that modes TE_1 and TE_3 will not couple to mode TE_0. The product of an even and odd function in a symmetric integral always yields a null result. Therefore two of the coupling coefficients equal zero. The only mode that the TE_0 mode can couple to is the even symmetry TE_2 mode. The coupling coefficient will be

$$\begin{aligned} \mathcal{K}_{02} &= 2\frac{\omega}{2} \int_{0.45h}^{0.5h} -2.647 \times 10^{-13} \cos(1335.12x)\ 287.7 \frac{\cos(3949.8x)}{\cos(3949.8h/2)} dx \\ &= -3.912\ \mathrm{cm}^{-1} \end{aligned}$$

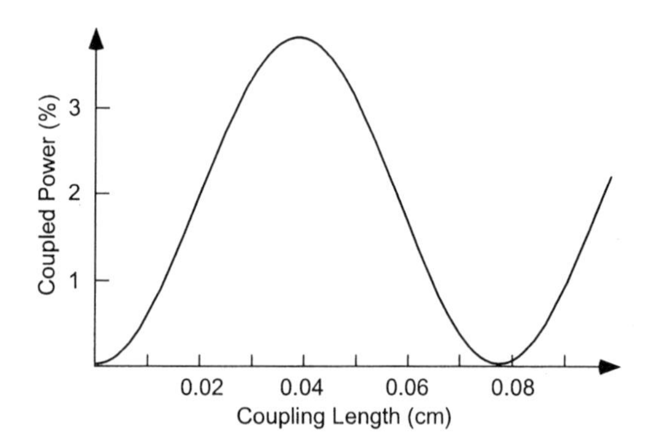

Figure 10.8. The power transfer from mode TE_0 to mode TE_2 as a function of the notch length in Fig.10.6. The coupling varies periodically with notch length.

The coupled mode equation relating the TE_2 mode to the TE_0 mode is then

$$\frac{\partial A_2^-}{\partial z} e^{j(\beta_2 z+\omega t)} - \frac{\partial A_2^+}{\partial z} e^{-j(\beta_2 z-\omega t)} + c.c. = j\mathcal{K}_{02} A_0^+ e^{-j(\beta_0 z-\omega t)}$$

Multiplying both sides by $e^{j(\beta_2 z-\omega t)}$ converts the equation to

$$\frac{\partial A_2^-}{\partial z} e^{2j\beta_2 z} - \frac{\partial A_2^+}{\partial z} + c.c. = j\mathcal{K}_{02} A_0^+ e^{-j(\beta_0-\beta_2)z}$$

Integrating over z from $-a$ to a, the first term on the left hand side averages to zero to first order. The amplitude of mode TE_2 at the end of the perturbation, assuming that $A_2^+(-a) = 0$, is

$$\begin{aligned} A_2^+(a) &= (2j\mathcal{K}_{02}/(\beta_0 - \beta_2))A_0^+ \sin(\beta_0 - \beta_2)a \\ &= j(7.83/40)A_0^+ \sin(\beta_0 - \beta_2)a \end{aligned}$$

Squaring this amplitude gives the actual power in the mode. The power in mode TE_2 as a function of perturbation length a is shown in Fig. 10.8. The coupling length ranges from 0 to 0.1 cm in the plot.

We can see that a small fraction ($< 4\%$) of the mode energy transfers into mode TE_2 at the optimum interaction length. One of the assumptions we made in solving for A_2^+ is that the amplitude of mode TE_0 stays constant. While this is not rigorously true, the amplitude, A_0^+, is reduced by less than 1% at the peak of the transfer, so this is an excellent assumption.

It is hoped that this example, with all its brutal detail, shows you how to normalize a mode, how to systematically apply the perturbation equations, and how to deal with the units. Also note the difference in magnitude between the phase constants in the terms of Eq. 10.28. One term oscillates at a rate of approximately 300,000 rads per cm, and the other oscillates at approximately 40 rads per cm. Our assumption that only one of the terms is significant is valid.

3.3 Reciprocity

It can be shown [2] that if energy is conserved the coupling will be reciprocal. This means that if mode i couples to mode j, then mode j couples with equal strength to mode i. Because of this reciprocity, the equations governing coupling between modes A and B can be written in the general form

$$\begin{aligned} \frac{dA}{dz} &= j\mathcal{K}_{ab}Be^{-j(\beta_a-\beta_b)z} \\ \frac{dB}{dz} &= j\mathcal{K}_{ba}^{*}Ae^{j(\beta_a-\beta_b)z} \end{aligned} \tag{10.29}$$

where A and B are the amplitudes of two modes, and $\mathcal{K}$ is a coupling constant. We will develop an expression for the coupling constant in the next section. The conservation of energy is expressed as

$$\frac{d}{dz}(|A|^2 + |B|^2) = 0 \tag{10.30}$$

4. Degenerate Mode Coupling

There are many examples of structures where two modes with identical propagation constants are coupled. This leads to degenerate coupling, and a different approach is required to solve the problem. Consider the structure involving two coupled waveguides shown in Fig. 10.9. This device is a coupler, where two waveguides are brought into proximity to each other for a short length a. The evanescent field of one guide extends out and partially overlaps the adjacent guide. Energy can tunnel from one guide to the other through the interaction of the evanescent tail. Being symmetric, energy can flow either way in this structure. Structures of this type serve as mode combiners for heterodyne receivers,and optical taps. In this example, we will calculate the amount of coupling between the two guides. We will assume that initially only one of the waveguides carries energy. At the conclusion of this section we will compare the coupling predicted by the Beam Propagation Method for a coupled waveguide in Chapter 9.8 with the coupling predicted by coupled mode theory.

In this device, the polarization perturbation arises from the presence of an external electric field in the waveguide, rather than from a dielectric defect. We assume that the waveguides are single mode structures. Fig. 10.10 shows the geometry of the structure.

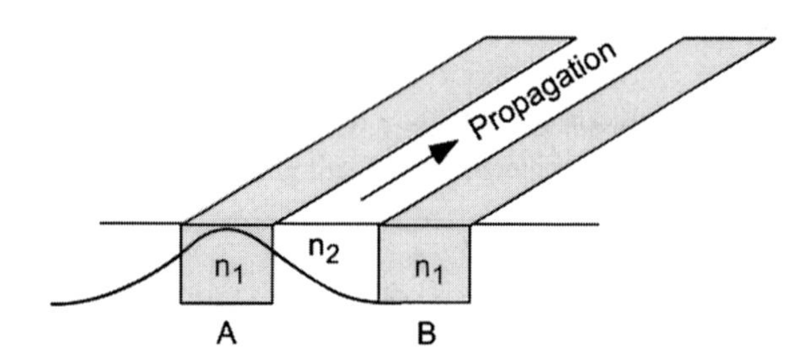

Figure 10.9. A directional coupler can be made by placing two waveguides in close proximity to one another for a finite distance. The evanescent field from one waveguide overlaps the core of the second waveguide, leading to coupling.

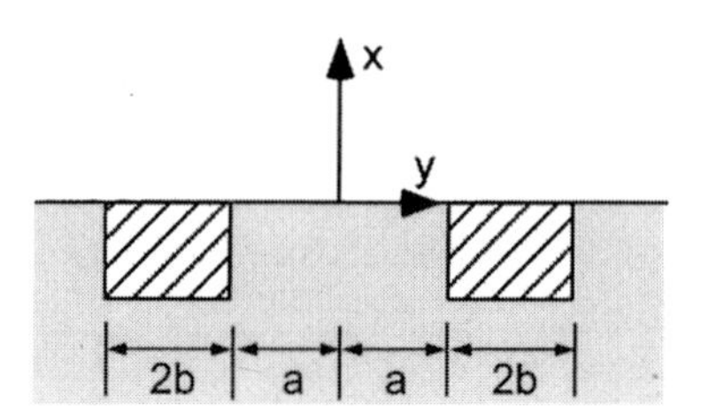

Figure 10.10. The coupled waveguides are separated in the y-direction by a distance of $2a$, and are $2b$ wide. The origin is on the surface between the two waveguides.

The fields in the cores of the ideal (uncoupled) waveguides will be of the form

$$\begin{aligned} E_A(x,y,z) &= A\cos(\kappa_x x + \phi_x)\cos(\kappa_y(y+(a+b))e^{-j\beta z} \\ E_B(x,y,z) &= B\cos(\kappa_x x + \phi_x)\cos(\kappa_y(y-(a+b))e^{-j\beta z} \end{aligned} \quad (10.31)$$

where both waveguides have the same transverse and longitudinal wavevectors, κ_x, κ_y, and β, respectively, and the center for each waveguide is a distance $a+b$ from the origin of the coordinate system. Outside the core, the fields decay exponentially.

The polarization perturbation in waveguide B arises from the presence of the evanescent tail of mode A. The perturbation is actually due to the difference in index the evanescent field sees when it is in the core of waveguide B, compared to the normal cladding index. The polarization induced by mode A acts as a source to excite mode B. From mode B's perspective, the perturbation is

$$\begin{aligned} P_{pert}(x,y,z) &= \epsilon_0(n_1^2(x,y) - n_2^2)E_A(x,y) \\ &= \epsilon_0(n_1^2(x,y) - n_2^2)\left[\frac{A}{2}\mathcal{E}_A(y)e^{-j(\beta z - \omega t)} + c.c.\right] \end{aligned} \quad (10.32)$$

The device shown in Fig. 10.9 operates with codirectional coupling, so we need not include the backward waves in the amplitude equation of motion. In fact, due to phase matching, it is impossible to couple energy into the backward wave of mode B. The general equation of motion (Eq. 10.11) for the structure will then be

$$\begin{aligned} -\frac{\partial B}{\partial z}e^{-j(\beta z-\omega t)} + c.c. &= -\frac{j}{2\omega}\frac{\partial^2}{\partial t^2}\int_S \mathcal{E}_B(x,y)P_{pert}dS \\ &= \frac{j\omega}{2}\int_{-\infty}^{\infty}\epsilon_0(n^2(x,y)-n_2^2)\mathcal{E}_B(x,y)\left[\frac{A}{2}\mathcal{E}_A(y)e^{-j(\beta z-\omega t)} + c.c.\right]dx\,dy \\ &= j\mathcal{K}Ae^{-j(\beta z-\omega t)} \end{aligned} \tag{10.33}$$

where the coupling constant $\mathcal{K}$ in the last equation represents the integral and all the fixed terms. By symmetry, the coupling between waveguide A and B will be identical to the coupling between B and A. (See prob.10.1). We can therefore write down an equivalent coupling equation for amplitude flowing from waveguide B to waveguide A.

$$-\frac{\partial A}{\partial z}e^{-j(\beta z-\omega t)} = j\mathcal{K}Be^{-j(\beta z-\omega t)} \tag{10.34}$$

Eqs. 10.33 and 10.34 are strongly coupled, and must be solved simultaneously. Since the propagation constants are identical for each waveguide, the formulae reduce to

$$\begin{aligned} \frac{\partial A}{\partial z} &= -j\mathcal{K}B \\ \frac{\partial B}{\partial z} &= -j\mathcal{K}A \end{aligned} \tag{10.35}$$

Taking the derivative of the first equation, and plugging it into the second equation reduces these two first order differential equations into a single (uncoupled) second order equation

$$\frac{\partial^2 A}{\partial z^2} = -\mathcal{K}^2 A \tag{10.36}$$

which can be directly solved. If we assume the initial conditions for the problem are $A(0) = 1$, and $B(0) = 0$, then the solutions become

$$\begin{aligned} A(z) &= \cos(\mathcal{K}z) \\ B(z) &= -j\sin(\mathcal{K}z) \end{aligned} \tag{10.37}$$

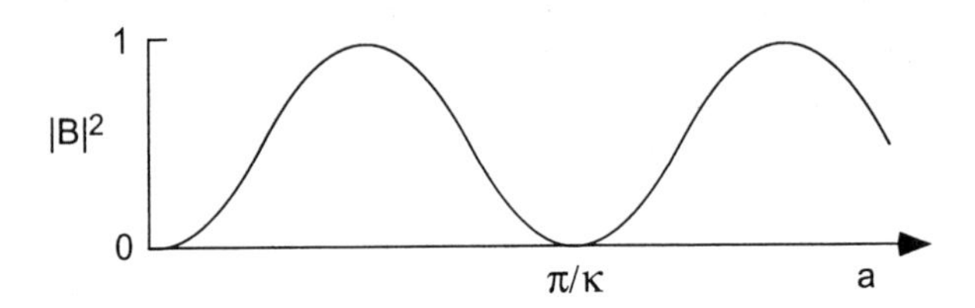

Figure 10.11. The power in waveguide B varies sinusoidally as a function of coupling length, a.

Note that the power can travel back and forth between the waveguides with 100% transfer efficiency. Fig. 10.11 plots the power in mode B as a function of coupling length, a. Also note there is a distinct phase difference between the driving and driven field. The phase in the driven field always lags by $90°$. This phase relationship results from the basic mechanism of coherent energy transfer: when polarization leads the field, the polarization does work on the field, and effectively increases the amplitude of the field. Likewise, when the material polarization lags the field, the field does work on the material, and dissipation of the field occurs. Thus we expect to have a leading polarization in the driven waveguide.

The energy transfer from one waveguide to the other will continue until the driving waveguide is totally depleted of energy. At that point, the driven waveguide suddenly becomes the source which perturbs the original waveguide, and the energy flows in the opposite direction. Also, because of this phase relation, no energy transfers into the backward wave direction. For this reason, such couplers are often called *directional couplers*.

The length of the interaction region determines the exact value of coupling. If coupling distance z_0 is set to be

$$z_0 = \frac{\pi}{2\mathcal{K}} + \frac{q\pi}{\mathcal{K}} \quad \text{(q integer)} \tag{10.38}$$

then complete energy transfer will occur. Other lengths produce values between 0 and 100%. This freedom allows the designer to make couplers of any desired strength. For example, to tap a broadcast signal, a coupler might extract only 1% of the signal, (-20dB coupling), passing the rest on. There are other applications, such as heterodyne detection, where a 50% (3dB) coupler would be desired. The freedom to choose the coupling strength comes with a price. Since the coupling constant depends on β, a change in wavelength can lead to a change in the coupling ratio. Thus, a 3dB coupler at 1.3 μm might not be a 3 dB coupler at 1.5 μm. And since the coupling constant depends critically on β,

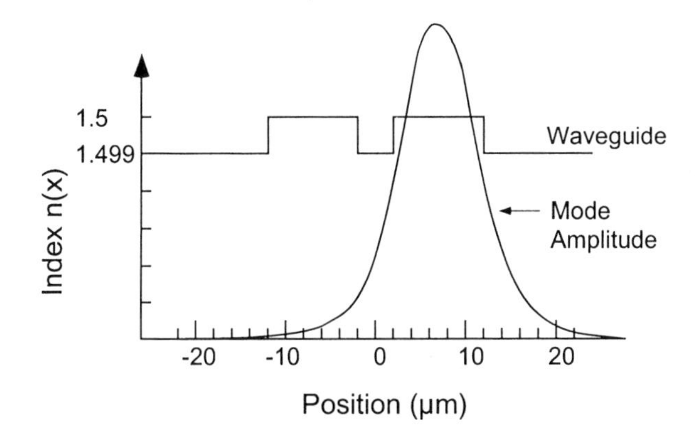

Figure 10.12. Two identical slab waveguides are separated by 4 μm. The eigenmode for one of the slabs is superimposed on the right waveguide.

multimode waveguides will not display the strong power contrast that a single mode coupler will.

The wavelength dependence allows new opportunities as well. It is possible to make couplers which couple strongly at one wavelength while letting another wavelength pass freely with no coupling. Such devices are used commonly to couple pump energy into fiber amplifiers, allowing the signal to be amplified remain on the amplifier waveguide.

What happens if the two waveguides have slightly different values of β? This is a realistic question, and it has practical implications. It turns out that it is difficult to fabricate waveguides which are exactly matched in characteristics. Power transfer will still occur, but it will not be complete[3]. The actual behavior [8] will fall between the 100% transfer predicted for the degenerate case, and the non-degenerate case graphically depicted in Fig. 10.5.

Example 10.2 Analysis of a co-directional waveguide coupler

In Chapter 9, we examined the operation of a coupled waveguide using the Beam Propagation Method. In this example, we will use coupled mode theory to calculate the performance, and compare theory to simulation. To review, the coupled slab waveguide structure is shown in Fig. 10.12. Plotted on top of one waveguide is the amplitude eigenmode solution for one of the uncoupled slabs.

To calculate the coupling between the waveguides, we must first determine the eigenmodes of the uncoupled waveguides. Using techniques that are by now familiar, β was found to be $\beta = 94227$ cm^{-1} ($\lambda = 1\mu$m), and the normalized mode for the right-hand waveguide is

$$\begin{aligned}\mathcal{E}_a(x) &= C\exp[-2840(x-0.0012)] \quad (x>0.0012\text{cm})\\ &= C\frac{\cos[1942(x-0.0007)]}{\cos[1942*0.0005]} \quad (0.0002<x<0.0012\text{cm})\\ &= C\exp[+2840(x-0.0002)] \quad (x<0.0002cm)\end{aligned}$$

The amplitude distribution for the left-hand mode is identical in form, requiring only appropriate offset of the coordinates. The modes are normalized according to Eq. 10.10. Two identical slab waveguides are separated by 4 μm. The eigenmode for one of the slabs is superimposed on the right waveguide.

$$\int_{-\infty}^{\infty}\mathcal{E}_i(x)\mathcal{E}_j(x)dx=\frac{2\omega\mu}{\beta_i}\delta_{ij}$$

where $\omega = 2\pi 300\times 10^{12}$ sec^{-1}, and $\mu = 4\pi\times 10^{-9}$ Henry/cm. Note that all units describing the amplitude distribution and physical constants are in centimeters. Eq. 10.10 was numerically evaluated using *Mathematica*, yielding $C = 433.56$.

The coupling coefficient, $\mathcal{K}$, is found using Eq. 10.21. In this example, we assume that initially the mode energy is completely contained in the right hand waveguide. The perturbation therefore only exists in the core of waveguide B, equaling

$$\begin{aligned}P_{pert} &= \mathcal{E}_a(x)\epsilon_0(n_1^2-n_2^2)\left[\frac{A}{2}e^{-j(\beta z-\omega t)}+c.c.\right]\\ &\qquad \text{for}(-0.0012<x<-0.0002)\\ &= 0 \qquad \text{elsewhere}\end{aligned}$$

Noting the the permittivity, ϵ_0, has units $\epsilon_0 = 8.85\times 10^{-14}$ Farad/cm, the coupling coefficient is found from

$$\begin{aligned}\mathcal{K} &= \frac{\epsilon_0\omega}{4}\int_{-0.0012}^{-0.0002}(1.5^2-1.499^2)\mathcal{E}_b(x)\mathcal{E}_a(x)dx\\ &= 3.6217\ \text{cm}^{-1}\end{aligned}$$

So how does this result compare to the BPM simulation shown of Chapter 9? Since this is a degenerate coupled system, there will be strong coupling. The amplitude will couple periodically back and forth between the waveguides according to Eq. 10.29

$$A(z)=A_0\cos(\mathcal{K}z)$$

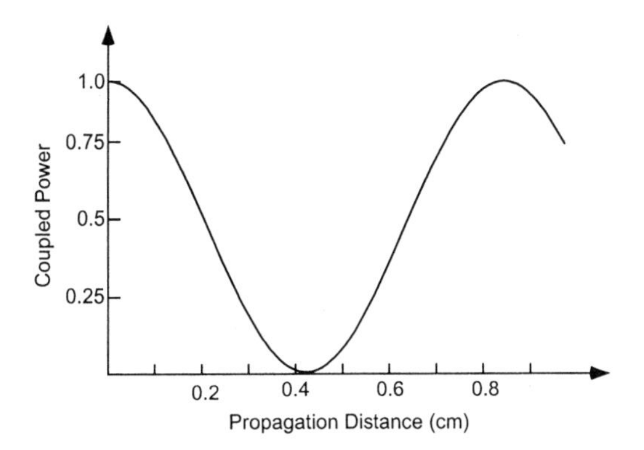

Figure 10.13. The power in the original waveguide couples back and forth between the two waveguides. The coupling period for the power is approximately 0.4 cm.

The power is proportional to the square of the amplitude. Therefore we would predict the power will couple back and forth as

$$A(z)^2 = A_0^2 \cos(\mathcal{K}z)^2$$

Fig. 10.13 shows how the power couples back and forth between the waveguides. By inspection, we see that the spatial coupling period is approximately 0.4 cm, which is exactly what we observed in the BPM example.

5. Coupling by a Periodic Perturbation: Bragg Gratings

As a last example of mode coupling based on the scalar theory of ideal modes, we will consider a very practical problem used in many semiconductor laser diodes and more recently in fiber Bragg gratings. Placing a corrugated index structure in a waveguide can provide strong coupling between forward and backward waves for selected wavelengths. Such wavelength dependence is used to advantage in lasers to control the output wavelength, or in Bragg gratings to make selective filters. The so-called DFB (Distributed FeedBack) laser utilizes a periodic structure on top of the waveguide to couple forward and backward waves of a specific wavelength, thereby restricting laser operation to a narrow and well-defined wavelength. Narrowline operation is a requirement for long-distance, low dispersion optical communication in fibers. In this section we will explore the operation of a similar example of periodic perturbation, the Bragg grating in an optical fiber. The example also serves to illustrate the concept of phase matching between the perturbation and coupled modes.

The fiber Bragg grating is widely used in optical fibers to create spectrally-defined reflectors in the optical fiber. Based on the length of the grating and the strength of the modulation, it is possible to create extremely narrow-band

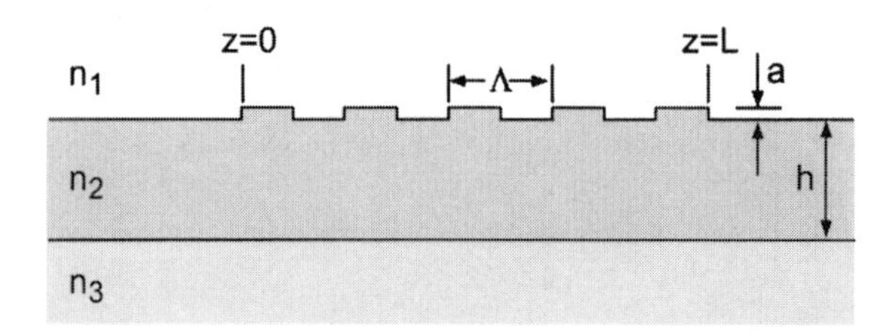

Figure 10.14. Two UV laser beams are crossed to form an interference pattern in the core of a single mode optical fiber. The period of the modulation is controlled by the angle of the crossed beams.

spectral reflections, or broad spectral reflections, or even periodic series of spectral reflections. Coupled mode theory allows the analysis and design of such structures.

The basic Bragg grating is made by exposing a single mode optical fiber to ultraviolet light with a wavelength of 248 nm. This wavelength causes defects in the Ge-doped core of the fused silica fiber to coalesce, effectively densifying the glass and hence increasing the index of refraction. This effect only occurs where there is Ge-doping, and hence only the core experiences this index modulation. The index change is essentially permanent, so it is possible to "write" index modulations into the core of a fiber with an external laser. To make a grating, it is necessary to establish an interference pattern of the UV light in the core. A simple way to create such a standing wave of UV light in a single mode fiber is shown in Fig. 10.14, where two beams from a UV laser (typically a KrF excimer laser) are crossed at an angle θ as they travel through the core. By adjusting θ it is possible to adjust the period of the index grating. Other techniques can be used, the most common being the use of a phase mask to create the two crossed beams.

Since the interference pattern of two crossed beams has a sinusoidal pattern, the index modulation in the core of the fiber has a sinusoidal modulation. Further, because the UV light can only increase the index, the core index will described as

$$\begin{aligned} n(r,z) &= n_{\text{core}} + \Delta n[1/2 + cos(2\pi z/\Lambda)/2] \quad \text{for } r < a \\ n(r,z) &= n_{\text{clad}} \quad \text{for} \quad r > a \end{aligned} \tag{10.39}$$

where Δn is the peak change in index following the exposure to UV light, and Λ is the spatial period of the modulation. Typical index values used in practice range from $\Delta n = 0.001$ to $\Delta n = 0.005$. The period Λ is chosen to provide reflectivity at a specified wavelength, as described below.

We will assume that the waveguide carries a single HE_{11} mode. Since this mode is degenerate in polarization we need only consider the scalar equations.

To make the calculations straightforward, we will assume the guided mode has a gaussian shape, and the forward travelling wave can be described by

$$E(r) = (1/2)A^{+}(z)\mathcal{E}_0 e^{-r^2/w^2} e^{-j(\beta z - \omega t)} + c.c \tag{10.40}$$

where A^{+} is the amplitude of the unperturbed waveguide, $\mathcal{E}_0$ is the normalized amplitude of the mode, and w is the mode radius. *We want to calculate the coupling between the forward wave and the backward wave of the same spatial mode.* The perturbation can be described as

$$P_{pert}(r, z, t) = \epsilon_0 \Delta n^2(r, z) \left[(1/2)A^{+}\mathcal{E}_0 e^{-r^2/w^2} e^{-j(\beta z - \omega t)} + c.c.\right] \tag{10.41}$$

where $\Delta n^2(r, z)$ contains all the spatial structure of the modulated region, and the perturbation only exists in the region between $z = 0$ and $z = L$. Substituting this into Eq. 10.10, the coupling between the forward and backward wave is described by

$$\frac{\partial A^{-}}{\partial z} - \frac{\partial A^{+}}{\partial z} e^{-2j\beta z} = \frac{j\omega\epsilon_0}{4} A^{+} e^{-2j\beta z} \int_0^{\infty} \Delta n^2(r, z)\mathcal{E}_0^2 e^{-2r^2/w^2} 2\pi r dr \tag{10.42}$$

where the integral is over the cross-section of the fiber.

As we observed in the previous examples, there will be negligible coupling between the forward (A^{+}) and backward (A^{-}) wave unless the index modulation $\Delta n(r, z)$ contains a periodic term with a spatial frequency of approximately 2β. It turns out that this is easy to arrange if we choose the period of the corrugated index modulation properly. The perturbation term can be expanded

$$\Delta n^2(r, z) = (\Delta n^2(r)/4)[1 + 2\cos(2\pi z/\Lambda) + \cos^2(2\pi z/\Lambda)] \tag{10.43}$$

If we set

$$2\pi/\Lambda = 2\beta \tag{10.44}$$

one of the phase terms in Eq. 10.42 will equal unity, and a reflection will form. Note that the condition on Λ is equivalent to saying the period of the index modulation should be one-half that of the incident wavelength, forming effectively what is known as a "half-wave" stack of reflectors. The reflection at each period is small because the perturbation is small ($\Delta n \approx 0.001$), but in a long grating, many small reflections add up in phase to build a significant value if the number of cycles, N is sufficiently large.

The forward and backward wave are coupled when this condition is satisfied. If we define $\delta = 2\beta - 2\pi/\Lambda$, and extract the middle term from Eq.10.43 which satisfies Eq. 10.44, the coupled mode equation becomes

$$\frac{\partial A^{-}}{\partial z} = \frac{j\omega\epsilon_0}{4} A^{+} e^{-j\delta z} \int_0^{\infty} \Delta n^2(r)\mathcal{E}_0^2(\frac{1}{2}) e^{-2r^2/w^2} 2\pi r dr \tag{10.45}$$

The r integral is straightforward to evaluate. In this case, the perturbation, $n(r)$ is constant and lies inside the core, so only the field in the region $r < a$ couples to the perturbation. The spatial integral evaluates to

$$\int_0^a \Delta n^2(r)\mathcal{E}_0^2(\frac{1}{2})e^{-2r^2/w^2}2\pi r dr = \Delta n^2\mathcal{E}_0^2[\frac{1}{4}(1-e^{-2a^2/w^2})\pi w^2] \quad (10.46)$$

The final coupled mode equation is thus

$$\begin{aligned} \frac{\partial A^-}{\partial z} &= \frac{j\omega\epsilon_0}{4}A^+e^{-j\Delta z}\Delta n^2\mathcal{E}_0^2[\frac{1}{4}(1-e^{-2a^2/w^2})\pi w^2] \\ &= \mathcal{K}A^+e^{-j\Delta z} \end{aligned} \quad (10.47)$$

We cannot directly solve Eq. 10.47 for $A^-(z)$ until we have an expression for $A^+(z)$. Fortunately, using reciprocity, we know that the backward wave will act as a source for the forward wave in exactly the same manner as we have just calculated for the forward to backward coupling. After some straightforward calculations, the coupled equation is derived

$$\begin{aligned} \frac{\partial A^+}{\partial z} &= \tfrac{j\omega\epsilon_0}{4}A^-e^{+j\Delta z}\Delta n^2\mathcal{E}_0^2[\tfrac{1}{4}(1-e^{-2a^2/w^2})\pi w^2] \\ &= \mathcal{K}A^-e^{j\Delta z} \end{aligned} \quad (10.48)$$

Under conditions of phasematching ($\Delta = 0$), we can solve these equations simultaneously to get

$$\begin{aligned} A^-(z) &= A^+(0)\tfrac{\sinh(\mathcal{K}(z-L))}{\cosh(\mathcal{K}L)} \\ A^+(z) &= A^+(0)\tfrac{\cosh(\mathcal{K}(z-L))}{\cosh(\mathcal{K}L)} \end{aligned} \quad (10.49)$$

These functions are plotted in Fig. 10.15. Here, by appropriate choice of length, L, we can adjust the reflectivity of the mirror in a corrugated waveguide. This is very useful for laser design. When the phasematching condition is not met, there can still be reflection, however there will be reduced efficiency over a limited bandwidth. Details of non-phasematched operation is described in Ref. [1].

Because the Bragg grating in an optical fiber has become a commodity item, numerous ways have been developed to calculate the transmission and reflection of fiber Bragg gratings. Recognizing that the reflection occurs from a periodic series of half-wave layers, matrix techniques developed for the design of thin-film coatings can be applied to the design of gratings, and are typically used instead of the technique shown above. Nevertheless, the coupling parametes needed for matrix techniques ultimately depends on a couped mode analysis as shown here.

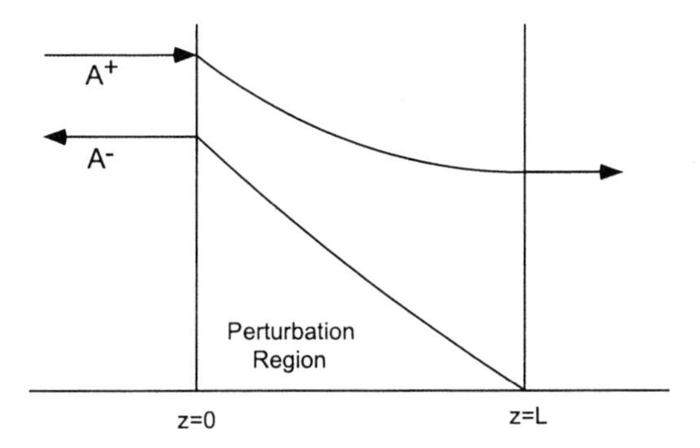

Figure 10.15. The forward and backward amplitudes of an HE_{11} mode in a fiber. Note that amplitudes, not power, are plotted.

6. Summary

The theory we have just developed works well on ideal waveguides with small localized perturbations. Some waveguide defects cannot be localized to a finite region, however. For example, a tapered waveguide has a perturbation which changes along the z-direction. The propagation along such waveguides can be described if the variations occur slowly. The description uses the concept of local modes. We will leave exploration of such techniques to more advanced texts, such as Snyder and Love[5], or Marcuse[6].

In this chapter we explored some examples of amplitude coupling between two waveguides based on a simple scalar theory. We found that coupling is maximized when the propagation coefficient, β, for the two modes is equal. We also found that coupling between dissimilar waveguides is enhanced by adding periodic structures such as gratings which alter the effective β of a waveguide. The coupling of radiation from a waveguide into free space is an important problem in optoelectronics. The calculation is difficult because the guided modes couple to free space modes, which are not normalizable. The design of such couplers is as much art as engineering. As more advance optoelectronic circuits are devised, coupled mode theory will become one of the key tools in the designers toolchest.

The theory developed here is based on a scalar analysis. More advanced theories are being developed to address modern problems such as coupled semiconductor lasers. References [7],[8],and [9] introduce the recent vector mode coupling theory of three research groups.

References

[1] A. Yariv, IEEE Jour. Quant. Elec. 9, 919 (1973)

[2] W. H. Louisell, *Coupled mode and parametric electronics*, Ch.1, John Wiley and Sons, New York (1960)

[3] S. Somekh, et.al, App. Opt. 13, 327 (1974)

[4] R. G. Hunsperger, *Integrated Optics: Theory and Technology*, Vol. 33 Springer Series in Optical Sciences, Chap. 7, Springer-Verlag, Germany (1982)

[5] A. W. Snider and J. D. Love, *Optical Waveguide Theory*, Chapman and Hall, London (1983)

[6] D. Marcuse, *Theory of Dielectric Optical Waveguides* Chapman & Hall, London (1983)

[7] A. Hardy and W.S. Streiffer, "Coupled mode theory of parallel waveguides," J. Lightwave Tech. Vol. LT-3, 1135-1146 (1985)

[8] S.L. Chuang, "A coupled mode formulation by reciprocity and a variational principle," J. Lightwave Tech. Vol. LT-5, 515 (1987)

[9] H.A. Haus, W.P. Huang, S. Kawakami, and N.A. Whitaker, "Coupled mode theory of optical waveguides," J. Lightwave Tech. Vol. LT-5, 16-23 (1987)

Practice Problems

1. Conservation of total power in a waveguide is expressed as

$$\frac{d}{dz}(|E_1|^2 + |E_2|^2) = 0$$

for codirectional coupling between two modes. Show that this can only be satisfied when

$$\kappa_{12} = -\kappa_{21}^*$$

2. Confirm that the normalization constant used in Eq. 10.10 for the TE orthogonal modes of a waveguide is in fact equal to $2\omega\mu/\beta_i$.

3. Show that the proper normalization coefficient for TM modes is

$$\int H_{y_i} H_{y_j} dx = \frac{2\omega\epsilon}{\beta}$$

4. Complete the missing steps in the derivation of Eq. 10.7.

5. Assume that two infinite slab waveguides are build on top of each other, separated by 4 μm. Each slab waveguide is 2 μm thick, and has an index of refraction of 1.52. The cladding indices of refraction are 1.5. The wavelength is 1 μm.

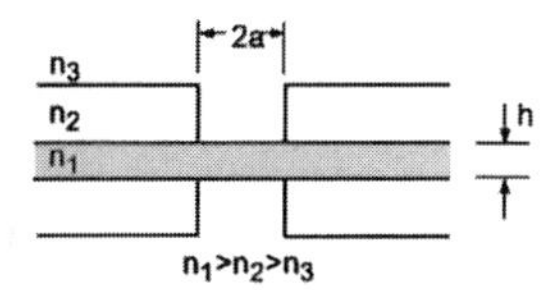

Figure 10.16. Notch coupled structure for Problem 6.

(a) Write down the perturbation polarization in waveguide 2 caused by the field in waveguide 1.

(b) Calculate the coupling coefficient, κ, for these two waveguides.

(c) Determine the length of the interaction region required to make the energy in one waveguide completely couple over to the other waveguide.

(d) Imagine that the interaction length is made to be infinite. Note that if one waveguide is initially excited, the energy will transfer back and forth between the two waveguides forever. It is possible to create "Supermodes" that consists of a superposition of two individual modes that will propagate down the system without any change in energy distribution. Find the two lowest order supermodes. This is a classic eigenvalue problem - you want to find an eigenmode of the total system.

6. Consider a symmetric planar waveguide as shown in Fig.10.16. The waveguide has a symmetric notch of length $2a$ in the cladding index surrounding the guiding layer. The waveguide is designed to only carry the two lowest order modes, which will have even (cosine) and odd (sine) symmetry. Assume the cladding is infinitely thick except in the region of the notch, and the core has thickness h.

   (a) Derive an expression for the phase shift observed for the lowest order forward waves as they propagate past the defect.

   (b) Will there any coupling between mode 1 and mode 2 due to the notch in the cladding? Justify your answer.

7. Consider the planar waveguide shown in Fig. 10.17 with a notch on one side. Assume $\lambda = 1\mu$m.

   $n_1 = 1.5$, $n_2 = 1.48$, and $n_3 = 1$. The guiding film is 5μm thick.

   (a) Determine the exact electric field amplitude distributions, for the first two lowest order TE modes for this structure.

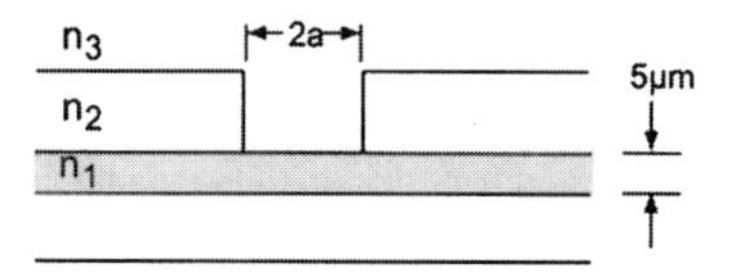

Figure 10.17. Notch coupled structure for Prob.6.b

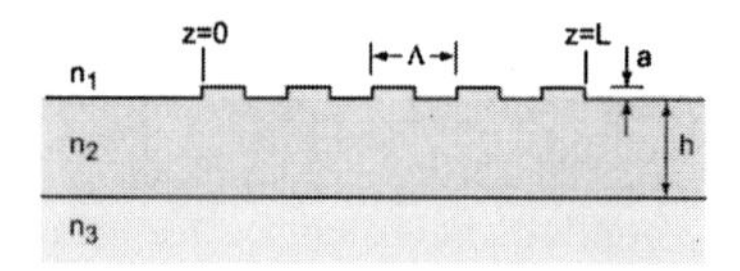

Figure 10.18. Semiconductor laser with a grating for Prob. 8.

(b) Calculate the coupling coefficient, κ_{12} that couples the forward wave of mode 1 to the forward wave of mode 2.

(c) What is the coupling coefficient that couples the forward and backward waves of mode 1?

(d) Develop an expression for the amplitude of the electric field for the second mode, $A_2(z)$, at the end of the notch (i. e., find $A(z = a)$.

8. A semiconductor laser is designed with a grating section on the top surface to act as a wavelength selective mirror. Fig. 10.18 shows the index profile of the waveguide. Assume the guiding layer has index $n_2 = 3.5$, the substrate has index $n_3 = 3.49$, the cladding layer has index $n_1 = 1.5$, and the thickness of the guiding layer is 1 μm.

   (a) What is the period, Λ, necessary to make the waveguide reflect light with vacuum wavelength $\lambda = 0.8\mu$m?

   (b) If a is set to a height of 0.1μm, what is the effective coupling coefficient for connecting the forward wave to the backward wave?

   (c) How long, L, should the grating structure be in order to reflect 90% of the power incident in the forward wave?

9. Design a tunable filter by putting a periodic step structure on top of a semiconductor waveguide. Design your waveguide to carry a single mode at 1.3 μm. Design the grating to selectively reflect signals at 1.31 μm (with 20 dB reflectivity) , but pass (with less than 3 dB loss) signals 50 nm away from this central wavelength. Using the fact that the index of refraction can be varied by injecting charge into the semiconductor, calculate the effect on the wavelength selectivity of this filter as the index

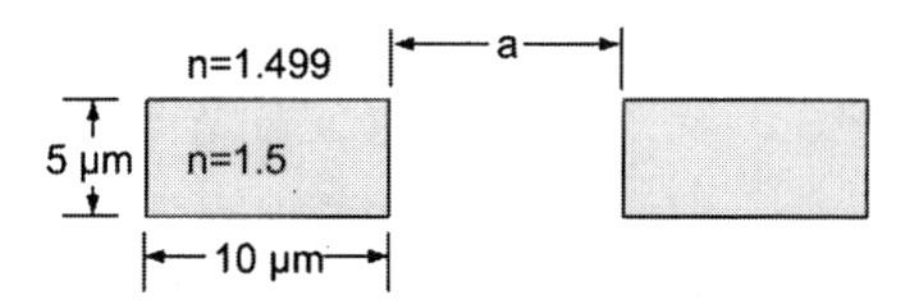

Figure 10.19. Waveguide structure for Prob. 10..

of refraction of the waveguide material decreases. How much should the index be changed to change the wavelength of the reflected signal by 50 nm?

10. Design a 2×2 coupler based on two single mode waveguides in close proximity to each other. Let the waveguides have rectangular cross section, 5μm by 10μm, with core index $n = 1.5$, and cladding index $n = 1.499$. Let the wavelength of operation be $\lambda = 1.55\mu$m.

    (a) Determine the distance between the two waveguides, a, such that 100% of the power in one waveguide is coupled to the other in a length of 500 μm.

    (b) For your design in part a, what is the effective coupling at $\lambda = 1.3\mu$m?

Chapter 11

COUPLING BETWEEN OPTICAL SOURCES AND WAVEGUIDES

1. Introduction

There can be significant loss in optical connections due to misalignment or mismatch of the modes between the two devices. Misalignment between a source and a single mode waveguide by dimensions of less than 1 μm can cause coupling loss exceeding 1 dB. Coupling problems are exaggerated by the small dimensions of typical optical waveguides and sources, which makes alignment a critical and challenging task. In this chapter, we will establish the fundamental rules for coupling optical power between two waveguides and between a source and waveguide. The coupling techniques are based on the concepts developed in previous discussion on coupled mode theory.

2. Coupling of Modes Between Waveguides

Calculating the coupling between two optical waveguides is based on a modal description of the waveguides, and depends on alignment, dimensional differences, and geometric shape. Following this section, you should be able to calculate the coupling efficiency between any two waveguides. Large core waveguides which have many modes are not well served by this theory, and require a different approach. Multimode coupling is discussed in the next section.

Consider the problem of coupling two single mode slab waveguides. Fig. 11.1 below shows the electric field distributions of the source and input waveguides. To efficiently excite the TE_0 mode of the input waveguide, the incident field should spatially overlap the mode profile in the waveguide as closely as possible. Any deviation between the input field and the guided mode shape will simply excite other waveguide and radiation modes.

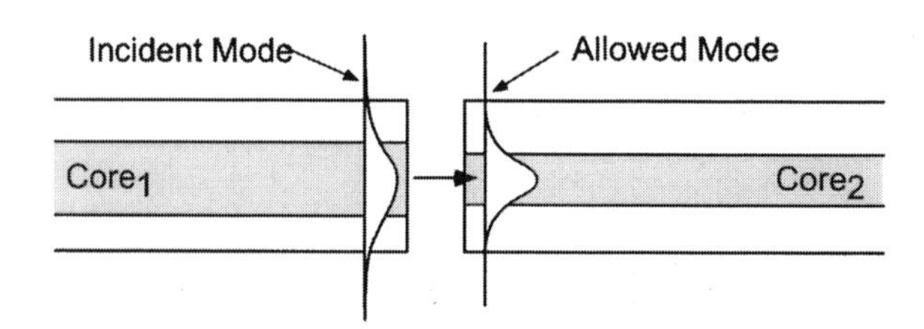

Figure 11.1. An incident mode can have a different spatial profile than the field in the input waveguide. The overlap of the two modes determines the degree of coupling between the input and guided mode.

To formally calculate the coupling efficiency between waveguides, we apply the requirement for continuity of the transverse electric and magnetic fields at a dielectric interface. There are several fields involved in the coupling: the forward waves of the waveguide; the reflected wave of the incident field; and radiation modes for both the reflected and transmitted field due to unguided propagation. If we define the normalized input spatial mode as $\mathcal{E}_i$(x), the normalized transmitted mode as $\mathcal{E}_t$(x), and radiation modes as $\mathcal{E}_\beta(x)$ the fields must satisfy the equation

$$\begin{aligned} \mathcal{E}_i(x) + r\mathcal{E}_i(x) + \int_0^\infty r(\beta)\mathcal{E}_\beta(x)d\beta &= t\mathcal{E}_t(x) + \int_0^\infty t(\beta)\mathcal{E}_\beta(x)d\beta \quad (11.1) \\ \mathcal{H}_i(x) - r\mathcal{H}_i(x) - \int_0^\infty r(\beta)\mathcal{H}_\beta(x)d\beta &= t\mathcal{H}_t(x) + \int_0^\infty t(\beta)\mathcal{H}_\beta(x)d\beta \end{aligned}$$

where r and t are the amplitude reflection and transmission coefficients for the guided modes, and $r(\beta)$ and $t(\beta)$ are the reflection and transmission coefficients for the radiation modes, respectively. If the phase of the mode is described by $exp(-j(\beta z - \omega t))$, then the guided power per unit length flowing in the z-direction is

$$\frac{1}{2}\int_{-\infty}^{\infty} \mathcal{E}_n(x)\mathcal{E}_m(x)dx = \frac{\omega\mu}{\beta_n}\delta_{nm} \quad (11.2)$$

This follows from the relation

$$\mathcal{H}_x = \frac{-j}{\omega\mu}\frac{\partial \mathcal{E}_y}{\partial z} = -\frac{\beta}{\omega\mu}\mathcal{E}_y \quad (11.3)$$

It is impossible to obtain exact solutions to the expressions in Eq. 11.1, because there are an infinite number of radiation modes. We can make an approximate solution by assuming that the amount of energy scattering into radiation modes is negligible, which will be true when the incident field and waveguide mode are not too dissimilar. Thus, we drop the integral terms from Eq. 11.1. We can isolate the transmission terms, t, by taking advantage of the orthonormality of the modes. We multiply both sides of the electric field equation by $\mathcal{E}_t^*$, and integrate over all space. We get

$$t = \frac{\beta_t}{\omega\mu}(1+r)\int_{-\infty}^{\infty} \mathcal{E}_i(x)\mathcal{E}_t^*(x)dx \tag{11.4}$$

We can repeat the process for the magnetic field equation, using Eq. 11.3, to get

$$t = \frac{\beta_i}{2\omega\mu}(1-r)\int_{-\infty}^{\infty} \mathcal{E}_i(x)\mathcal{E}_t^*(x)dx \tag{11.5}$$

Now we have two equations and two unknowns. Solving for t and r respectively yields

$$\begin{aligned} t &= \frac{2\beta_r\beta_t}{2\beta_i+\beta_t}\frac{1}{2\omega\mu}\int_{\infty}^{\infty} \mathcal{E}_i(x)\mathcal{E}_t^*(x)dx \\ r &= \frac{\beta_i-\beta_t}{\beta_i+\beta_t} \end{aligned} \tag{11.6}$$

It is important to realize that $|t|^2+|r|^2 \neq 1$ for most situations, because the incident power can couple into modes other than $\mathcal{E}_i$ and $\mathcal{E}_t$. From Eq. 11.6 we see an overlap integral is used to determine the coupling between the incident mode and exiting mode. Based on the coupled mode theory that was discussed in the last chapter, this should seem reasonable to you. The overlap integral simply calculates the fraction of the incident field that "looks like" the desired mode. Also note that the reflection coefficient, r, is very similar to the Fresnel reflection formula derived in Chapter 2 when the wave is incident normal to the surface. Replacing β by $k_0 n_{eff}$ in r in fact produces the Fresnel expression.

The power coupling efficiency is equal to the square of the amplitude coupling,

$$\eta = |t|^2 \tag{11.7}$$

Often, mode amplitudes will not be formally normalized. In such a case, a working formula for coupling efficiency can be derived from Eq. 11.6 to be

$$\eta = \left[\frac{4\beta_i\beta_t}{(\beta_i+\beta_t)^2}\right]\frac{\left[\int E_t(r,\phi)E_i^*(r,\phi)\,dr\,d\phi\right]^2}{\int E_t(r,\phi)E_t^*(r,\phi)\,rdr\,d\phi\int E_i(r,\phi)E_i^*(r,\phi)\,rdr\,d\phi} \tag{11.8}$$

where $E_i(r,\phi)$ is the input field amplitude, and $E_t(r,\phi)$ is the transmitted field amplitude.

Example 11.1 Power coupling between two fibers with different MFDs

Let's calculate the coupling between two single mode fibers as a function of their relative mode field diameters (MFD). To simplify the calculation, we assume that the spatial HE_{11} modes can be approximated by Gaussian profiles. Define one fiber as having a beam radius w_1 and amplitude A_1, and the second fiber has beam radius w_2 and amplitude A_2. We will assume that these fibers have nearly identical propagation coefficients, so we will neglect the effect of β mismatch.

The power coupling efficiency is given by Eq.11.7

$$\eta = \frac{\left[\int_0^\infty \int_0^{2\pi} A_1 e^{-r^2/w_1^2} A_2 e^{-r^2/w_2^2} r\, dr\, d\phi\right]^2}{\int_0^\infty \int_0^{2\pi} A_1^2 e^{-2r^2/w_1^2} r\, dr\, d\phi \int_0^\infty \int_0^{2\pi} A_2^2 e^{-2r^2/w_2^2} r\, dr\, d\phi}$$

$$= \frac{\left[2\pi A_1 A_2 \int_0^\infty r e^{-r^2\left(1/w_1^2+1/w_2^2\right)} dr\right]^2}{(2\pi)^2 A_1^2 A_2^2 \int_0^\infty r e^{-2r^2/w_1^2} dr \int_0^\infty r e^{-2r^2/w_2^2} dr}$$

This integral can be evaluated in closed form

$$\eta = 4\frac{w_1^2 w_2^2}{(w_1^2 + w_2^2)^2}$$

Note that the amplitudes of the modes drop out of the equation.

The coupling efficiency is plotted in Fig. 11.2 below as a ratio of w_1/w_2. Coupling is maximized when the fibers have identical mode field diameters, and decrease for all other ratios. It is especially interesting to note that the coupling is reciprocal, meaning that it does not matter which fiber has the larger mode. Geometrically, one might expect that a smaller mode would couple more efficiently into a large mode, while it seems intuitively obvious that a large mode will not couple as well to a small mode. Physically, the coupling symmetry arises because we are discussing coupling between two *modes*, not waveguides. A small mode may couple very efficiently into a large core fiber, but it does

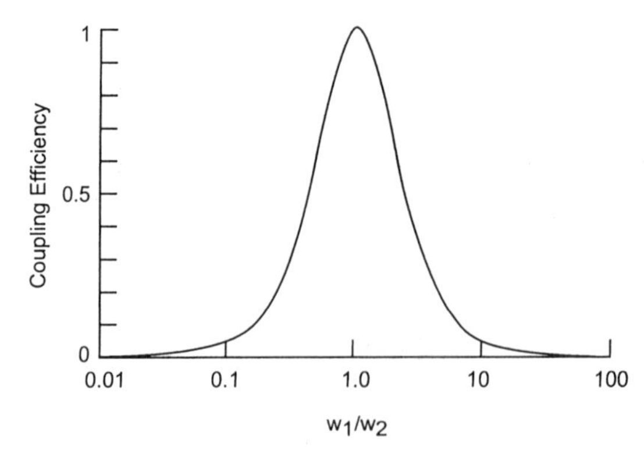

Figure 11.2 Coupling efficiency between two fibers. Notice that the horizontal axis is the log of the ratio of w_1/w_2.

so by exciting a superposition of modes in the larger core fiber. The coupling between the two fundamental HE_{11} modes is reciprocal in this and every case. It makes no difference which way the energy flows. If the mode field diameters are not identical, there will be a reduced overlap integral.

Example 11.2 Coupling a single mode fiber to a Gaussian beam

The 1.52 μm output from a HeNe laser is to be coupled onto a commercial fused silica single mode optical fiber. The laser beam has a Gaussian amplitude profile, with a characteristic beam radius, $w_0 = 0.70$ mm. The fiber has a mode field diameter of 10 μm. What focal length lens should be used to focus the beam onto the fiber to maximize coupling efficiency? What is the maximum possible coupling efficiency?

To answer this, we must know some features of Gaussian optics[1]. When a Gaussian beam is focussed using a high quality lens with focal length f, the focal spot size is approximately described by the formula

$$w_1 = \frac{2\lambda}{\pi}\frac{f}{2w_0}$$

Plugging in numbers, we find a relation between the spot size at the focal point of the lens

$$w_1 = 6.9 \times 10^{-4} f$$

To maximize the coupling between the laser and the fiber, the spot size of the focussed beam should exactly match the mode field radius of the fiber.

$$(6.9 \times 10^{-4} f) = 5 \times 10^{-4}\ \text{cm}$$

Solve for f

$$f = 0.723\ \text{cm}$$

This is a moderately short focal length lens, and is difficult to achieve with a simple lens. Examination of several optics catalogs reveals that a 10x microscope objective has an effective focal length of 1.48 cm, a 20x objective has a 0.83 cm focal length, and a 40x objective has a 0.43 cm focal length. The 20x objective would be the best choice for this fiber. Using the results of the last section, the power coupling efficiency, η would equal 0.98. The Fresnel reflection at the input surface will further reduce the coupling efficiency. In this case the reflection coefficient will be given by

$$R = \left(\frac{\beta_1 - \beta_2}{\beta_1 + \beta_2}\right)^2 = \left(\frac{k_0 n_{eff} - k_0}{k_0 n_{eff} + k_0}\right)^2 = \left(\frac{n_{eff} - 1}{n_{eff} + 1}\right)^2$$

Since n_{eff} is not exactly defined, we cannot proceed. But note that this is in fact the Fresnel reflection that would be expected when crossing a dielectric interface, and $n_{eff} \approx n_{SiO_2} \approx 1.5$. For fused silica with index $n = 1.5$, the power reflection is on the order of 4%, so the maximum coupling will be approximately $0.96 \times 0.98 = 0.94$. Practically speaking, this limit is rarely reached. Achieving 80% coupling efficiency between a laser and a single mode fiber is considered excellent by most practitioners.

3. Coupling From an Optical Fiber to an Integrated Waveguide

The previous examples illustrate the difficulty in connecting two different waveguides efficiently. For this reason industry has standardized to a few mode sizes in order to allow fiber-based instruments and generic devices to be added to a network without a great concern over coupling loss. But coupling a fiber to a chip is still a challenge. In this section we will describe how a knowledge of mode profiles can be used to enhance coupling between different types of single mode waveguide.

Silica optical fibers are the *de facto* standard for transmitting light over any significant distance (anything greater than a few centimeters). Many integrated photonic devices are constructed using thin films of high-index material, so a problem arises when coupling from a fiber to an integrated device due to the dramatic difference in mode size between the two waveguide systems. Silica optical fibers rely on a very small Δn to provide waveguiding, and so the mode size can be relatively large (on the order of 10 μm diameter). In contrast, integrated waveguides are typically made from high index materials such as Si. The high index material provides strong confinement of the mode, allowing tight bends on the chip which conserves space, but the strong confinement also requires that the waveguides have very small dimensions (on the order of 1μm) to sustain single mode operation. This leads to a serious interconnect problem at the chip-fiber interface. We are left with the problem of coupling dramatically different modes (1 μm mode versus a 10 μm mode) and dramatically different effective indices.

In microwave electronics, such mis-matches can be corrected by creating an impedance matching circuit between the two waveguides. A similar solution can be applied to optical waveguides, as is shown in the next example.

Example 11.3 Coupling between a fiber and a high confinement waveguide

Consider the coupling of a fiber with a 10 μm mode field diameter to a square waveguide, 0.5μm x 0.5μm, composed of a Si core and a SiO_2 cladding. Because of the high index contrast between the core and cladding of the Si waveguide, we can safely assume the mode is tightly confined within the Si, and the approximate mode field diameter of the mode must be 0.5μm. To

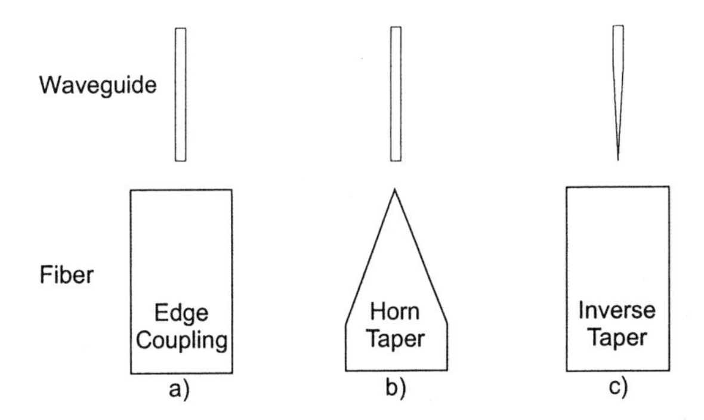

Figure 11.3. Top view of different schemes for fiber-to-waveguide coupling. (a) simple butt coupling, (b) using a horn taper, and (c) using an inverse taper.

expedite a calculation of the coupling, we will assume both guided modes are gaussian. If we simply butt-couple the two waveguides, the coupling efficiency can be estimated ssing Eqs.11.7 and 11.8 and the results from Example 11.2

$$\eta = 2\left[\frac{n_{eff}^{i} n_{eff}^{t}}{n_{eff}^{i} + n_{eff}^{t}}\right] \frac{4w_1^2 w_2^2}{(w_1^2 + w_2^2)^2}$$

Taking $w_1 \approx 0.25\mu$m and $n_{eff}^{i} \approx 3.5$ for the integrated waveguide, and $w_2 \approx 5\mu$m and $n_{eff}^{t} \approx 1.5\mu$m for the fiber, the coupling efficiency turns out to be approximately 10^{-2}, or 1%. Clearly simple butt-coupling is not a viable scheme.

Inverse taper coupling

Fig.11.3 shows some examples fiber-to-waveguide coupling schemes. Simple butt-coupling 11.3(a) leads to poor coupling as we just calculated. Horn-type tapers as shown in 11.3(b) improve coupling efficiency, however in order to avoid excessive coupling to radiation modes in the taper, the required taper length must be on the order of millimeters. This is usually impractical, and it would also require the manufacture and packaging of special tapers for each connection. Ideally one would like to simply attach a commercial fiber to the chip and make the connection without a great deal of fuss. Consider the structure in 11.3(c), which is an inverse taper. To see how this might work, consider a mode on the waveguide travelling toward the fiber. Recall that coupling is maximized when the overlap between the two waveguide modes is maximized, so we want to transform the mode in one of the waveguides so to look similar to the other. The inverse taper provides this mode transformation.

Consider a mode in the Si waveguide heading toward the taper. As the Si waveguide becomes smaller, the mode will become less well confined within the

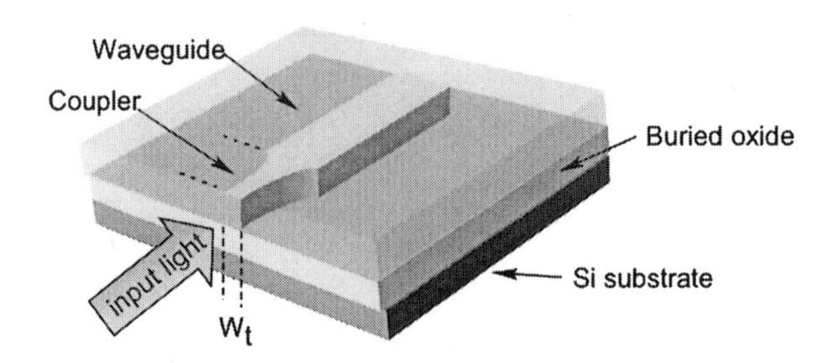

Figure 11.4. Schematic of a waveguide with a nano-taper coupler.

Si, and will actually start to grow in size and become more and more decoupled from the Si core. At some point the size of the decoupled mode will increase to match that of the fibers mode. Placing the fiber at that spot will maximize the overlap and ensure strong coupling. Example 11.4 shows how this occurs.

Example 11.4 Mode size as a function of waveguide dimension

A simple way to see that the mode enlarges as it leaves the inverse taper is to calculate the size of the mode as a function of waveguide dimension. We can do this simply using the normalized parameters developed in Chapter 3. Consider a Si/SiO_2 waveguide operating at 1.5 μm. What is the mode size if the waveguide has a thickness of 0.15 μm? and with 0.02μm?

Solution: We can quickly estimate the size of the mode by estimating the penetration depth, $1/\gamma$, in the cladding using Eq. 3.6 and noting that $\beta = k_0 n_{eff}$

$$1/\gamma = \frac{1}{\sqrt{k_0^2 n_{eff}^2 - k_0^2 n_c^2}} = \frac{\lambda}{2\pi} \frac{1}{\sqrt{((n_f^2 - n_s^2)^{1/2} b)}}$$

From Eq. 3.42 the normalized frequency is

$$V = \frac{2\pi}{\lambda} h (n_f^2 - n_s^2)^{1/2} = \frac{2\pi}{1.5} h (3.5^2 - 1.5^2)^{1/2} = 20.77h$$

From Fig. 3.14 for $h = 0.15\mu$m the normalized effective index for the TE_0 mode is $b_1 \approx 0.64$, and for $d = 0.02\mu$m it is $b_2 \approx 0.02$. Using these values in the expression for $1/\gamma$ we get

$$\text{For} \quad h = 0.15\mu\text{m} \quad 1/\gamma = \left(\tfrac{1.5}{2\pi}\right) \frac{1}{\sqrt{(3.5^2 - 1.5^2)^{1/2} 0.64}} = 0.11\mu\text{m}$$

$$\text{For} \quad h = 0.02\mu\text{m} \quad 1/\gamma = \left(\tfrac{1.5}{2\pi}\right) \frac{1}{\sqrt{(3.5^2 - 1.5^2)^{1/2} 0.02}} = 3.7\mu\text{m}$$

It is easy to see how reducing the size of the guiding layer causes the mode to expand. Simulations based on FDTD and BPM methods show that the coupling efficiency can reach 95% for properly tapered waveguides.

4. Coupling to an Optical Source

None of the information we have discussed in this text so far would be very interesting if it were not possible to somehow couple light onto the waveguide in the first place. In this section we will describe the relative parameters that are essential for estimating coupling efficiency. The discussion in this section is primarily directed at multimode structures, where exact modal calculations are impractical.

A useful measure of an optical source is its *brightness*, B. Brightness is defined as the optical power radiated into a unit solid angle per unit surface area. Brightness is specified in terms of watts per square centimeter per steradian. Brightness is the critical parameter for determining source-to-waveguide coupling for multimode waveguides.

Consider the case shown in Fig.11.5, showing an optical source end-fire coupling onto the end of a waveguide. The source emits light into a cone, or solid angle, that partially overlaps the numerical aperture of the waveguide. Any light falling outside either the numerical aperture or the physical core dimension obviously will not couple to the waveguide. The total power coupled to the waveguide will be given by [2]

$$\begin{aligned} P &= \int_{A_f} dA_s \int_{\Omega_f} d\Omega_s B(A_s, \Omega_s) \\ &= \int_0^{r_{min}} \int_0^{2\pi} \left[\int_0^{2\pi} \int_0^{\theta_{max}} B(\theta, \phi) \sin\theta d\theta d\phi \right] d\theta_s r dr \qquad (11.9) \end{aligned}$$

where the subscript f refers to the fiber or waveguide, subscript s refers to the source, and r_{min} is the smaller radius of either the fiber or the source. In this expression, the Brightness, $B(\theta, \phi)$, is integrated over the acceptance solid angle of the fiber. The maximum acceptance angle, θ_{max}, is defined through the numerical aperture, $\sin\theta_{max} = (n_{core}^2 - n_{clad})^{1/2}$. We have implicitly assumed circular symmetry for the waveguide, but this is not essential to the arguments. You should use common sense when choosing the spatial and angular limits for the integral in Eq. 11.9.

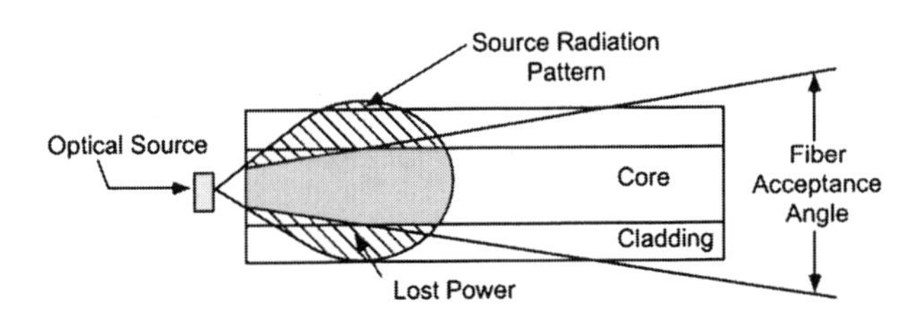

Figure 11.5. An optical source couples light to a multimode waveguide by launching light within the numerical aperture of the guide.

4.1 Optical Source Characterization

It is clear from the coupling equation (Eq. 11.9) that the brightness of the source is critical to coupling, and must be known. The radiation pattern from a source can be fairly complex. There are two extremes of source characterization: Lambertian, and spatial single mode. Most sources fall somewhere in-between these extremes.

A Lambertian source has emission which is uniform across the surface area. A sheet of paper is a good example of a Lambertian source: the light falling on it from a light source is uniformly scattered away in all directions. For a Lambertian source of fixed area, the power delivered from the source measured relative to the normal decreases as $\cos\theta$ because the projected area of the source decreases as $\cos\theta$. The brightness is therefore

$$B(\theta,\phi) = B_0 \cos\theta \tag{11.10}$$

where B_0 is the brightness normal to the surface. The other extreme source, the single spatial mode, is best illustrated by lasers, especially gas lasers such as the HeNe laser. Lasers can emit power into a single spatial mode, and can have extremely narrow beams with angular divergence on the order of a milliradian. These sources are extremely bright, even when the total power is only a few milliwatts.

A surface-emitting LED is a Lambertian source, while edge-emitting LEDs and laser diodes have more complicated angular structure. In general, one must refer to the manufacturer's data sheet accompanying such devices in order to describe the brightness. For example, a typical commercial semiconductor laser is specified to have radiation angle of 11° parallel to the junction, but a radiation angle of 33° perpendicular to the junction. This difference in divergence angle arises because the spatial mode of the laser is smaller in the perpendicular direction than in the vertical direction. Such sources can be approximated by the generalized formula[3]

$$\frac{1}{B(\theta,\phi)} = \frac{\sin^2\phi}{B_0\cos^T\theta} + \frac{\cos^2\phi}{B_0\cos^L\theta} \tag{11.11}$$

The integers T and L are the transverse and lateral power distribution coefficients, respectively. A Lambertian source has $T = L = 1$, while a laser might have values in the hundreds. Fig. 11.4 shows the radiation pattern for a Lambertian ($\cos\theta$) and a laser ($\cos^{100}\theta$). The distance from the origin to the particular solid curve represents the magnitude of the power emitted in that direction.

The values of several types of light source are listed in Table 11.1. Commonly available sources include the tungsten filament bulb, an LED, a semiconductor laser, and a gas laser. A filament bulb will radiate according to blackbody

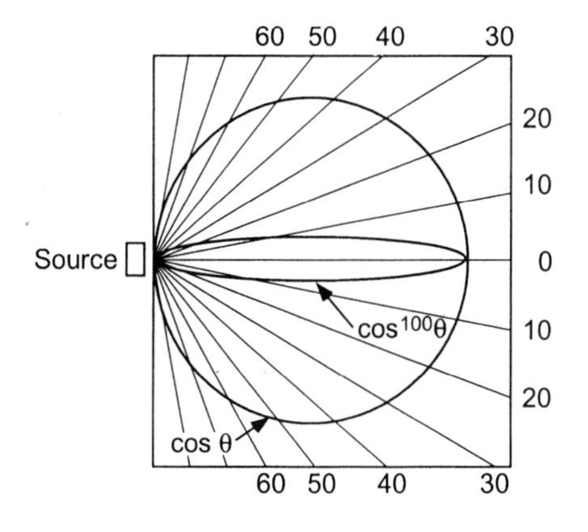

Figure 11.6 The radiance pattern of a Lambertian ($\cos\theta$) source and of a laser ($\cos^{100}\theta$). The plots are normalized to unity B_0.

Table 11.1. Brightness of common optical sources

Source	Power	$\Delta\lambda$ (nm)	Area	I (W/m^2)	Sterads	Brightness (W/cm^2/sr)
Filament	14.4 W	20	1 cm^2	14.4	2π	2.3
LED	20 mW	20	(50 μm)^2	800	2π	125
Diode Laser	10 mW	10	5 by 10μm	2×10^4	0.1	2×10^5
HeNe laser	1 mW	0.1	1 mm^2	0.1	10^{-7}	1×10^6

radiation laws over a large spectral bandwidth. For the purposes of this table, we restricted the source bandwidth to be 20 nm or less, centered at 1 μm.

Inspection of the data in Table 11.1 shows the reason why lasers and LEDS are the preferred sources for optical waveguide communication. The common filament light source, calculated in this case for a filament temperature of 2700°K, is relatively low in brightness compared to the other sources. That, along with the poor overall optical conversion efficiency and relatively slow modulation rate, effectively rules out the use of blackbody sources for efficient waveguide excitation.

The semiconductor LED is also a Lambertian source which radiates over 2π sterads, but due to its small size, and relatively narrow spectral emission bandwidth, its brightness is sufficient for many purposes, especially in low frequency (less than 100 MHz) communication links. The semiconductor laser, due to its much smaller area, and smaller solid angle of emission, has a brightness several orders of magnitude greater than the LED. This is the reason single mode waveguides are almost exclusively excited by laser sources rather than LEDs.

Finally, a low power gas laser is included for comparison. The HeNe laser has relatively low power and intensity, but the extremely clean spatial mode of the gas laser allows for extremely well collimated beams which have a small solid angle. Due to this, the HeNe laser is the brightest source listed in the table. Due to its bulky size, and the need for high voltage discharges, gas lasers are not preferred sources for most waveguide applications. It is clear from Table 11.1 that the semiconductor LED and semiconductor laser are well suited for waveguide excitation. It is truly fortunate that the emission spectrum of the semiconductor devices can be made to coincide with the minimum dispersion and minimum attenuation wavelengths (1.3 μm and 1.55 μm, respectively) of fused silica fibers. This overall compatibility has been one of the key reasons for the dramatic success of optical fiber communications in the last 20 years.

4.2 Coupling an LED to a Step-Index Waveguide

The most efficient coupling between a waveguide and an LED occurs when the LED is butted up against the cleaved end face of the waveguide. This is called "butt coupling", or "end-fire coupling." Since the surface emitting LED is a Lambertian source, we can apply Eq. 11.9 directly to the problem of calculating the power coupled onto a waveguide. In this case, lets assume the waveguide is a circular step-index fiber with core radius a. Eq. 11.9 becomes

$$\begin{aligned} P &= \int_0^{r_s}\int_0^{2\pi}\left(2\pi B_0\int_0^{\theta_{max}}\cos\theta\sin\theta d\theta\right)d\theta_s r dr \\ &= \pi B_0\int_0^{r_s}\int_0^{2\pi}\sin^2\theta_{max} d\theta_s r dr \\ &= \pi B_0\int_0^{r_s}\int_0^{2\pi}(\text{NA})^2 d\theta_s r dr \end{aligned} \tag{11.12}$$

Notice that the numerical aperture of the waveguide is the critical parameter in the coupling. If the optical source is smaller than the core of the fiber, then the power coupled becomes

$$P = \pi^2 r_s^2 B_0(\text{NA})^2 \tag{11.13}$$

The total power emitted by an LED, P_s, is simply

$$\begin{aligned} P_s &= A_s\int_0^{2\pi}\int_0^{\pi/2} B(\theta,\phi)\sin\theta d\theta d\phi \\ &= \pi^2 r_s^2 B_0 \end{aligned} \tag{11.14}$$

Combining the results of Eq. 11.13 and 11.14, we get a simple formula for calculating the power coupled to a step-index multimode fiber

$$P = P_s(\text{NA})^2 \tag{11.15}$$

If the source radius is larger than the core, then Eq. 11.15 can be easily modified to become

$$P = \left(\frac{a}{r_s}\right)^2 P_s(\text{NA})^2 \tag{11.16}$$

4.3 Coupling an LED to a Graded Index Waveguide

The coupling models we have been using are based on ray tracing. For a step-index waveguide, it is simple to characterize an acceptance angle based on the maximum angle that will support total internal reflection inside the core. What do we do if the core is graded? In such a case, the numerical aperture becomes a local function of the distance from the axis of the waveguide. For a circular core graded index fiber with cladding index n_2, the numerical aperture is defined as

$$\text{NA}(r) = \sqrt{n^2(r) - n_2^2} \tag{11.17}$$

Since the maximum acceptance angle is defined as $\theta_{0,max} = \sin^{-1}\text{NA}$, the acceptance angle depends on radius. The coupled power is simply derived by modifying Eq. 11.12 for the step-index fiber, using the fact that $\sin\theta_{0,max} = \text{NA}$

$$P_{graded} = 2\pi^2 B_0 \int_0^r [n^2(r) - n_2^2] r dr \tag{11.18}$$

This integral cannot be evaluated until the index profile is defined. The integration limit is the smaller of the core or source radius.

Example 11.5 Coupling power to a step-index fiber

Consider coupling a large area surface emitting LED to a step-index fiber with core diameter of 50 μm. If the brightness of the LED is 125 W/cm^2/sr, how much power is coupled onto the fiber if the LED is butt-coupled to the end of the fiber? The fiber has an NA = 0.12.

Solution: Since the core is smaller than the LED, we simply need to let $r = 25\mu$m in Eq. 11.13.

$$P = \pi^2 (25 \times 10^{-4})^2\ 125\ (0.12)^2 = 1.1 \times 10^{-4}\ W$$

4.4 Using a Lens to Improve Coupling

Lenses can be used in certain circumstances to improve coupling between a waveguide and a source. A lens can be thought of as an optical transformer: it trades off divergence angle for area. A large collimated beam can be transformed into a tightly focussed, but strongly diverging beam by passing through a lens. The brightness of the source is not changed by the lens.

If the emitting area of the *source* is smaller than the guiding core of the waveguide, then a lens can improve the coupling efficiency compared to simple butt-coupling. The function of the lens is to magnify the image of the source by a factor M to match exactly the core area of the waveguide. In the process, the solid angle in which the source emits is reduced by the magnification factor, M, so that more of the emitted light falls in the NA of the waveguide.

If the Lambertian source area is larger than the core, then the lens will not help improve the coupling. In the case of a single spatial mode beam, such as a laser beam, where the solid angle of the beam is much smaller than the NA of the waveguide, a lens may be useful for converting the large collimated beam into a small focal point that matches the core dimension and NA of the waveguide. Optimum coupling of a source to a waveguide requires matching the dimension of the optical beam, and its numerical aperture, to that of the waveguide.

5. Surface Coupling a Beam to a Slab Waveguide

There are two basic methods for coupling an external beam to a slab or surface waveguide. The most direct is to "end-fire" couple, where the beam is focussed onto the end of the waveguide. This method works well if a clean edge of the waveguide is optically accessible, and if one has the ability to tightly focus and position the external beam. Unfortunately this is often not the case. The second method relies on surface coupling of a beam to one of the waveguide modes. Since an unguided beam (effectively, a radiation mode) cannot directly couple to a guided mode of a waveguide, some coupling mechanism must be invoked to transfer energy between the field and the mode. The most common ways are to 1) use a prism to evanescently couple the optical fields, or to 2) use a grating on the waveguide to couple radiation and guided modes. Both surface methods are examined in this section.

5.1 Prism coupling

The prism coupler is widely used for characterizing thin films on substrates. The process is totally non-intrusive, and requires no clean, exposed edges of the waveguide for end-fire coupling. Large planar substrates can be characterized quickly and nondestructively for their mode structure.

Consider the problem of coupling a radiation mode to a guided mode inside a waveguide. Fig. 11.7 illustrates a radiation mode travelling parallel to a guided mode in a waveguide. Both fields have the same frequency (and vacuum wavevector, k_0). The external field will extend into the waveguide, either as an evanescent field, or as a travelling wave passing through the structure, and it will induce a polarization perturbation within the waveguide. So why doesn't this external field couple energy into the guided mode? The reason is due to the lack of "phase matching," i.e. the phase velocities of the two waves are different. The field in the waveguide propagates with a spatial wavevector β, which is nearly equal to $\beta \approx k_0 n_1$. The radiation field propagates with spatial wavevector k_0. The polarization perturbation created by the radiation field within the waveguide does in fact excite the guided field. However, due to the phase velocity difference between the induced guided wave and the radiation field, they rapidly get out of phase. In a distance, L, such that

$$(\beta - k_0)L = \pi \tag{11.19}$$

the field induced at $z = 0$ will be exactly out of phase with the field being induced at $z = L$. For fused silica, this distance is about one wavelength. The net effect is that no energy is transferred to the guided mode.

Effective coupling between two fields requires that they be phase matched, i.e. the two waves travel at the same phase velocity in the waveguide. One method of accomplishing this is to use a prism to effectively slow down the radiation field. Consider the optical structure in Fig. 11.8, where a prism is located a distance h from the surface of a waveguide. The prism has an index of refraction n_p, while the waveguide on the surface has guiding film index n_f, on a substrate with index n_s. The cover index, n_c, is assumed to be unity, although it is only critical that it's value be less than that of the waveguide or the prism.

The incident beam is directed into the prism at an angle such that *total internal reflection* occurs at the n_p–n_c interface. This is satisfied if the angle of incidence, θ as defined in Fig. 11.9, is greater than the critical angle

$$\theta > \theta_c = \sin^{-1}(n_a/n_p) \tag{11.20}$$

Satisfying this condition requires that the prism have a higher index of refraction than the substrate. Inside the prism, the incident and reflected waves form a standing wave pattern. The k-vector for the field in the prism can be described

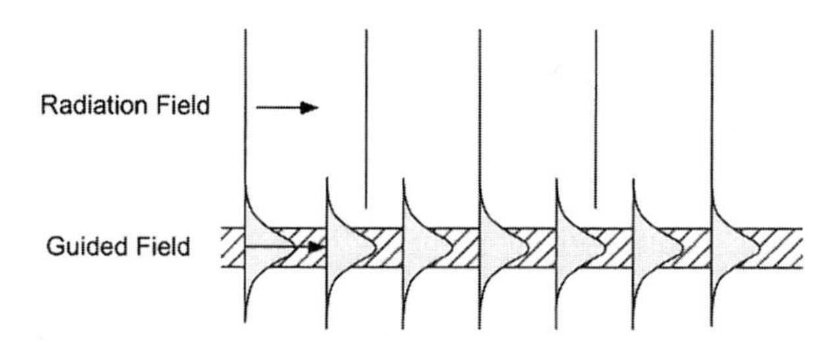

Figure 11.7. The lines of constant phase for a radiation mode propagating parallel to a guided mode are shown. Because of the difference in propagation coefficients, the radiation mode accumulates phase at a slower rate than does the guided mode, and therefore rapidly becomes out of phase with the guided field.

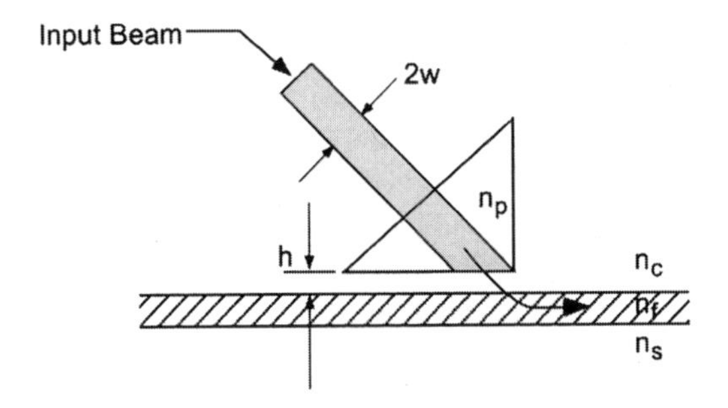

Figure 11.8. The prism with index n_p is located a small distance above the guiding film, n_f. A beam is incident in the prism, and under proper conditions can couple into the waveguide.

in terms of its components

$$\begin{aligned}\hat{k}_o n_p &= n_p(\pm k_x\hat{x} + k_z\hat{z}) \\ &= n_p(\pm k_0\cos\theta\hat{x} + k_0\sin\theta\hat{z})\end{aligned} \tag{11.21}$$

Below the prism–air interface, the field decays exponentially with increasing distance. The x-component of the propagation coefficient is imaginary in this region, but the z-component remains the same as inside the prism. Since the z-component of k depends on the angle of incidence, it is possible to adjust the angle so that the waves travel at the same velocity as those in the waveguide. When this happens, strong coupling occurs. Specifically, to maximize the interaction between the fields in the prism and the waveguide, the angle of incidence of the beam with respect to the bottom surface should be

$$k_0 n_p \sin\theta = \beta_f \tag{11.22}$$

This is called the phasematching condition. Note that this angle is measured inside the prism. The external angle at which the beam enters the prism must be adjusted to account for refraction at the prism-air interface to satisfy this equation. The utility of prism coupling is that the angle of incidence can be adjusted to satisfy Eq. 11.22 for each and every mode in the waveguide (not simultaneously, however), allowing the selective coupling of energy to individual modes, and allowing the experimental determination of mode structure.

5.2 The Coupling Constant for Prism Coupling

We cannot easily use coupled mode theory to describe the interaction of between the prism and the waveguide, because the radiation field is difficult to normalize. In view of this, one usually resorts to a full-field description of the interaction. Such an analysis is not conceptually difficult, but it is beyond the scope of this book, especially in view of the limited application prism couplers have to current devices.

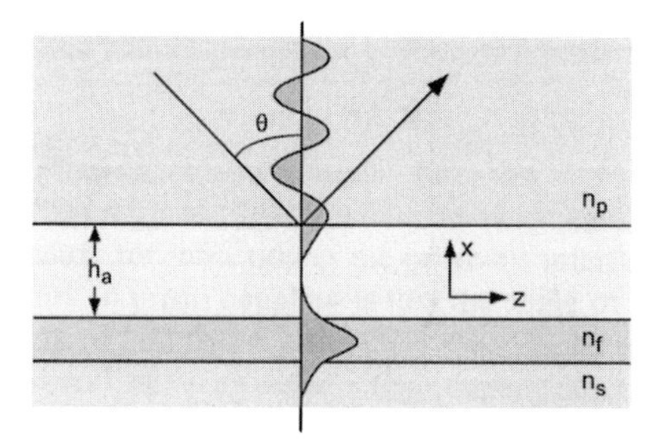

Figure 11.9. A side-view of the prism coupler. The evanescent wave from a wave that is total internal reflected by the glass-air interface couples to the guided mode in the slab waveguide. The coupling depends strongly on the distance, h, separating the prism and substrate.

Coupling depends critically on the separation between the prism and guiding film. As the gap decreases, weak fields from each region can extend across the gap and couple to the opposite region. This evanescent overlap is the source of the weak coupling that occurs between the prism and waveguide. The energy exchange can be considered a tunneling event. It should be intuitively obvious based on the discussion above that the coupling will become stronger as the distance between the prism and waveguide decreases. Exact analysis of the full wave problem is required to find a coupling coefficient for the structure. The interested reader is referred to the literature for a full wave description of the coupling[6, 7, 8]. Typical separations are less than one-half wavelength, which in practice is difficult to achieve over a broad area. Generally one must resort to clamps which press the prism onto the surface. The clamping pressure is experimentally adjusted to give the desired coupling.

Example 11.6 Mode analysis using a prism coupler

A step index thin film waveguide is constructed on a glass substrate. The guiding film has an index of 1.53, and the guiding film has an index of 1.6, with thickness 2 μm. The waveguide is excited by a HeNe laser operating at $\lambda = 0.6328\mu$m. We can assume that all possible spatial modes in the waveguide have been excited by the source. If a 45-45-90° prism made from SF-14 glass with an index 1.73 is placed on the surface of the waveguide, at what angle will each mode couple out of the prism?

Solution: The waveguide and prism are shown in Fig. 11.10. We can assume that the prism is gently pushed into the surface and that optical coupling between the guide and prism occurs.

The first step is to calculate the values of β for the allowed modes in the waveguide. Using a numerical routine, we find that there are three allowed TE

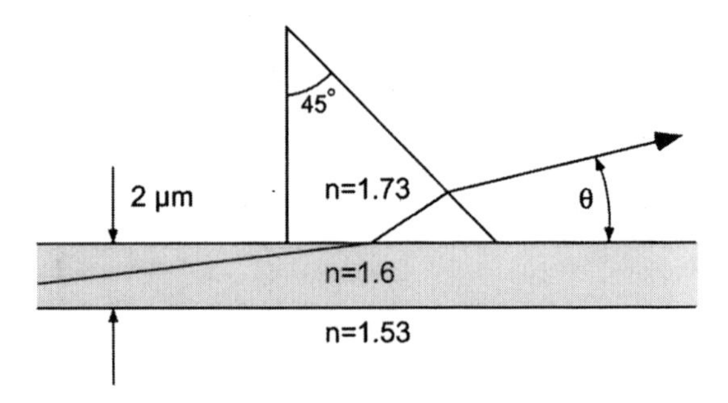

Figure 11.10. A prism coupler is used to couple radiation from a waveguide.

modes, with $\beta_0 = 15.826\mu\text{m}^{-1}$, $\beta_1 = 15.651\mu\text{m}^{-1}$, and $\beta_2 = 15.369\mu\text{m}^{-1}$. There are also three TM modes, but we will assume that the excitation source was polarized to only excite the TE modes.

Using Eq. 11.22, we can calculate the angle of each coupled mode within the prism, using $k_0 = 9.93\mu\text{m}^{-1}$.

$$\begin{aligned}\theta_0 &= \sin^{-1}\frac{\beta_0}{k_0 n_p} = \frac{15.826}{1.73\cdot 9.93} = 67.13^\circ\\ \theta_1 &= \sin^{-1}\frac{\beta_1}{k_0 n_p} = \frac{15.651}{1.73\cdot 9.93} = 65.65^\circ\\ \theta_2 &= \sin^{-1}\frac{\beta_2}{k_0 n_p} = \frac{15.369}{1.73\cdot 9.93} = 63.46^\circ\end{aligned}$$

Straightforward geometry allows us to determine the angle of incidence of each ray with respect to the hypotenuse of the prism. Refer to Fig. 11.11 The incident angle on the interface is given by

$$\theta_{inc} = \theta_i - 45^\circ$$

Using Snell's law, we can solve for the exit angles of the beams from the prism.

$$\theta_{exit} = \sin^{-1}(1.73 \sin \theta_i)$$

For the three angles determined above, the exit angles with respect to the prism hypotenuse are $40.67^\circ, 37.64^\circ$, and 33.25° for β_0, β_1, and β_2, respectively. Since the hypotenuse makes a 45° angle with respect to the substrate, we should subtract these angles from 45° to find the angle, θ as indicated in the figure for each of the modes. In this case we find that the lowest order modes travels at an angle of 4.33° from the substrate, while the other modes travel at 7.36° and 11.75° from the substrate. These modes can be easily distinguished from each other on a card placed a small distance from the prism coupler.

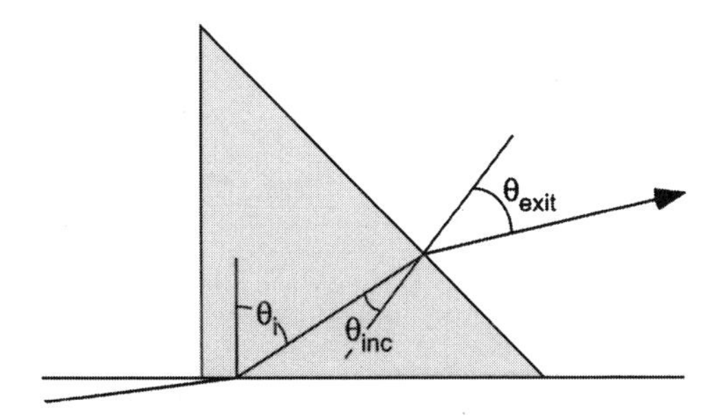

Figure 11.11. Determining the final exit angle of a ray involves evaluating the refraction of the beam at several surfaces.

5.3 Disadvantages of Prism Coupling

Prism coupling is useful for characterizing large thin film structures, but it has several disadvantages that have limited its application in integrated optics. First, prism coupling is inconvenient. It requires precision location of a relatively large bulk optic over a waveguide. This effectively precludes the use of prism coupling in any sort of integrated structure due to manufacturing difficulties. Second, the air gap adjustment is critical for controlled coupling. This can be circumvented to some extent by putting down a low index cover over the waveguide, to which the prism can be directly glued with index matching cement. Nevertheless, there is little room for error, and no simple adjustment is possible to accommodate tolerances. Finally, the index of the prism must be larger than that of the waveguide. For fused silica waveguides, this is not a major problem. But for waveguides made on Si or GaAs substrates with indices on the order of 3.5, or even on a material like Lithium Niobate with index on the order of 1.8, it is very difficult to find optically transparent material of sufficiently large index. Such prisms, if they do exist, are extremely expensive and not particularly rugged.

6. Grating Couplers

A second technique for coupling an optical beam onto a thin film waveguide is to use corrugations in the waveguide. We examined one example of this in Chap.10, where a corrugation was used as a wavelength-dependent reflector, coupling forward waves to backward waves in a waveguide. In the case of a *surface coupler*, we want to couple a guided wave into a radiation mode of the field. While we used coupled mode theory in the last chapter to describe the effect of a waveguide corrugation, we do not have the luxury of having normalized modes when dealing with radiation modes, so other techniques must be used to determine the effective coupling. As with the prism coupler,

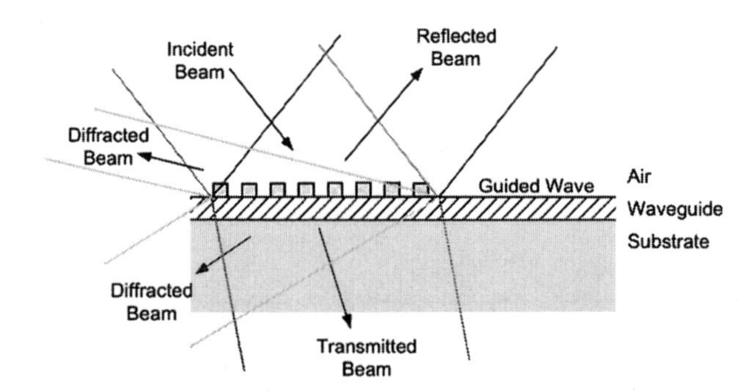

Figure 11.12. A grating coupler illuminated with an incident beam. The incident beam is broken into a transmitted, reflected, and diffracted beams. Under proper conditions, power can also be coupled into the guided mode of the waveguide.

such analysis is beyond the scope of this book. In this section we will describe qualitatively the operation of the coupler, and refer interested readers to selected papers.

The grating coupler is shown in Fig. 11.12. An incident beam strikes the grating on the waveguide. This grating can be created through lithography, holographic development, or volume index variations through ion implantation, to mention only a few techniques. Here, we are only concerned with the fact that there will be a grating. The incident wave strikes the grating, and is broken into several other beams. There is usually reflected wave, and if the substrate is transparent, there will be a transmitted beam. The directions of these beams follow Snell's law. There can also be several diffracted waves, where the direction of the beam is dramatically altered from what would be expected through reflection or refraction. If conditions are correct, a portion of the wave can couple into the guided mode of the waveguide. We will show that the effect of the grating is to modify the longitudinal component of the k vector of the wave.

6.1 Basic Grating Physics

If the grating structure is oriented along the z-direction, then it can modify the z (longitudinal) component of the incident wavevector. Specifically, if the grating has a period Λ, the vector relation between incident and diffracted light is

$$\hat{k}_{out} = \hat{k}_{in} \pm \frac{2q\pi}{\Lambda}\hat{z} \tag{11.23}$$

where q is an integer, subject to the condition that the magnitude of the wavevector does not change

$$|k_{in}| = |k_{out}| \tag{11.24}$$

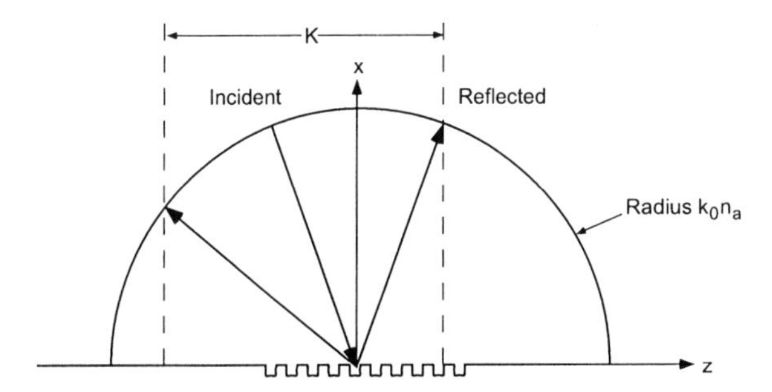

Figure 11.13. k-space diagram of a wave diffracting off of a grating. The grating subtracts a value K from the z component of the input wavevector, resulting in a backward directed wave. Note the backward wave has the same magnitude, $|k|$

when the diffracted wave stays in the same index medium. This is most easily viewed graphically. Fig. 11.13 shows the input, reflected, and diffracted wave from a surface grating. The period of the waveguide satisfies $K = 2\pi/\Lambda$. The incident wave has a wavevector k_0n_a where n_a is the index of the medium above the grating. Upon striking the grating, a reflected and diffracted wave are generated. The reflected wave has the same longitudinal wavevector, $k_z = k_0n_a \sin\theta$ as the input wave, but the transverse (x-directed) component of the wavevector is reversed. The magnitude of the reflected wavevector is identical to the input. The grating can add or subtract integer units of K to the z-component of the incident wave. The radius k_0n_a shows the locus of allowed wavevector values based on Eq. 11.24. Any diffracted wavevector must fall on this radius. Graphically, we add or subtract the vector K from the z-component of the incident wave, and see where it intersects the radius. In this case, it is impossible to add a vector K and remain on the proscribed radius. Only when K is subtracted from the incident wave is there an allowed solution. This is shown by the shaded arrow.

Now consider what would happen if the beam is incident on a grating placed in a thin film waveguide structure, sitting on a transparent substrate. The same type of k-space diagram can be drawn, as shown in Fig. 11.14. In this case, the magnitude of the transmitted wave is larger due to the increased index. The lower radius in the figure represents the allowed values of the diffracted or transmitted wavevectors in the substrate. The allowed k values of the waveguide are represented by the small section of a radius horizontally located at k_0n_f. In this case we can see that there can be one diffracted beam backward into the substrate, and one beam diffracted forward into the waveguide.

The horizontal diffracted beam corresponds to the grating scattering a wave into the waveguide, which has an effective wavevector, k_0n_f. If the incident

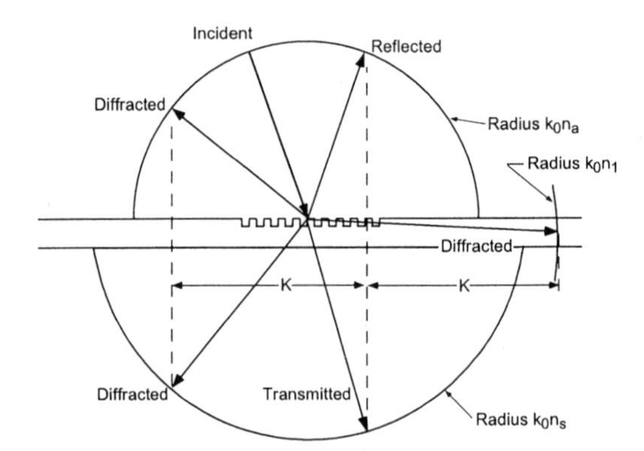

Figure 11.14. k-space diagram of a grating on a high index thin film, placed on a moderate index substrate. In this special case, the angle of incidence is exactly right for the grating to couple light into the waveguide.

angle is θ, then by geometry, the guided wave propagation coefficient is

$$\beta = k_0 n_a \sin\theta + K \tag{11.25}$$

From these diagrams, we can see how the longitudinal value of the wavevector is converted by the grating into new values. The value k_0 stays the same in all media, but the influence of the dielectric constant, n_i, is seen to scale the wavevector in each media. However, the effective wavenumber of the grating, K, is independent of the index, and simply adds or subtracts to the longitudinal component. If total wavenumber can be preserved after adding in the effect of the grating, then a diffracted order can occur.

6.2 Output Coupling with a Grating Coupler

The waveguide grating is advantageous over prism coupling because it can be manufactured using lithographic techniques that are standard in the semiconductor industry. This means that it is possible to mass-produce integrated optical devices with waveguide couplers. In this section we want to consider some of the details of such a coupler. Consider the case where a guided mode in a waveguide is incident upon a section of waveguide with a grating. The grating will act as a coupler, and coherently scatter some of the light out of the waveguide. This technique is becoming popular for semiconductor laser output coupling, because it is not necessary to cleave the substrate in order to create an output mirror. An example of such a structure is shown in Fig.11.15. The top figure shows the waveguide configuration, with the incident and transmitted wave, and the diffracted waves going into the substrate and the air.

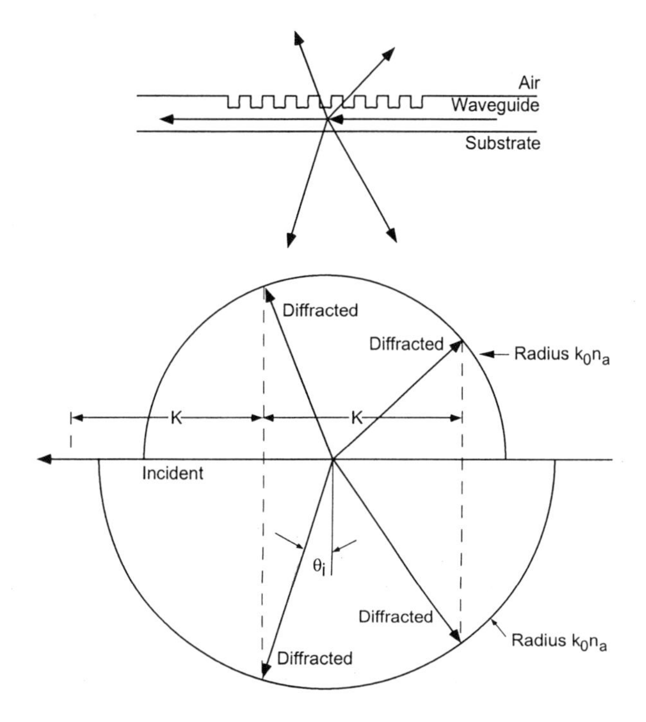

Figure 11.15. A guided wave incident upon a grating will couple some energy into a diffracted waves. The angle at which the diffracted beams leaves is determined by the phase matching conditions shown in the lower diagram.

The incident wave is a guided mode with propagation coefficient β, which by definition is the z-component of the wavevector. The grating can add or subtract to this z-component. By inspection of Fig. 11.15, we can see that it is impossible for a photon to scatter off the grating in a forward direction and have $k_z = \beta + K$. We must consider only cases where the z-component is reduced. For the case shown, we see that there are four cases where

$$k_0 n_i \sin\theta = \beta - qK \tag{11.26}$$

can be satisfied. These are represented by two rays going into the substrate, and two into the air. This illustrates one difficulty with grating couplers: they tend to couple light in both directions out of the waveguide. If we were trying to efficiently couple light from a waveguide for an application, we might try to adjust the grating period so that only one beam was coupled into the air. But there would still be a beam coupled into the substrate, which represents a potential power loss of 50 %. This can be combatted by placing a reflector under the active waveguide, reflecting the power back out of the structure,

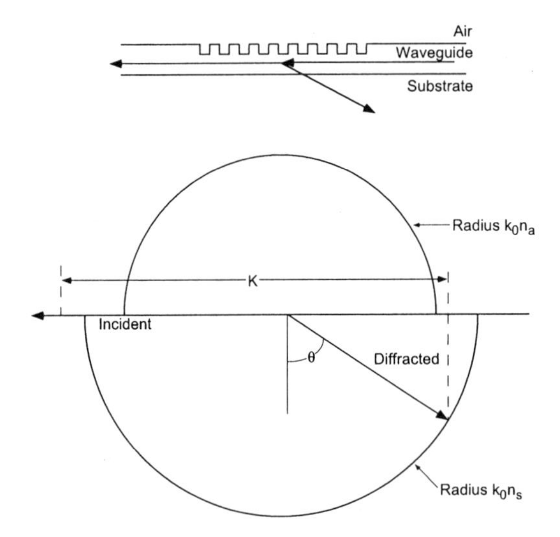

Figure 11.16. A single beam coupler can be built by making the grating wavenumber so large that the phase matching condition cannot be met in the cover region. The only coupled power radiates into the substrate.

although one has to be very careful about interference effects between the two coherent waves combining above the waveguide. A second method is to design the grating such that only one beam can couple out of it. If the index of the substrate is greater that the cover index, it is possible to make it impossible for the phase matching conditions to be realized in the cover region. Fig. 11.16 shows the phase diagram of such a structure.

In this case the grating wavenumber is chosen to almost retroreflect the beam back down the grating. Graphically, we can see that this condition will be met if

$$\beta + k_0 n_s > K > \beta + k_0 n_c \tag{11.27}$$

Unfortunately, there are several problems with this scheme. First, the exiting beam will strike the lower substrate-air interface above the critical angle, and will not couple into free space. To get light out, it is necessary to deform the lower substrate surface by adding a prism or another grating. This defeats many of the reasons for using gratings in the first place. However, there are situations where it could useful to couple light out of the waveguide and into other regions of the substrate, such as into a detector. The second problem has to do with the fact that most semiconductors operate in a standing wave. The laser light travels back and forth along the waveguide axis. Using a coupler as shown in Fig. 11.16, two light beams would be coupled out, in nearly opposite directions.

This is essentially the same problem as in the air coupling, when the power was divided between two beams.

Another approach to the multiple beam problem is to use a blazed grating, where the dielectric grating has an asymmetric profile[7]. The effect of the blaze (i.e. the asymmetry) is to cause certain diffraction orders to preferentially receive more of the power. This technique has been used for years in the manufacture of diffraction gratings for scientific instruments such as spectrophotometers and laser tuners.

6.3 The Coupling Coefficient

As noted above, we cannot directly use coupled mode theory to derive the coupling coefficient for a diffraction grating. There have been many calculations of these structures, using Green's functions, Bloch waves, and variational techniques. There is currently a great deal of research being done on gratings couplers for semiconductor lasers. The grating can serve as both an output coupler and as a combination mirror-tuned filter for providing the necessary optical feedback to sustain laser operation. The interested reader is directed at much of the current literature to see the latest in design methodology[8]. The overview presented here is intended only to show the qualitative aspects of grating coupling.

7. Summary

In this chapter, we introduced a set of very important rules for calculating coupling efficiency between waveguides and free space optical beams. Calculations can be done for determining the coupling efficiency between two waveguides that are slightly tilted with respect to each other, or slightly offset to each other. As problem 11.5 will show, misalignment of less than 1μm can result in 1 dB of loss. This is a major problem for the manufacturers of single mode fiber connectors. It is very difficult to mass produce connector ferrules to tolerances of less than 1 μm. Even if it can be done, thermal expansion caused by everyday temperature variation can easily introduce creep and distortion of noticeable magnitude.

Multimode waveguides are much more tolerant of misalignment, because there are more guided modes available to create the necessary superposition.

We found that most coupling problems can be described by an overlap integral between the incident and exiting field. Maximum coupling arises when the spatial fields are similar. We explored two examples of coupling for which coupled mode theory was not applicable, namely prism and grating coupling. Coupled mode theory does not work in these cases because it is difficult to normalize the radiation modes. We described qualitatively how these couplers worked, hoping that the reader can explore more involved solutions with greater insight and understanding.

References

[1] J. Verdeyen, Laser Electronics, 2nd ed., Prentice-Hall , USA, chaps.2-6 (1989)

[2] G. Kaiser, *Optical Fiber Communications, 2nd ed.*, McGraw-Hill, Inc., USA (1991)

[3] Y. Uematsu, T. Ozeki, and Y. Unno, "Efficient power coupling between an MH LED and a taper-end multimode fiber," *IEEE J. Quantum Electronics*, QE-15, 86-92 (1979)

[4] J. E. Midwinter, *IEEE J. Quantum Electronics* 6, 583-589 (1970)

[5] R. Ulrich, "Theory of the prism-film coupler by plane-wave analysis," *JOSA* 60, 1337-1350, (1970)

[6] P. K. Tien and R. Ulrich, "Theory of prism-film coupler and thin-film light guides," *JOSA* 60, 1325-1337 (1970)

[7] K.C. Chang and T. Tamir, "Simplified approach to surface-wave scattering by blazed dielectric gratings," *Applied Optics* 19, 282 (1980)

[8] J. H. Harris, R. K. Winn, and D. G. Dalgoutte, *Applied Optics* 10, 2234 (1972)

Practice Problems

1. Consider coupling a laser beam into the slab waveguide described in Example 11.1. Use the data in Table 11.1 to calculate the exact wavefunctions for the first three TE modes of the unperturbed structure. A laser beam is focussed in one direction using cylinder lens so that the y-component of the beam remains wide (several mm's), but the x-dimension of the beam is focussed down to a Gaussian profile with characteristic radius $w_x = 10\mu$m. (The characteristic radius is the distance from the beam center at which the amplitude is reduced to e^{-1} of the peak value.) Assume the wavelength of the incident radiation is 1.3 μm, as in the example.

   (a) If the x-axis of the beam is perfectly centered on the waveguide, how much power is coupled to the TE_0 mode? The TE_1 mode? the TE_2 mode?

   (b) If the beam is lowered so that the center of the optical beam strikes 5 μm below the center of the waveguide axis, determine the coupling to the TE_0 and TE_1 modes.

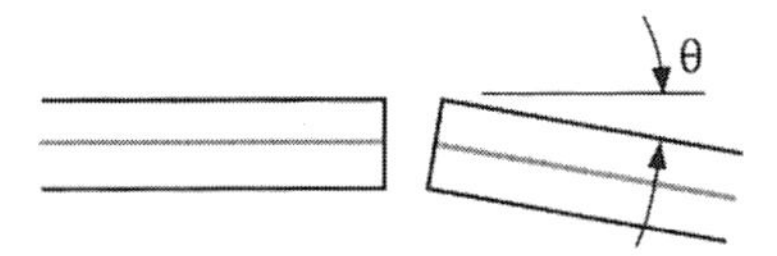

Figure 11.17. Two identical fibers are angularly misaligned by an angle θ.

2. Extend the derivation of Eq. 11.6 to derive Eq.11.8 for unnormalized modes.

3. An LED with circular radius of 40 μm and a Lambertian emission pattern of 125 W/cm^2/sr is used to excite two optical fibers. The first has a core radius of 25 μm, and an NA=0.15. The second fiber has a core radius of 31 μm and an NA=0.12. How much optical power is coupled into both fibers if the LED is butted against the end of the cleaved fiber. You may neglect Fresnel loss.

4. Consider the problem of coupling two single mode step–index fibers together as a function of bending angle between the two fibers. Assume that the fibers are identical, with electric field amplitude described by a Gaussian function
$$E(r) = A\exp(-r^2/w_0^2)$$
where $w_0 = 5\mu$m is the Gaussian beam radius. The wavelength is 1.3μm. Fig. 11.17 illustrates the geometry of the problem.

5. If the two fibers have the same beam radius, numerically calculate the effect of relative tilt between the core axis on the coupling efficiency. Calculate and plot the coupling efficiency between the two fibers as the tilt angle between the fibers increases from 0 to 10° in steps of 2 °.

6. For the same fiber as described in Problem 11.4, calculate the coupling efficiency as a function of translational offset, δr, between the two fibers. See Fig. 11.16. Assume that fibers are parallel to one another, but that the core axes are separated radially by a distance r. Calculate the coupling efficiency as a function of r for $r = 0 \to 2w_0$.

7. A major problem in optical fiber systems is the reflection that can couple back into a waveguide from the end of the fiber. The dielectric surface will display a Fresnel reflection unless it is carefully index-matched with the outside world.

8. One way to fix this problem is to polish the end of the fiber at a slight angle with respect to the fiber axis. If the core index is 1.45, the cladding

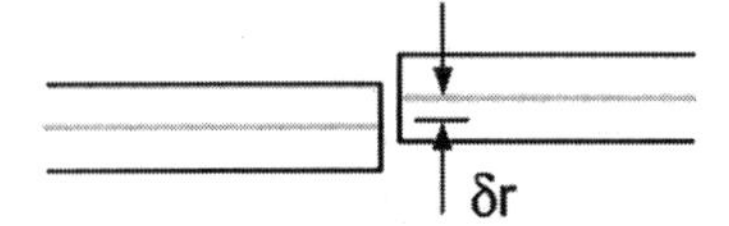

Figure 11.18. Two identical fibers translationally offset.

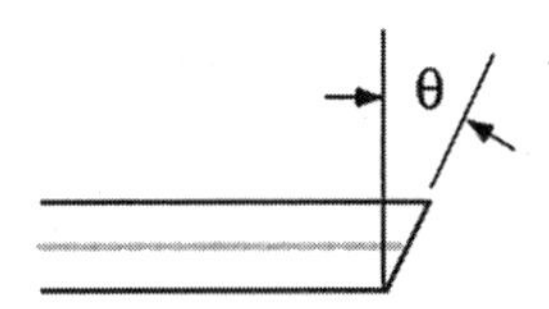

Figure 11.19. A fiber is cut and polished at an angle θ with respect to the plane perpendicular to the core axis.

index is 1.44, the core diameter is 10μm, and the outside index is unity, at what angle should the end of the fiber be polished in order to reduce the coupling of back-reflected light to less than 20 dB? (Hint: you don't need all of these parameters.)

9. Another way to fix this problem is to "index-match" the end of the fiber with its mate, and leave the angle of the fiber perpendicular to the axis. What is the maximum allowable difference index of refraction between the core and interface that can be tolerated if reflections are to be kept below 30 dB?

10. Calculate the coupling efficiency of a Gaussian beam to a symmetric slab waveguide. Assume the waveguide is 3 μ m thick, has a core index $n_{core} = 1.5$, and a cladding index $n_{clad} = 1.485$. Assume the incident Gaussian beam has a characteristic radius, $w_0 = 3\mu$m.

11. Determine the accuracy of the Gaussian approximation for the HE_{11} mode by calculating the coupling of a Gaussian beam to a single mode fiber with core radius $a = 5\mu$m, $\lambda = 1.3\mu$m, and normalized frequency $V = 1.8$. Use Eq. 5.53 to determine the input Gaussian beam size. Calculate the coupling efficiency by numerically evaluating the overlap integral of the true mode of the fiber.

12. Prove that if Eq. 11.38 is satisfied, the beam coupled from the diffraction grating into the substrate will be totally internally reflected by the substrate-air interface on the bottom of the substrate.

13. A diffraction grating is designed for operation in the head of a CD player for coupling light from a laser onto the probe head. Assume the head is

made from a polymer waveguide on glass, with substrate index $n_s = 1.45$, waveguide index $n_f = 1,5$, and cover index $n_c = 1$. The laser has a design wavelength of 750 nm.

(a) What grating period, Λ, should be chosen if the laser is to be coupled into the waveguide from directly above, so the angle of incidence is $90°$?

(b) Assume the laser beam has a diffraction angle of $2°$. Over what range of wavelengths will the waveguide still couple light from the laser, assuming that all light enters within $\pm 2°$ of normal?

(c) If the grating has a spectral acceptance width of 5 nm, meaning that it will couple light at normal incidence over a spectral range of 5 nm, what is the maximum angular mis-alignment that the laser can be mounted at, and still couple light to the waveguide?

14. A prism is to be used to couple light onto a waveguide with thickness 5μm, $n_s = 1.55$, $n_f = 1.65$, and $n_c = 1$. If the prism has index $n_p = 1.7$, what angle(s) should the light be sent into the prism in order to couple to the TE mode(s)?

Chapter 12

WAVEGUIDE MODULATORS

1. Introduction

There are two common methods for encoding a signal onto an optical beam: either directly modulate the optical source, or externally modulate a continuous wave optical source. Direct modulation is the most widespread method of modulation today, but it introduces demanding constraints on the semiconductor lasers. For example, it is difficult to directly modulate a semiconductor laser at frequencies above a few GHz. Furthermore, it is difficult to maintain single mode operation of these pulsed lasers. Non-single-mode lasers have a larger spectral bandwidth which leads to increased pulse spreading due to dispersion. External modulators offer several advantages over direct modulation. First, one can use a relatively simple and inexpensive continuous wave laser as the primary optical source. Second, since a modulator can encode information based on a number of externally controlled effects, it is not compromised by the need to maintain a population inversion or single mode control. Finally, direct phase modulation (for FM or PM systems) is possible in external modulators, but is nearly impossible to achieve in a laser.

Fig. 12.1 shows a modulator in an optical system . A continuous wave laser couples through the modulator onto an optical fiber. The laser can be a simple and inexpensive source, since the burden of information encoding is placed on the modulator. Separating the generation and the modulation functions between two devices often makes the system work better, although it adds to the system complexity.

In this chapter, we will examine two methods for modulation. For communications and high speed links, *electro-optic* thin-film modulators are today's choice. For signal processing and detection, *acousto-optic* modulation is used. Most physical processes that can be exploited to modulate light require a sig-

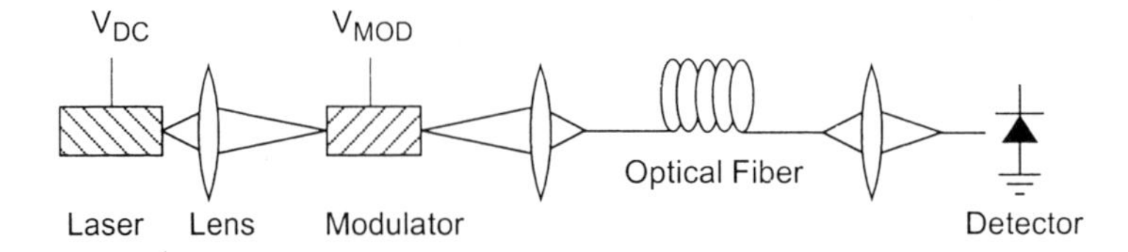

Figure 12.1. An intensity modulator installed between a continuous wave laser and an optical fiber.

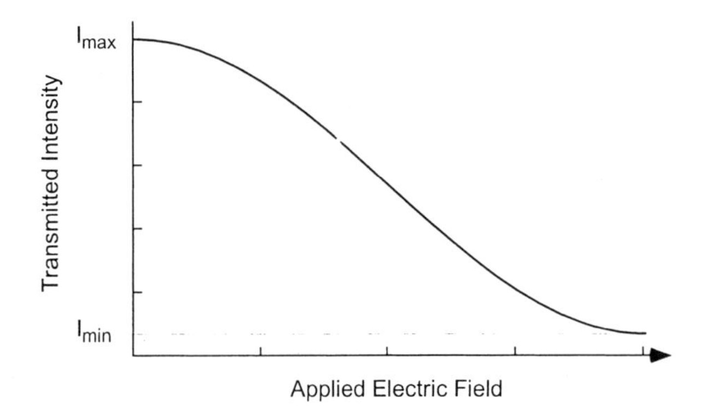

Figure 12.2. The transmission of a modulator can be described in terms of the minimum and maximum intensity transmitted through the device. The transmission of an electro-optic modulator is a function of applied field.

nificant power per unit volume. By miniaturizing the interaction volume using thin films and guided wave optics, the required modulation power can be significantly reduced. Integration often also leads to increased speed.

2. Figures-of-Merit For a Modulator

There are five basic parameters used to characterize a modulator: modulation efficiency, bandwidth, insertion loss, power consumption, and isolation between different channels.

The *Modulation Efficiency*, η, depends on the form of modulation. For intensity modulation, η is defined as

$$\eta = \frac{I_{max} - I_{min}}{I_{max}} (\times\ 100\%) \tag{12.1}$$

where I_{max} is the maximum transmitted light, and I_{min} is the intensity transmitted when the modulator is adjusted for minimum transmission. An example is shown in Fig12.2.

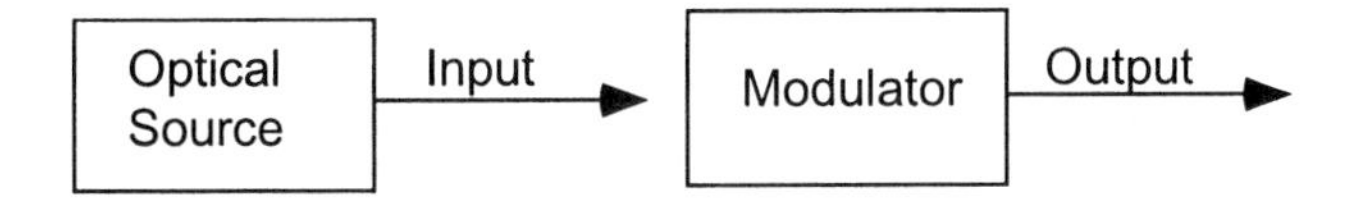

Figure 12.3. The modulator will introduce some passive losses between the optical source and the end user of the light. This is due to reflections, absorption and scattering in the modulator, and mode mismatch between the source, modulator, and waveguide, if used.

Often the modulation depth is described in decibels, using the term *Contrast Ratio*

$$\text{Contrast Ratio} = 10 \log \frac{I_{max}}{I_{min}} \tag{12.2}$$

Electro-optic modulators can be configured for either intensity or phase modulation. Eq. 12.1 is relevant to intensity modulation. For phase modulators, the modulation index is $\eta = \sin^2(\Delta\phi/2)$, where $\Delta\phi$ is the extreme value of the phase modulation. This form of η describes the intensity contrast derived from an interferometric measurement of the phase shift.

Modulation bandwidth, $\Delta\nu$, is defined by the 3 dB points of the frequency transfer function for the modulator (i.e. the frequencies where the modulation index is reduced to 50% of its maximum value). Bandwidth establishes the maximum information transfer rate for a modulator. If the switching time, τ, is defined instead of a frequency bandwidth, then the equivalent bandwidth is

$$\Delta\nu = \frac{0.35}{\tau} \quad \text{Hz} \tag{12.3}$$

where τ is the *10-to 90-percent rise time*.

Insertion loss, L, describes the fraction of power lost when the modulator is placed in the system. The insertion loss does not include the additional modulation losses induced by the modulator. The definition is

$$L = 10 \log \frac{P_{out}}{P_{in}} \tag{12.4}$$

where P_{out} is the transmitted power of the system when the modulator is not in the beam, and P_{in} is the transmitted power when the modulator is placed in the beam and adjusted to provide maximum transmission.

Do not confuse *insertion loss* with modulation index. The insertion loss is a passive loss, arising from reflections, absorption, and imperfect mode coupling between the modulator and source. Insertion loss must be compensated with either a higher power optical source, a more sensitive detector, or an optical amplifier. All these schemes are less than optimum, so insertion loss should be minimized. Insertion loss does provide a crude isolation between the optical source and any reflections coming back from the destination of the light, however there are better ways to eliminate this coupling.

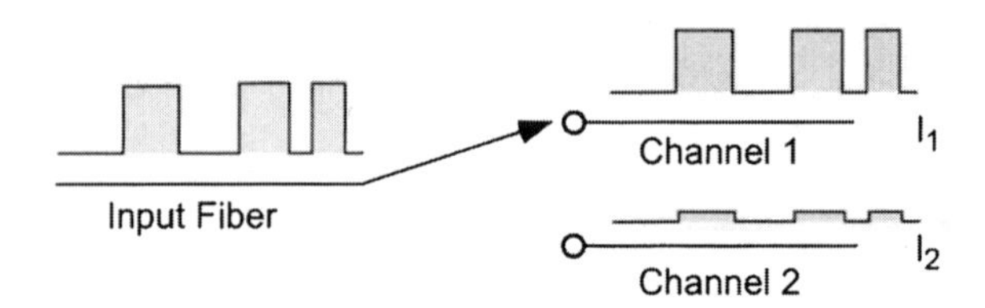

Figure 12.4. An optical switch can connect one of several ports. Isolation is the degree to which unconnected ports are coupled.

Power consumption is determined by the power per unit bandwidth required for intensity modulation, or in the case of phase modulators, the power per unit bandwidth per unit radian of modulation. This value depends on the electro-optic or acoustooptic properties of the material, but is most affected by the volume of the modulator. Waveguide modulators with small effective volumes have better power consumption performance than bulk modulators. Total power consumption determines how many devices can be put on a single substrate before thermal loading or power supply loading becomes a serious problem.

The final issue, *isolation*, describes how effectively a signal is isolated between two unconnected channels. Fig. 12.4 shows a switch with one input and two outputs. The signal from the Input connects to Channel 1 of the output. Isolation in this case describes how much of the Input signal appears on Channel 2. Ideally there would be no feedthrough from the input to the unselected channel. Unfortunately, there is always some coupling due to evanescent fields, scattering, or unwanted reflections. Isolation is specified in decibels,

$$\text{Isolation [dB]} = 10 \log \frac{I_2}{I_1} \tag{12.5}$$

where I_1 is the intensity in the driven channel, and I_2 is the intensity in the unselected, or *off* channel. A switch which coupled 0.1% energy between the two channels would have an isolation of 30 dB. The degree to which isolation is required depends on the application. Local Area Networks (LANs) sometimes specify isolation in excess of 40 dB.

3. Electrooptic Modulators and the Electrooptic Effect

In 1875, Kerr discovered that amorphous, optically isotropic material became birefringent in a strong electric field, with the optic axis parallel to the applied field [1]. You may recall the optical Kerr effect that we discussed in Chapter 7 concerning self phase modulation in optical fibers. About 20 years later, Pockel discovered a similar but much weaker effect in certain crystals[2]. Isotropic crystals became uniaxial in the presence of electric fields. Uniaxial crystals become biaxial. In the Kerr effect, the polarization depends quadratically on

E. The Pockel's effect is linear in **E**, and is therefore better suited for making modulators. Symmetry arguments can be applied to show that the crystal must *not* possess inversion symmetry in order to display a Pockel's effect.

3.1 The Propagation of Electromagnetic Waves in an Anisotropic Medium

Before we describe the Pockel's effect, and how it is used in modulators, it is necessary to understand how light travels through an anisotropic crystal. This all depends on the index of refraction, and the key point we will make is that the index of refraction is a function of both the propagation direction and the wave polarization.

In an anisotropic crystal, the electric displacement vector, **D**, is related to the electric field by a dielectric tensor,

$$\mathbf{D}_i = \epsilon_{ij}\mathbf{E}_j \tag{12.6}$$

where the subscripts represent cartesian coordinates. It can be shown through energy conservation arguments[3] that

$$\epsilon_{ij} = \epsilon_{ji} \tag{12.7}$$

so there are only six possible values of ϵ_{ij}. Unlike the isotropic media, the electric field, **E**, and displacement vector, **D**, are not necessarily parallel.

This difference in direction between the electric field and displacement vector has a major effect in the propagation of a wave through a crystal. The power in a beam follows the Poynting vector

$$\mathbf{S} = \mathbf{E} \times \mathbf{H} \tag{12.8}$$

and keep in mind that the Poynting vector, **S**, travels perpendicular to both **E** and **H**. Maxwell's equations for a single frequency plane wave can be written as

$$\begin{aligned} \mathbf{k} \times \mathbf{E} &= \omega\mu_0\mathbf{H} \\ \mathbf{k} \times \mathbf{H} &= -\omega\mathbf{D} \end{aligned} \tag{12.9}$$

Eliminating **H** from the equations yields

$$\mu_0\omega^2\mathbf{D} = k^2\mathbf{E} - (\mathbf{k}\cdot\mathbf{E})\mathbf{k} \tag{12.10}$$

where the term $(\mathbf{k}\cdot\mathbf{E})$ does not necessarily equal zero. Therefore, since the power flows perpendicular to **E**, the power does not travel in the same direction as **k**. The angle between **k** and **S** is a complicated function of the susceptibilities. This effect leads to "walk-off" of the power from the phasefront direction, and is responsible for the distortion seen in double-refracting crystals.

To describe electro-optic modulation we want to develop a general formalism for describing plane wave propagation through an anisotropic crystal. It is our goal to describe the effective index of refraction that an arbitrary plane will "see" inside a crystal. The next few paragraphs will derive the expression for the *Index Ellipsoid* which is the essential tool for wave calculations. The dis-interested reader can jump ahead to the gist of the discussion, Eq. 12.16.

The stored electric energy in the medium is

$$w = \frac{1}{2}\mathbf{E} \cdot \mathbf{D} = \frac{1}{2}E_i \epsilon_{ij} E_j \tag{12.11}$$

where ϵ_{ij} is the dielectric tensor for the medium. Using cartesian coordinates, and expanding the stored energy term yields

$$2w = \epsilon_{xx}E_x^2 + \epsilon_{yy}E_y^2 + \epsilon_{zz}E_z^2 + 2\epsilon_{yz}E_yE_z + 2\epsilon_{xz}E_xE_z + 2\epsilon_{xy}E_xE_y \tag{12.12}$$

This expression can be simplified if we use the *principal dielectric axes*, which depend on the crystal structure. For many crystals these axes lie along the familiar x, y, and z axes, while others lie in non-orthogonal directions. Examples that follow will illustrate some of these orientations. The principal axes are the orientations where an applied electric field, E, produces a *parallel* displacement, D. The principal axes are found by diagonalizing the dielectric tensor. In terms of the principle axes, x', y', and z', the energy is defined as

$$2w = \epsilon_{x'}E_{x'}^2 + \epsilon_{y'}E_{y'}^2 + \epsilon_{z'}E_{z'}^2 \tag{12.13}$$

Recasting this in terms of the displacement vector,

$$2w\epsilon_0 = \frac{D_{x'}^2}{\epsilon_{x'}/\epsilon_0} + \frac{D_{y'}^2}{\epsilon_{y'}/\epsilon_0} + \frac{D_{z'}}{\epsilon_{z'}/\epsilon_0} \tag{12.14}$$

Using the little-known identity, $\mathbf{D}/\sqrt{2\omega\epsilon_0} = \mathbf{r}$, Eq. 12.14 becomes

$$\frac{x'^2}{n_{x'}^2} + \frac{y'^2}{n_{y'}^2} + \frac{z'^2}{n_{z'}^2} = 1 \tag{12.15}$$

This expression describes the *Index Ellipsoid*. The index ellipsoid is a locus of points which form a 3-dimensional ellipse. The distance from the origin to the surface of the ellipse is equal to the index of refraction for an electric field polarized along that direction.

To find the effective indices of refraction for a beam we follow this recipe: draw a line parallel to the **k** vector through the origin of the index ellipsoid. A plane wave travelling along this direction will have an electric field polarized perpendicular to the **k** vector, so it will lie somewhere in the plane perpendicular to **k**, as shown in Fig. 17.6. The index of refraction experienced by the wave depends on the orientation of the polarization.

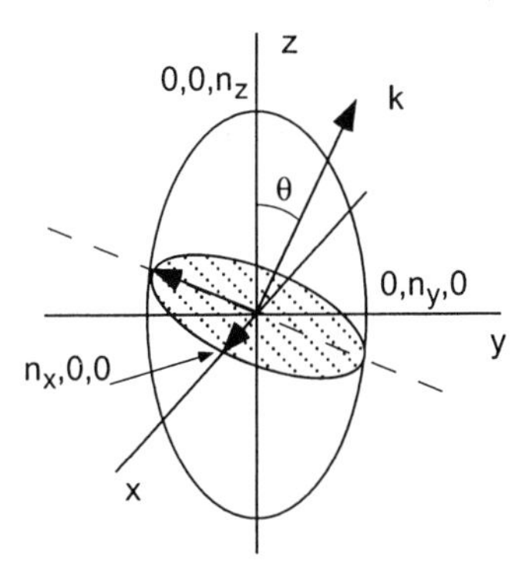

Figure 12.5. The index ellipsoid is a three dimension ellipse. A wave propagating in the yz plane makes an angle θ to the z-axis. A plane perpendicular to the ray intersects the walls of the ellipse.

A special case occurs if the field is polarized along one of the principle axes. Consider the case of a wave polarized in the x-direction, travelling in the yz plane. As the angle θ varies, the width in the x-direction remains constant, indicating that the index, n_x, is independent of angle θ. This is called the *ordinary wave*. If the polarization lies in the yz plane, then the index depends on the angle θ, ranging from n_y when $\theta = 0$, to n_z when $\theta = 90°$. This is the *extraordinary wave*. The extraordinary index that such a wave will experience is given by

$$\frac{1}{n_{ext}^2} = \frac{\cos^2\theta}{n_y^2} + \frac{\sin^2\theta}{n_z^2} \tag{12.16}$$

For the ellipse shown, the ellipse major and minor axes lie along the x, y, and z directions. The most common example of an anisotropic medium is called a *uniaxial* crystal, where the index of refraction is identical along two axes. This is called the *ordinary index*, n_o. The index of refraction along the third axis is called the *extraordinary index*, n_e. Common examples of uniaxial crystals include quartz and sapphire. Biaxial crystals have three unique indices of refraction. Examples include minerals such as calcite, tourmaline, and forsterite.

So what does all of this have to do with wave propagation? A plane wave in a crystal has two polarization eigenstates. An eigen-polarization is one that does not change as it propagates. In this case, if the field is linearly polarized along either axis of the ellipse, it will remain linearly polarized. Fields with polarizations that do not lie along the major or minor axis will not remain linear polarized. The electric field will be decomposed into two linear polarized components oriented along each axis. These components will travel separately,

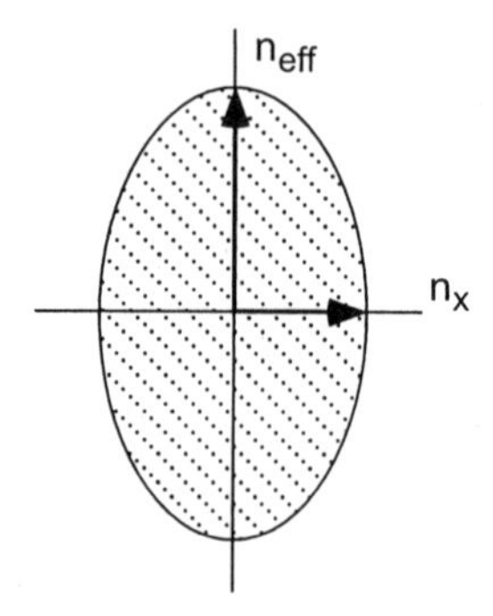

Figure 12.6. The cross-section depicted in Fig. 12.5 is plotted in 2-dimensions. The two directions correspond to eigenstates for the polarization: a wave polarized along either axis will remain in that axis. Polarizations that are not parallel to the major or minor axis will change as they propagate through the crystal.

accumulating phase according to the index of the axis. The general field will be elliptically polarized as it propagates through the crystal.

3.2 The Pockel's Effect

The linear electro-optic effect, or Pockel's effect, is the change in the index of refraction that occurs when an external electric field is applied to a crystal. The magnitude of the change is critically dependent on the orientation of the electric field and crystal. Since we are interested in how the Pockel's effect will alter the propagation through a crystal, it is most convenient to describe the effect in terms of the general modified index ellipsoid

$$\left(\frac{1}{n^2}\right)_1 x^2 + \left(\frac{1}{n^2}\right)_2 y^2 + \left(\frac{1}{n^2}\right)_3 z^2 + \left(\frac{1}{n^2}\right)_4 2yz + \left(\frac{1}{n^2}\right)_5 2xz + \left(\frac{1}{n^2}\right)_6 2xy = 1 \qquad (12.17)$$

where $(1/n^2)_i$ represents the appropriate dielectric tensor terms along the regular cartesian coordinates. If x', y', and z' are chosen to be the principle axis, the terms of Eq.12.17 reduce (for $E = 0$) to those of Eq.12.15, so

$$\left(\frac{1}{n^2}\right)_1 = \frac{1}{n_x'^2}, \quad \left(\frac{1}{n^2}\right)_2 = \frac{1}{n_y'^2}, \quad \left(\frac{1}{n^2}\right)_3 = \frac{1}{n_z'^2}$$
$$\left(\frac{1}{n^2}\right)_4 = \left(\frac{1}{n^2}\right)_5 = \left(\frac{1}{n^2}\right)_6 = 0 \qquad (12.18)$$

Since the applied electric field modifies the index of refraction, we define the change of the index ellipsoid in terms of *electro-optic coefficients*, r,

$$\begin{bmatrix} \Delta\left(\frac{1}{n^2}\right)_1 \\ \Delta\left(\frac{1}{n^2}\right)_2 \\ \Delta\left(\frac{1}{n^2}\right)_3 \\ \Delta\left(\frac{1}{n^2}\right)_4 \\ \Delta\left(\frac{1}{n^2}\right)_5 \\ \Delta\left(\frac{1}{n^2}\right)_6 \end{bmatrix} = \begin{bmatrix} r_{11} & r_{12} & r_{13} \\ r_{21} & r_{22} & r_{23} \\ r_{31} & r_{32} & r_{33} \\ r_{41} & r_{42} & r_{43} \\ r_{51} & r_{52} & r_{53} \\ r_{61} & r_{62} & r_{63} \end{bmatrix} \cdot \begin{bmatrix} E_1 \\ E_2 \\ E_3 \end{bmatrix} \tag{12.19}$$

The matrix, r_{ij}, is called the electro-optic tensor. Unlike the dielectric tensor, even if the axes are aligned along the principle axes, the cross terms (elements 4, 5, and 6) are not necessarily zero. Crystals with an inversion symmetry will have all r coefficients identical to zero. In fact, due to the high symmetry of most crystals, most values of the electro-optic tensor will be equal to zero. For example, GaAs, which is a cubic crystal with $\bar{4}3m$ symmetry, has an electro-optic tensor in the form

$$r_{ij} = \begin{bmatrix} 0 & 0 & 0 \\ 0 & 0 & 0 \\ 0 & 0 & 0 \\ r_{41} & 0 & 0 \\ 0 & r_{41} & 0 \\ 0 & 0 & r_{41} \end{bmatrix} \tag{12.20}$$

We will illustrate how to use these in the following example. The form of the electro-optic tensor can be determined strictly from a knowledge of the crystal symmetry. The magnitudes of the coefficients are determined through molecular polarizability calculations or experimental measurement. Table 17.1 lists the non-zero electro-optic coefficients for some relevant materials. Note that the values depend on wavelength. More extensive tables are available in the references [5, 6].

Example 12.2 The electro-optic effect in GaAs

GaAs is a popular substrate for active and passive optical devices. Consider the effect of an electric field oriented along the propagation direction of a waveguide oriented along the [001] axis (z-axis) of the crystal. Let's determine how the electric field will affect the propagation of light being carried by the waveguide.

Table 12.1. Linear Electro-optic Coefficients for Some Relevant Crystals

Material	Symmetry	Wavelength (μm)	Electro-optic coefficient (10^{-12} m/V)	Index of Refraction
$LiNbO_3$	3m	0.632	$r_{13} = 9.6$	$n_O = 1.8830$
			$r_{22} = 6.8$	$n_e = 1.7367$
			$r_{33} = 30.9$	
			$r_{51} = 32.6$	
$LiIO_3$	6	0.633	$r_{13} = 4.1$	$n_O = 1.8830$
			$r_{41} = 1.4$	$n_e = 1.7376$
GaAs	$\bar{4}3m$	0.9	$r_{41} = 1.1$	$n = 3.60$
		1.15	$r_{41} = 1.43$	
KDP	$\bar{4}2m$	0.633	$r_{63} = 11$	$n_o = 1.5074$
			$r_{41} = 8$	$n_e = 1.4669$
ADP	$\bar{4}2m$	0.633	$r_{63} = 8.5$	$n_o = 1.52$
			$r_{41} = 28$	$n_e = 1.48$
Quartz	32	≈ 0.632	$r_{41} = 0.2$	$n_O = 1.54$
			$r_{63} = 0.93$	$n_e = 1.55$
$BaTiO_3$	$4mm$	≈ 0.632	$r_{33} = 23$	$n_O = 2.437$
			$r_{13} = 8$	$n_e = 2.180$
			$r_{42} = 820$	
$LiTaO_3$	$3m$	≈ 0.632	$r_{33} = 30.3$	$n_O = 2.175$
			$r_{13} = 5.7$	$n_e = 2.365$

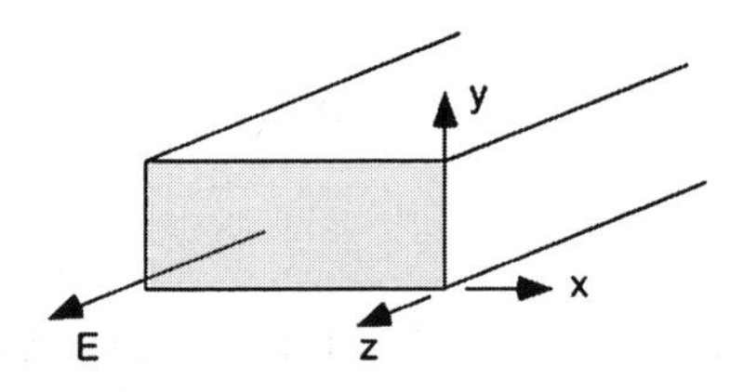

Figure 12.7. GaAs waveguide, where the modulating electric field is parallel to the axis of the waveguide.

Using Eqs. 12.17 and 12.19, and the information in Eq. 12.20, the index ellipsoid from GaAs can be written as

$$\frac{x^2}{n^2} + \frac{y^2}{n^2} + \frac{z^2}{n^2} + 2r_{41}E_xyz + 2r_{41}E_yxz + 2r_{41}E_zxy = 1$$

where the first three terms are independent of the applied field. For a z-directed field, the expression reduces to

$$\frac{x^2 + y^2 + z^2}{n^2} + 2r_{41}E_zxy = 1$$

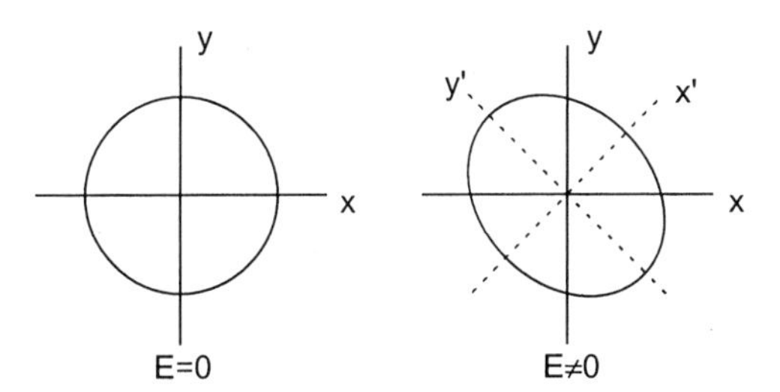

Figure 12.8. The index ellipsoid with zero field is simply a circle in the x-y plane. When a field is added in the z-direction, the ellipsoid constricts along the xy axis, and expands along the $x\bar{y}$ axis.

The effective index ellipsoid has a mixed term. This means that in the presence of a z-directed electric field, the cartesian x, y, and z coordinates are no longer the principle axes of the system (in fact, they never were, but because GaAs is isotropic with zero field, it is a matter of convenience to define the coordinate axes to lie parallel to the crystal axes). The coupled term can be removed (diagonalized) by finding a new coordinate system that lies parallel to the principle axes, x', y', and z'. The index ellipsoid in the presence of a field is shown in Fig. 12.8. We see that the ellipsoid changes from a circle when no field is present, to an ellipse rotated 45° from the x and y axes when the z-directed field is applied.

By inspection, we can identify the principal axes from Fig. 12.7, to be 45° rotated from the original axes. Remember that when the principal axes are properly chosen, there will be no terms in the ellipsoid which couple two directions. You can verify by substitution that the proper transformation to form the principle axes is

$$\begin{aligned} x &= x' \cos 45^\circ + y' \sin 45^\circ \\ y &= -x' \sin 45^\circ + y' \cos 45^\circ \end{aligned}$$

The transformed equation becomes

$$\left(\frac{1}{n^2} - r_{14}E_z\right) x'^2 + \left(\frac{1}{n^2} + r_{14}E_z\right) y'^2 + \frac{z^2}{n^2} = 1$$

Since there are no cross terms, the principle axes of the perturbed system are indeed x' and y'. The propagating field will see the index structure as shown in Fig. 12.9.

The length of the index ellipsoid along the two axes yields the effective indices of refraction:

$$\frac{1}{n_{x'}^2} = \frac{1}{n^2} - r_{14}E_z$$

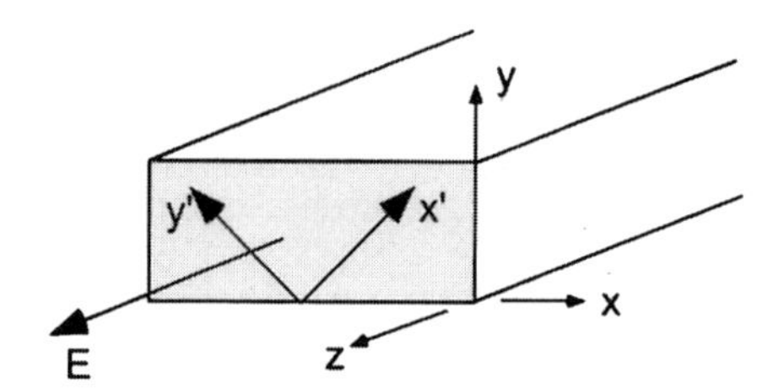

Figure 12.9. The principle axes of the perturbed waveguide are the x' and y' directions, which are rotated by 45° from the crystal axes.

$$\frac{1}{n_{y'}^2} = \frac{1}{n^2} + r_{14}E_z$$

If the magnitude of $r_{14}E_z$ is small compared to n^2, which is usually the case, the expression can be inverted and simplified using a simple binomial expansion of the quadratic terms to yield

$$\begin{aligned} n_{x'} &= n + \frac{n^3}{2} r_{14}E_z \\ n_{y'} &= n - \frac{n^3}{2} r_{14}E_z \\ n_z &= n \end{aligned}$$

We see that the principle indices are linearly modified by the applied field. We will examine how this index change affects propagation in the next section.

4. Phase Modulators

The Pockel effect makes it possible for an electric field to alter the index of refraction of a material. It is possible to construct devices which use this index change to directly modulate the phase, the intensity, or the polarization of the light. All of these effects rely on optical retardation. We will consider phase modulation first.

Consider the case of a TE wave in a GaAs waveguide. As we saw in Example 12.2, the application of an electric field in the z-direction will alter the index of refraction along the x' and y' axes. These axes lie along the [110] and [101] directions in the crystal. Because the r-coefficients in GaAs are identical, applying an electric field along any single axis will alter the indices along the remaining axes. Keeping this in mind, let's try to make a GaAs based phase modulator. To make an efficient phase modulator, we will orient the waveguide along the z'-axis, or in the [101] direction (Fig. 12.10), and place an electrode over the waveguide so that the applied electric field is oriented along the y-axis. The conductive substrate act as a ground plane. The applied field

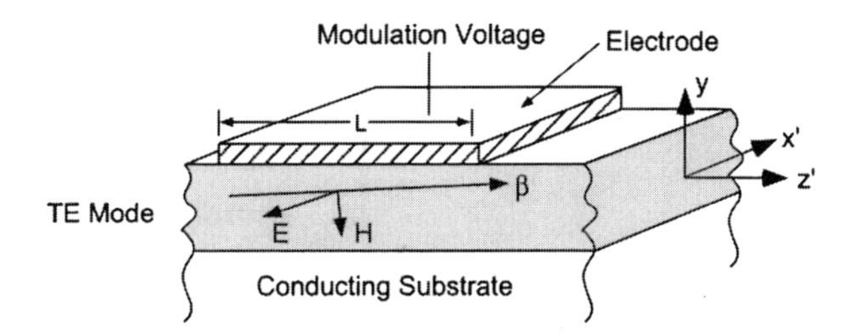

Figure 12.10. An electro-optic modulator based on GaAs. The waveguide on the doped substrate is made of low conductivity material, so essentially all of the applied field appears across the thickness t_g. The substrate is moderately doped so that it has a high conductivity. A metal electrode is applied over the waveguide for a distance, L, to create the modulator zone.

will cause the index to change in the z'- and x' directions. The index change in the z'-direction has no effect on the phase of the guided wave, because the electric field polarization does not lie in the z'-direction. However, any changes in n'_x will be felt by the TE mode. The TM mode, with its electric field predominantly polarized in the y-direction (there is a small component along the z'-direction, but in the weakly guided mode approximation, this component is almost negligible) will not be effectively modulated by the applied field.

A TE wave will experience a phase shift approximately as

$$\Delta\beta = \Delta n'_x k_0 \tag{12.21}$$

The total phase shift due to the interaction of an applied field, E_0, over a length L in this GaAs structure will be (using the results from Ex. 12.2)

$$\begin{aligned}\Delta\phi = \Delta\beta L &= k_0 L \Delta n'_x \\ &= \frac{2\pi}{\lambda}\frac{Ln^3 r_{14} E_y}{2}\end{aligned} \tag{12.22}$$

The electric field required to achieve a phase modulation of $\pi/2$ can be directly solved from Eq. 12.22 to be

$$E_{\pi/2} = \frac{\lambda}{2}\frac{1}{Ln^3 r_{14}} \tag{12.23}$$

This can also be written in terms of the applied voltage to the electrode, using $E = V/t_g$. The half-wave voltage is then

$$V_{\pi/2} = \frac{\lambda}{2}\frac{t_g}{L\, n^3 r_{14}} \tag{12.24}$$

The longer the modulation length, the lower the required voltage. We have implicitly assumed that the guided mode is entirely confined within the modulated thin film. In fact, some of the mode will exist as an evanescent wave in the substrate, and will not experience phase modulation. The effective modulation will be reduced proportional to the confinement factor, Γ, of the mode.

What are the limitations of such a phase modulator? There are two major problems. First, the bandwidth is limited by the capacitance of the electrode on the waveguide. GaAs has a large DC dielectric constant, so this capacitance can be significant. Second, the modulation only works for the TE mode. If the modulator were connected to a circular fiber, where polarization is uncontrolled, both TE and TM modes would be excited in the modulator. The absence of modulation on the TM mode would reduce the total modulation efficiency of the device. Finally, whenever the index of refraction of a material is modified, the imaginary component of the index also changes. This is a statement of the Kramers-Kronig relations[4]. The change of the imaginary component causes a change in the intrinsic attenuation of the waveguide. An applied electric field will in practice modulate both the phase and intensity of the transmitted light. Ideally, a phase modulator would not modulate the amplitude of the carrier wave. The coupling between phase and intensity must be treated as an additional source of noise on the signal. Fortunately the magnitude of the change in the imaginary component of n is small, and decreases as the wavelength gets further from the absorption edge of the material, so proper selection of materials and operating wavelengths can reduce the magnitude of this problem (see Section 12.8 on electro-absorption).

5. Power Required to Drive a Phase Modulator

The power required to drive a phase modulator can be directly calculated with some simple approximations. Assume that a digital signal is being transmitted. Modulation of the signal to send one bit requires an amount of energy, W, which is stored in the capacitor formed by the electrode and substrate. The power required to send a signal will depend on the fraction of marks and spaces, but is approximately

$$P = W \cdot \Delta f \tag{12.25}$$

where Δf is the data rate, or bandwidth of the signal. If the electro-optic modulator is ideal (no ohmic losses), the energy will be stored in the electrostatic field

$$W = \frac{\epsilon}{2} \int_V E_0^2 \, dV \tag{12.26}$$

where V is the volume of the device. For simple structures, E_0 will be constant over the volume of the modulator. The integral is then simple to evaluate

$$W = \frac{\epsilon}{2} H L t_g \, E_0^2 \tag{12.27}$$

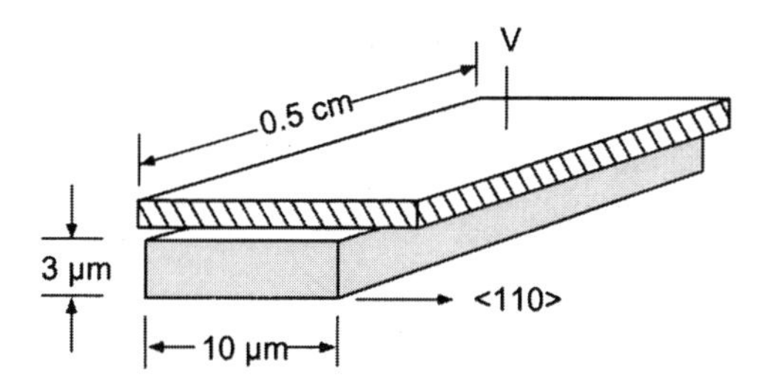

Figure 12.11. A simple phase modulator made by placing an electrode over a 3 by 10 μm waveguide.

where H, L, and t_g are the height, length, and thickness of the active region of the modulator. The required power is then

$$P = \frac{\epsilon}{2} H L t_g E_0^2 \Delta f \tag{12.28}$$

To illustrate in a GaAs modulator, if we assume that sending a mark or space requires changing the phase by $\pi/2$, then in terms of the $\pi/2$ field of Eq.12.24, the power per unit bandwidth is

$$\frac{P}{\Delta f} = \frac{\epsilon}{8} \frac{H\, t_g}{L} \frac{\lambda^2}{n^6 r_{14}^2} \tag{12.29}$$

Example 12.3 Electro-optic phase modulator: voltage and power considerations

To illustrate the power required to operate a modulator, let's reexamine the TE rectangular waveguide described above, oriented along the [110] axis on a GaAs substrate. Fig. 12.11 shows the structure. The waveguide has the following characteristics:

$$\begin{aligned} n &= 3.6 \frac{\epsilon}{\epsilon_0} = 12 \\ r_{41} &= 1.1 \times 10^{-12} \text{ m/V} \\ H &= 10 \mu\text{m} \\ t_g &= 3 \mu\text{m} \\ L &= 0.5\text{cm} \\ \lambda &= 0.9 \mu\text{m} \end{aligned}$$

The half-wave voltage required to achieve a $\pi/2$ phase-shift in this modulator is given by Eq.12.24

$$
\begin{aligned}
V_{\pi/2} &= \frac{\lambda}{2}\frac{t_g}{L\, n^3 r_{14}} \\
&= \frac{0.9\times 10^{-6}m}{2}\frac{3\times 10^{-6}m}{(5\times 10^{-3}m)\,(3.6)^3\,(1.1\times 10^{-12}m/V)} \\
&= 5.26V
\end{aligned}
$$

The applied field would need a peak amplitude of 5.26 V to delay or advance the phase of the TE carrier wave by $\pi/2$ radians. This corresponds to an electric field of $E = 1.73\times 10^6$ V/m across the 3 μm film. The power required to create this modulation is given by Eq. 12.28

$$
\begin{aligned}
\frac{P}{\Delta f} &= \frac{\epsilon}{8}\frac{H\,t_g}{L}\frac{\lambda^2}{n^6 r_{14}^2} \\
&= \frac{12\cdot 8.85\times 10^{-12}}{8}\frac{10\times 10^{-6}\cdot(3\times 10^{-6})}{5\times 10^{-3}}\frac{(0.9\times 10^{-6})^2}{3.6^6(1.1\times 10^{-12})^2} \\
&= 2.45\times 10^{-11}\ \text{W/Hz} = 24.5\ \mu\text{W/MHz}
\end{aligned}
$$

This value should be compared to that required to drive a bulk (non-waveguide) modulator. We keep the modulator length the same as in the example ($L = 0.5$ cm). To minimize the size of the bulk modulator, the optical beam should be focussed through the crystal. The smallest that the beam can be made, and still get through the crystal without excessive losses due to vignetting, is determined by the confocal parameter, $2z_0$ of the beam.(The *confocal parameter* is the distance over which an optical beam doubles its area due to diffraction from a focal spot. It represents the region where the optical beam is the most collimated.) The beam radius, ω_0, at the focus is determined directly from the confocal parameter:

$$2z_0 = \frac{2\pi n\omega_0^2}{\lambda}$$

Setting this equal to 0.5 cm, and solving for ω_0 yields $\omega_0 = 14\mu m$. At the face of the crystal, the beam will have it's largest radius, $\omega = \sqrt{2}\omega_0 = 20\mu m$. To avoid serious aperture loss, the crystal should have a lateral dimension at least *twice* the *diameter* of the beam at the input/output faces, or 80 μm. Let's set the dimensions of the bulk modulator as being $80\times 80\times 5000\mu\text{m}^3$. Using these dimensions we find a modulation voltage (Eq. 12.24) to be

$$V_{\pi/2} = \frac{0.9\times 10^{-6}}{2}\frac{80\times 10^{-6}}{(5\times 10^{-3})\,(3.6)^3\,(1.1\times 10^{-12})} = 140V$$

The applied electric field is still the same, namely $140/80 \times 10^{-6} \approx 1.73 \times 10^6 V/m$. The power per unit bandwidth to drive the bulk modulator is then

$$\begin{aligned}\frac{P}{\Delta f} &= \frac{12 \cdot 8.85 \times 10^{-12}}{8} \frac{80 \times 10^{-6} \cdot (80 \times 10^{-6})}{5 \times 10^{-3}} \frac{(0.9 \times 10^{-6})^2}{3.6^6 (1.1 \times 10^{-12})^2} \\ &= 5.22 \times 10^{-9} \text{ W/Hz} = 5.2 \text{ mW/MHz}\end{aligned}$$

which is approximately 200 times greater than the power required for an integrated modulator. This contrast illustrates perhaps the greatest virtue of integrated modulators.

6. Electro-optic Intensity Modulators

The phase modulation introduced by the electro-optic effect can be used to create an intensity modulation via changes in polarization, and through interferometric effects.

6.1 Polarization Modulation

Polarization modulation can be achieved using the differential retardation between two orthogonal polarizations of the optical wave. Fig.12.12 illustrates a wave linearly polarized at 45° to the x-axis propagating into a birefringent crystal. The crystal has two indices of refraction, the *ordinary* index, n_o, and the *extraordinary* index, n_e, oriented along the x and y axis, respectively. The incident wave is broken into an ordinary and extraordinary component, each of which travels with a different phase velocity. As they propagate a distance L, the waves will accumulate a relative phase difference,

$$\Delta\phi = k_0 L(n_x - n_y) \tag{12.30}$$

The superposition of the two waves will in general describe an elliptically polarized wave. When the phase difference is $\Delta\phi = \pi, 3\pi, 5\pi, \ldots$, the superposition will result in a linear polarization that is rotated by 90° relative to the input polarization. This is illustrated in Fig. 12.12, where the two waves emerge with the proper delay such that the superposition of the two waves produces a polarization that is orthogonal to the input polarization. A relative delay of $\Delta\phi = 2\pi, 4\pi, 6\pi, \ldots$ will restore the wave to its original state of polarization.

If the crystal is electro-optic, then application of an electric field along one axis of the crystal can lead to an electronically-controlled relative phase shift between the ordinary and extraordinary waves. This effect can be used to control (modulate) the polarization through the Pockel's effect. The crystal need not be birefringent to begin with (for example, GaAs), but the applied field must introduce a relative retardation between two components of the field. Note that it is essential that the input light excite both the ordinary and extraordinary

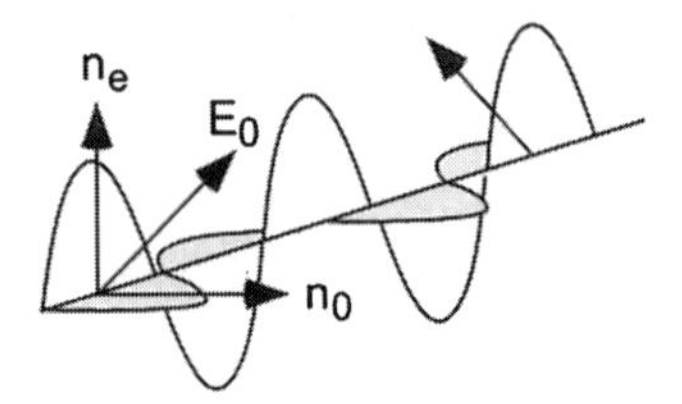

Figure 12.12. A linearly polarized wave at 45° to the polarization axes of the crystal will travel as two waves, an *ordinary wave*, and an *extraordinary wave*. They travel with different phase velocities. At certain lengths, a half-wave retardation exists where the two fields add to form a linear polarized wave rotated by 90 ° from the input.

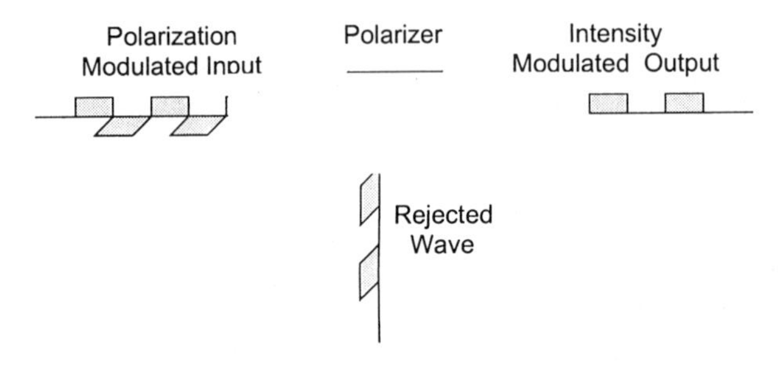

Figure 12.13. A polarization-modulated input beam is split into vertically and horizontally polarized components by passing through a linear polarizer.

fields. If only one of these is excited, there will be a phase delay introduced by the external field, but there will be no polarization rotation.

To convert this polarization rotation into an intensity modulation, it is necessary to run the output through a linear polarizer, also known as an *analyzer*. The analyzer transmits only one polarization component, either ejecting the other component into another direction or attenuating it. As the polarization is electro-optically rotated over 90°, the transmitted intensity will continuously vary from maximum to zero. Fig. 12.13 illustrates an optical system that converts polarization modulation into intensity modulation.

Polarization modulation has not been widely adopted in optical communication systems for a variety of reasons. A major problem is that most materials that are electro-optic are also naturally birefringent ($n_x \neq n_y$). This natural birefringence usually overwhelms the induced birefringence of the electro-optic effect ($\Delta n_{pockels}$ is on the order of 10^{-5} for reasonable fields, while $\Delta n_{birefringence}$ is typically 10^{-2}), and this makes choosing the absolute length of the modulator extremely critical. Second, the indices of these crystals are temperature dependent, and in general do not track each other. Any small change in temperature can lead to a polarization drift. Third, there are not many integrable polarizers

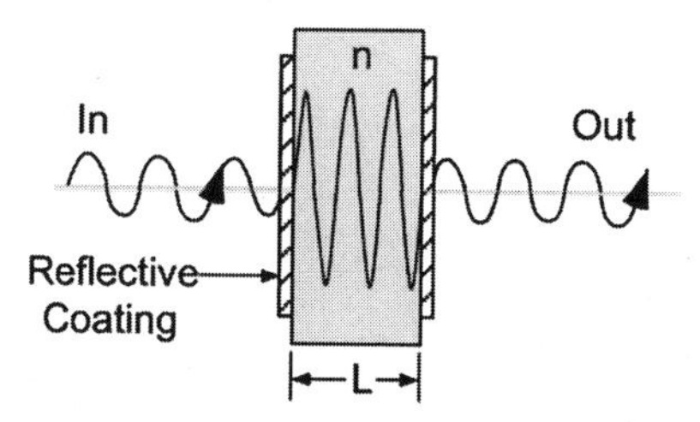

Figure 12.14. A solid Fabry-Perot etalon is made by putting a reflective coating on the facets of the substrate. At resonance, the wave intensity builds inside the mirrors.

yet available. Certain types of polarization maintaining fiber can now be used in fiber optic systems, but for pure integrated optics applications, the lack of a simple polarizer remains a problem.

7. Interferometric Modulators

Phase modulation can be converted into an intensity modulation through constructive interference between two waves. The Fabry-Perot interferometer and the Mach-Zender interferometer represent two examples for converting phase modulation into intensity modulation.

7.1 Fabry-Perot Modulators

The Fabry-Perot interferometer is commonly used in high resolution spectroscopy and for tuning of lasers. Fig. 12.14 shows the construction of a typical Fabry-Perot interferometer, also known as an *etalon*. Two partially transmitting mirrors are aligned parallel to one another, separated by a distance L. An etalon is constructed using either two mirrors separated by an air gap, or by coating reflective coatings on the parallel faces of a dielectric. Fig. 12.14 shows the latter case. The material has an index of refraction n. Transmission through the Fabry-Perot is maximum when the optical path length between the two mirrors is equal to an integer number of half-wavelengths, $L = m\lambda/2n$, where m is an integer, and the effective wavelength in the material is λ/n.

For non-integer wavelength separations, the transmission is given by

$$T = \frac{1}{1 + \frac{4R}{(1-R)^2}\sin^2\left(\frac{4\pi}{\lambda}nL\right)} \tag{12.31}$$

A plot of the transmission function for several reflectivities is shown in Fig. 12.15. The selectivity of the Fabry-Perot increases with reflectivity, R. For a reflection equivalent to the Fresnel reflection from fused silica (approximately 4% per surface) the modulation depth is approximately 8% between the minimum and maximum transmission points. For reflectivity of 30% (equivalent

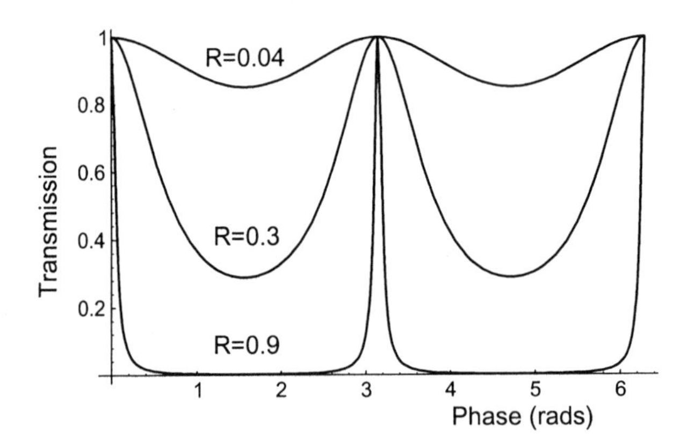

Figure 12.15. Transmission peaks for R=0.04, 0.3, and 0.9. The horizontal axis is the phase shift in radians that a wave accumulates per round trip.

to the Fresnel reflection from GaAs), the modulation depth is close to 80%. Generally, the reflectivity is adjusted using dielectric coatings to achieve the desired contrast ratio. As the reflectivity increases, the contrast between maximum and minimum transmission increases, while the width of the transmission peak becomes narrower.

This interferometer can operate as an intensity modulator by making the material between the mirrors electro-optic. When an electric field is applied to the material, the index is modified, which changes the effective optical path length between the mirrors. The goal in making a modulator is to switch the device from sitting at T_{max} to T_{min} by application of an external field. Fig.12.16 shows a schematic of one such device. The Fabry-Perot is made of a semiconductor material (GaAs) that is electro-optic. The entire structure is made through epitaxial growth of semiconducting material. First, a stack of alternating high/low index layers of $\lambda/4$ thickness is grown on the substrate. A bulk layer of GaAs is then grown on top of the dielectric stack to the desired thickness to achieve the proper spacing between transmission maxima. A second dielectric stack is grown on top of the bulk layer to form the final reflector. The entire structure can be fabricated in a epitaxial growth facility.

If an electric field is applied across the bulk layer, the index of the material changes, which shifts the transmission peak to a different wavelength. The spectral distance between two adjacent transmission maxima is called the *Free Spectral Range*, (FSR). A properly designed modulator must have a FSR that is larger than the bandwidth of the modulated signal. Equally important, the spectral transmission at resonance must be broad enough to transmit the entire signal, and not reject a significant portion of it due to the spectral filtering properties of the Fabry-Perot. The FSR of a Fabry-Perot etalon is straightforward to

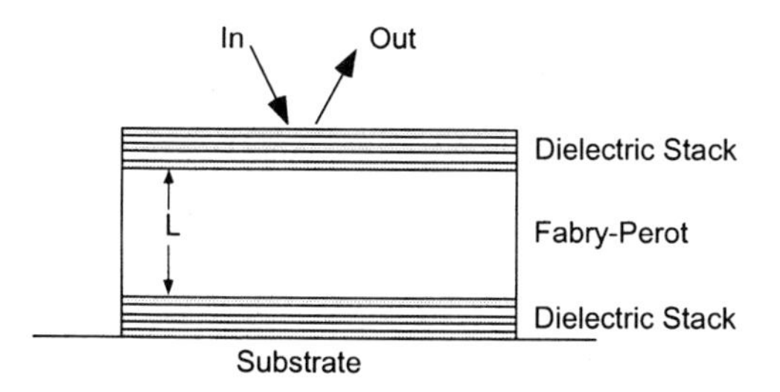

Figure 12.16. A Fabry-Perot modulator.

derive based on the requirement that the optical path length between the mirrors be an integer number of half-wavelengths. The FSR is defined as

$$\text{FSR} = \frac{c}{2nL} \tag{12.32}$$

The spectral bandwidth of an etalon depends on the FSR and on the reflectivity of the mirrors, as is apparent from Fig. 12.15. The resolution is often defined in terms of *finesse*, which is

$$F = \frac{\text{FSR}}{\text{Full width at Half Maximum}} = \frac{\pi(R_1R_2)^{(1/4)}}{1-(R_1R_2)^{1/2}} \tag{12.33}$$

To reduce the insertion loss for this type of modulator, the output from this device is usually taken from the reflection off the front surface. In anti-resonance, the reflection is nearly 100%, while at resonance the reflection is minimum. Devices to date have demonstrated contrast ratios up to 10 dB, at frequencies exceeding 10 GHz.

7.2 Mach-Zender Modulators

In a waveguide structure, the Mach-Zender interferometer can use interference between two waves to convert phase modulation into intensity variation. Fig. 12.17 shows a schematic Mach-Zender interferometer. The single mode waveguide input is split into two single mode waveguides by a 3 dB Y-junction. The split beams travel different paths of length l_1 and l_2, and then recombine at another Y-junction. If the optical path lengths of the two arms have an integer number of optical wavelengths, the two waves will arrive at the Y-junction in-phase, and constructively interfere. They will combine into a guided mode which then propagates down the remaining waveguide. If the optical path lengths are unequal, and the relative phase difference between the two combining beams is $\pi/2$, the two beams will destructively interfere and not couple into the following single mode waveguide.

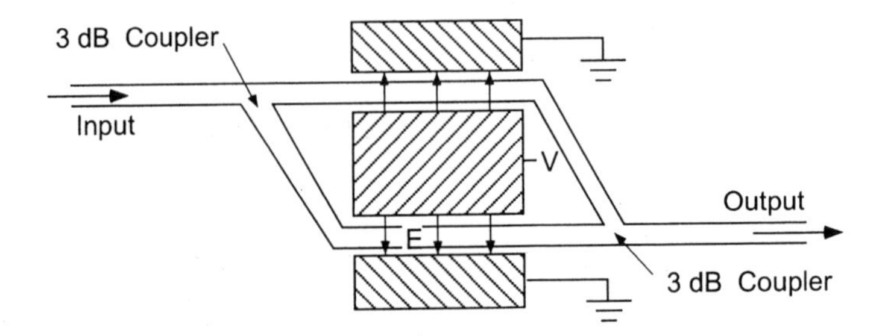

Figure 12.17. The Mach-Zender interferometer is made with two 3 dB couplers which split and recombine the beam, and two arms which can be modulated by application of an electric field.

The relative phase difference of the two beams can be electro-optically controlled by applying a voltage to the center electrode in the structure shown in Fig. 12.17. Because the change in index, Δn, depends on the direction of the crystal and the applied electric field, appropriate choice of the crystal axes will cause the applied field to increase the index in one arm, and decrease it in the other arm. This differential change in index is used to alter the relative phase of the recombining fields.

Results with Mach-Zender interferometers made from Lithium Niobate have demonstrated contrast ratios exceeding 20 dB with bandwidths exceeding 5 GHz. Much of the challenge in getting good contrast rations is caused by non-ideal 3 dB couplers. If the intensities of the two beams are not exactly balanced, the interference will not cancel entirely. Since there is only one output beam, isolation and contrast ratio are the same parameter for these devices.

8. Electro-Absorption Modulators

Another way to modulate an optical field with an electric field is through electro-absorption. This type of modulator is based on the Franz-Keldysh effect, in which the absorption edge of a semiconductor shifts in the presence of an electric field. Applying a large field to a semiconductor shifts the absorption profile toward the long wavelength direction. Fig. 12.18 shows the absorption spectrum for a GaAs sample with an applied field and with no field. With no field, the absorption coefficient shows the typical increase for optical energies that equal or exceed the bandgap energy of the material. Over a range of about 50 nm, the absorption coefficient, α, increases from 20 cm^{-1} to over 10^3 cm^{-1}.

An optical signal at wavelength $\lambda_0 = 0.9\mu$m will experience an absorption coefficient of approximately 10 cm^{-1} absorption with no electric field on the sample. If an electric field of 10^5 V/m is applied, the absorption coefficient will increase to approximately 600 cm^{-1}. The total change in intensity will depend

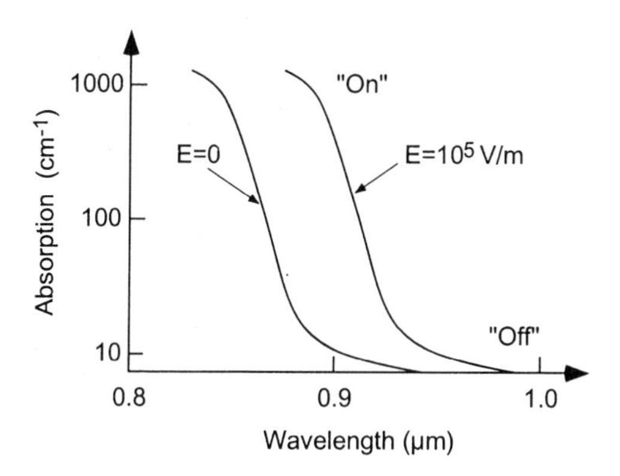

Figure 12.18. The Franz-Keldysh effect leads to a shift in the absorption bandedge for semiconductors. GaAs is shown in this example, with an applied field of 10^5 V/m.

on the path length through the modulator. Since the transmitted signal goes as

$$I(z) = I(0)e^{-\alpha z} \tag{12.34}$$

the contrast ratio will be

$$\frac{I_{max}}{I_{min}} = \frac{e^{-\alpha_1 z}}{e^{-\alpha_2 z}} \tag{12.35}$$

Devices to date have demonstrates modulation depths of up to 20 dB using this effect.

The Franz-Keldysh effect arises due to band-bending near the surface of the semiconductor. Fig. 12.19 shows the energy band diagram of a semiconductor exhibiting the Franz-Keldysh effect. The left-hand side of the diagram corresponds to the Schottky barrier contact on the surface. Application of a reverse bias causes a charge depletion layer to form. A non-uniform field is formed as the charge depletion decays away from the surface, ending in flat bands well within the material. In the flat band region, a photon can only be absorbed if its energy exceeds the bandgap potential. In the band bending region, absorption can occur for a lower energy photon. The photon lifts an electron far enough into the bandgap that it can horizontally tunnel into the conduction band. Horizontal movement in this figure requires no energy, so energy conservation is not violated. The effective change in the bandgap energy is

$$\Delta E = \frac{3}{2}(m^*)^{-1/3}(q\hbar\mathcal{E})^{2/3} \tag{12.36}$$

where E is the applied field strength, and m^* is the effective mass of the electron in the semiconductor.

A basic electro-absorption modulator is shown in Fig. 17.23. A lightly doped waveguide layer is grown on a conductive substrate, and a Schottky

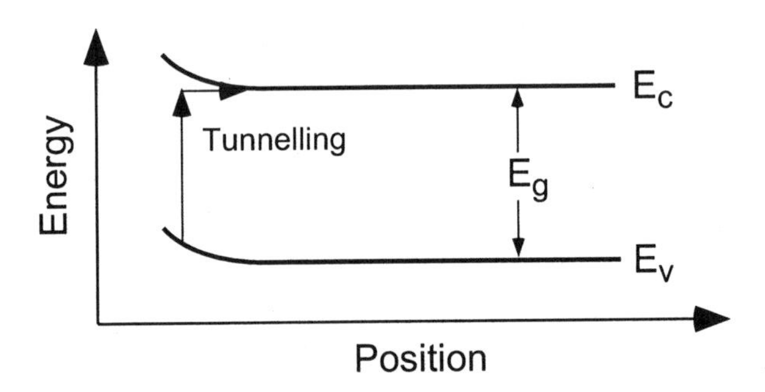

Figure 12.19. The band structure of the semiconductor becomes bent near the surface. This bandbending allows electrons to tunnel into the conduction band.

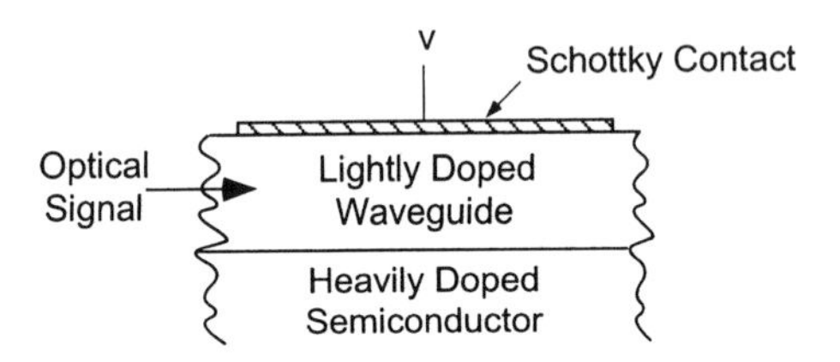

Figure 12.20. A simple electro-absorption modulator. The optical signal passes through the surface waveguide. The applied field is developed across the lightly doped waveguide.

barrier is placed on the top surface of the waveguide. Light which is slightly sub-bandgap in energy propagates down the waveguide. The length of the electrode is chosen to maximize the contrast experienced by the guided light when the field is applied. Because the absorption tails of the semiconductor die off slowly, these modulators have moderate insertion loss. 3 dB is not unusual, however when active, the attenuation can increase by 20 dB.

9. Acousto-optic Modulators

Like electro-optic modulators, acousto-optic modulators control the transmission of light by local changes in the index of refraction of the transmission medium. However, acousto-optic modulators differ from electro-optic devices in three important respects. First, the modulation occurs by means of a travelling sound wave which induces a stress-related modification of the local index. The sound wave can be transverse, longitudinal, or a combination of both, as in the case of a *surface acoustic wave* (SAW). The second difference is that acoustic interactions travel at the speed of sound in the material, while electro-optic interactions can occur at nearly the speed of light (actually they are limited by RC circuit time constants). Finally, while electro-optic interactions can be established with DC fields, acousto-optic modulation, being based on sound waves, always involves interacting with a travelling or standing wave in the solid. Good reviews of acousto-optic interaction and devices can be found in references [6, 8, 10, 11, 12].

9.1 The Photoelastic Effect

The photoelastic effect involves reflecting light off of the change of the index of refraction in a dielectric due to strain. Formally, the effect is characterized by a fourth rank tensor, p_{ijkl}, called the *photoelastic tensor*, via

$$\Delta\left(\frac{1}{n^2}\right)_{ij} = p_{ijkl}S_{kl} \tag{12.37}$$

where $1/n_{ij}^2$ is the term in the index ellipsoid, and S_{kl} is the strain, defined as

$$S_{kl}(r) = \frac{1}{2}\left[\frac{\partial u_k(r)}{\partial x_l} + \frac{\partial u_l(r)}{\partial x_k}\right] \tag{12.38}$$

The photoelastic effect is nonlinear, as it depends on the product of two interacting fields. Being nonlinear, we expect that the frequency of the scattered light can be different from either the frequency of the strain field or of the incident optical field. The frequency shift is usually attributed to a Doppler shift of the scattered light from a traveling acoustic wave. For a thorough discussion on the fourth rank photoelastic tensor, see *Nye*[9].

The acousto-optic strain interacts with an electric field component E_j to generate a polarization ΔP_i. The strain therefore causes a change in the index of refraction [10]. The change in index of refraction, Δn, is related to the acoustic power, P_a, through the relation

$$\Delta n = \sqrt{n^6 p^2 P_a / 2\rho v_a^3 A} \tag{12.39}$$

where n is the index of refraction in the unstrained medium, p is the appropriate element of the photoelastic tensor, P_a is the acoustic power (in Watts), ρ is the mass density, v_a is the acoustic velocity, and A is the cross-sectional area that the acoustic wave travels through. This expression is somewhat unwieldy, so it is often rewritten in terms of a *figure of merit*, M, defined as

$$M = n^6 p^2 / \rho v_a^3 \tag{12.40}$$

in the simplified form

$$\Delta n = \sqrt{M P_a / 2A} \tag{12.41}$$

The value of M depends on the material and, in the case of crystals, on the orientation. Table 12.2 lists some values of M for common acousto-optic materials

Table 12.2. Materials commonly used in Acousto-optic Modulators [10]

Materials	$\lambda(\mu m)$	n	$\rho(g/cm^3)$	$v_s(10^3$ m/s	M
Fused Quartz	0.63	1.46	2.2	5.95	1.51×10^{-15}
GaAs	1.15	3.37	5.34	5.15	104×10^{-15}
$LiNbO_3$	0.63	2.20	4.7	6.57	6.99×10^{-15}
YAG	0.63	1.83	4.2	8.53	0.012×10^{-15}
As_2S_3	1.15	2.46	3.20	2.6	433×10^{-15}
$PbMO_4$	0.63	2.4		3.75	73×10^{-15}

Acousto-optic modulators used in integrated optics generally use travelling wave acoustic fields. The acoustic field creates a grating structure which can diffract the incident optical field. Light reflected from the moving grating is Doppler shifted in frequency by an amount equal to $\pm mf_0$ where f_0 is the acoustic frequency, and m is the order of the reflection. Plugging numbers into the expression for Δn, using fused quartz as an example and an acoustic intensity of 100 W/cm^2, the magnitude of Δn is on the order of 10^{-4}, which is not a very large change. However, the interaction of an optical field with the strain field can be significant because the acoustic field has many periods of oscillation, so the small reflections at each crest can accumulate constructively (or destructively) if proper phase matching is arranged. Reflections approaching 100% can be generated from the grating.

Optical wave interaction can be produced by either bulk acoustic waves travelling in the volume of the material, or by surface acoustic waves (SAW) which propagates on the surface within approximately one acoustic wavelength of the surface. SAW devices are well suited to integrated optics applications, because the energy of the acoustic field is concentrated in the region of the optical waveguide.[11]

There are two basic configurations used in acousto-optic modulation. If the optical field propagates transverse to the acoustic beam, and the interaction length of the two beams is relatively short so that multiple diffraction does not occur, then Raman-Nath type diffraction occurs. If the acoustic field is large so that multiple refraction can occur, the interaction is called *Bragg* modulation. Bragg modulation tends to provide larger modulation depth, and is more commonly implemented in integrated optic devices.

9.2 Raman-Nath modulators

The basic structure of a Raman-Nath modulator is shown in Fig. 12.21. An optical field travels through a thin region of acoustic waves with spatial wavelength Λ. The phase that the optical beam accumulates on passing through

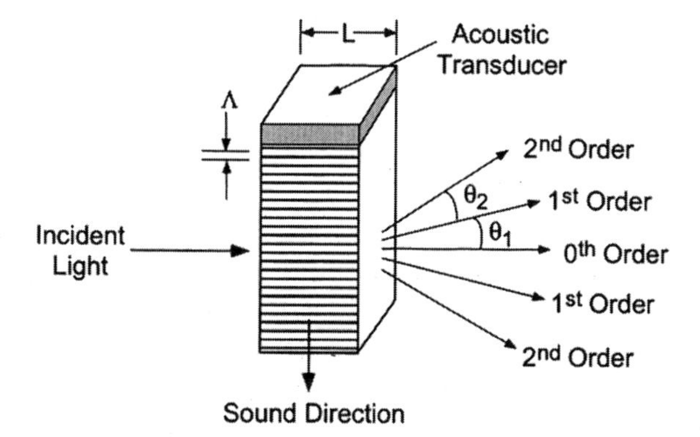

Figure 12.21. A Raman-Nath type interaction. Incident light travels through a thin region of acoustic energy, and is diffracted into a number of different orders.

the acoustic beam is

$$\Delta\Phi = \frac{\Delta n 2\pi l}{\lambda_0} \sin 2\pi y/\Lambda \tag{12.42}$$

Notice that this has a spatial component ($\sin 2\pi y/\Lambda$) which adds a corrugated structure to the phase front of the transmitted light.

There are two ways to view the scattering process. If the index structure created by the acoustic wave is considered to be a diffraction grating, the transmitted light reflects off the grating into many orders. In such a case, the grating analysis shows that the orders will leave the crystal at angles

$$\sin\theta = \frac{m\lambda}{\Lambda}, \quad m = 0, \pm 1, \pm 2, \ldots \tag{12.43}$$

The second point of view is to consider the modulated wavefront that emerges from the modulator as being a superposition of several plane waves, each directed in a different direction.

The Raman-Nath condition will hold so long as the interaction length is short enough to ensure that multiple refractions do not occur. This is the case when the interaction length, l, satisfies

$$l \ll \frac{\Lambda^2}{(\lambda/n)} \tag{12.44}$$

The intensity of the diffracted orders are described by Bessel functions[12]

$$\begin{aligned} \frac{I_m}{I_0} &= [J_m(\Delta\Phi_{max}]^2/2, \quad m\,0 \\ &= [J_0(\Delta\Phi_{max}]^2, \quad m = 0) \end{aligned} \tag{12.45}$$

where Φ_{max} is the peak amplitude of the index change, given in Eq. 12.42.

For use as a modulator, the output is usually taken as the zeroth order beam. The modulation index is then

$$\eta_{RN} = \frac{I_0 - I(m=0)}{I_0} = 1 - [J_0(\Delta\Phi_{max}]^2 \qquad (12.46)$$

A problem with the Raman-Nath modulator is that due to the short interaction length, the modulation depth is not as great as can be obtained from the Bragg modulator. So it is not typically used as a modulator in information systems. Also, since the output is spread out over several orders, it is not useful as a switch. In contrast, the Bragg modulator has been widely used as a modulator, beam deflector, and as a switch.

9.3 Bragg Modulators

The Bragg condition for scattering is satisfied when the interaction length of the acoustically-generated diffraction grating is large enough to allow multiple diffractions. Quantitatively

$$l \gg \frac{\Lambda^2}{\lambda} \qquad (12.47)$$

For optimum performance, the input angle of the optical beam should satisfy the Bragg condition

$$\sin\theta_B = \frac{\lambda}{2\Lambda} \qquad (12.48)$$

as illustrated in Fig. 12.21. For modulators, the output of the zeroth order beam is generally taken as the output. The modulation depth is then [12]

$$\frac{I_0 - I}{I_0} = \sin^2\left(\frac{\Delta\Phi}{2}\right) \qquad (12.49)$$

Devices demonstrating modulation depths exceeding 95% have been demonstrated.

Waveguide acousto-optic devices have been developed for applications including switching, modulation, and spectrum analysis. The complication of non-uniform optical and acoustic fields in thin film structures modifies the analysis presented here slightly. To accurately determine phase shift of an optical beam, an overlap integral between the optical and acoustic fields must be calculated. Details can be found in ref. [13].

10. Applications of Acousto-Optic Waveguide Devices

One of the most significant applications of acousto-optic modulation in integrated optics is in spectral analysis of radio frequency signals. The direct application of this device is to allow a pilot to obtain an instantaneous spectrum analysis of a radar signal, in order to determine if his plane is being tracked by a ground-based station, air-to-air missile, or other vehicle. The signature of the radar signal can be deciphered to extract this information.

Fig. 12.22 shows a schematic representation of a hybrid integrated optical spectrum analyzer. A laser is butt-coupled into a planar slab waveguide, and the beam is expanded and collimated by a pair of integrated lenses on the waveguide. These lenses can take the form of simple domes on top of the waveguide, or can look like cross-sections of conventional lenses placed on top of the guiding layer. The collimated beam then passes through a region where a surface acoustic wave (SAW) is established.

The SAW is generated by the incoming electrical signal. An antenna collects the RF signal, and sends it to an amplifier. The amplified signal is applied to an interdigitated array of electrodes on the surface of the planar waveguide. If the waveguide is constructed using a piezo-electric material, such as X-cut $LiNbO_3$, the electric field established between the fingers of the electrode will periodically constrict and expand the surface material, establishing an acoustic wave that propagates across the waveguide. The spatial period of the acoustic wave depends on the frequency of the applied RF signal.

The collimated optical beam is Bragg scattered off of this SAW structure. The angles of the scattered beams are described by Eq. 12.48, so each RF frequency will scatter the optical beam in a different direction. The scattering efficiency can approach 50%/W of applied electrical signal. The scattered beams are again passed through a lens to focus them on a detector array. Here, the fact that the lens acts as a Fourier transformer is exploited. The optical beam leaving the SAW region contains several distinct beams travelling in different directions, depending on the spectral content of the applied electrical signal. But the beams are essentially fully spatially overlapped. To separate the beams would require propagation over a long distance. A lens solves this problem by converting the angular variation of the incoming rays (k-space description) into a spatial variation (x-space) at the focus of the output beam. This is what is meant when we say a lens performs a Fourier transform. The focussed output of this lens is directed onto an array of detectors. Each pixel of the array corresponds to a specific frequency of the incoming electrical signal. All frequencies present on the incoming signal are simulataneously detected at the array, so the signature of an incoming signal can be readily determined. Performance of such devices have shown 5 MHz resolution over a bandwidth of 400 MHz. [14]

11. Summary

We reviewed the two major methods used for modulating light. Electro-optic effects are currently being explored for high speed, high performance modulators that will be suitable for the communication industry. These devices

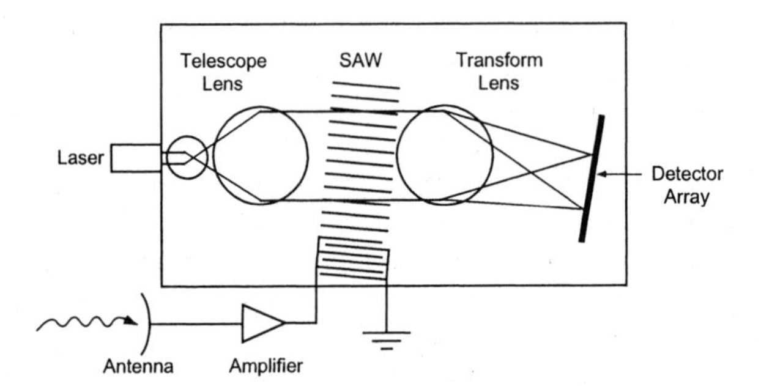

Figure 12.22. An integrated optic spectrum analyzer consists of a cw laser source, a collimating lens, a SW region, a transform lens, and a detector array.

are built using either semiconductor materials, or nonlinear dielectrics such as $LiNbO_3$. Only a fool would make a prediction about the future in this field, but it appears that there will be more interest in integrated sources and modulators, which can be made using semiconductors, than there will be for hybrid systems consisting of semiconductor lasers, optical interfaces, and dielectric modulators. However, the hybrid devices at present provide many of the best results. Manufacturing cost is the pressure pushing toward integrated systems.

Acousto-optic modulators, because of their inherently slow modulation speed, will probably never be serious competitors for communication modulation. However, acousto-optic devices are playing critical roles in optical computers, where convolutions, beam switching, and frequency shifting are used to perform various signal processing functions. We have discussed only a few of the possible implementations of electro-optic and acousto-optic modulators in this chapter. There are many other possible schemes. Interested readers should consult many of the references for overviews of some of these techniques. Entire texts have been devoted to single aspects of the discussion presented here.

References

[1] R. W. Wood, *Physical Optics, 3rd ed.*, Optical Society of America, 1988, USA

[2] A. Yariv and P. Yeh, *Optical Waves in Crystals*, Wiley Interscience,1984, USA

[3] A. Yariv, *Quantum Electronics, 3rd ed.*, Ch. 5, John Wiley and Sons, 1989, USA

[4] S. G. Lipson and H. Lipson, *Optical Physics, 2nd ed.*,Cambridge University Press, USA, 1981

[5] R. J. Pressey, ed. *CRC Handbook of Lasers*, Chemical Rubber Co., Cleveland, Ohio, (1971)

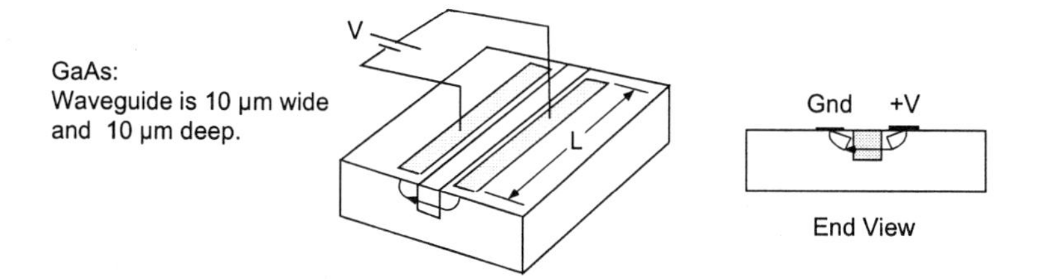

Figure 12.23. Waveguide structure for problem 2.

[6] A. Yariv, *Optical Electronics, 4th ed.* Ch. 9, Holt, Rinehart, and Winston, New York (1991)

[7] A. Yariv, *Quantum Electronics, 3rd ed.*, John Wiley and Sons, USA (1989)

[8] R. G. Hunsperger, *Integrated Optics: Theory and Technology, Vol. 33 Springer Series in Optical Sciences*, Berlin (1982)

[9] J. F. Nye, *Physical Properties of Crystals,* Oxford Clarendon Press, London, pp. 241 (1957)

[10] R. W. Dixon, "Photoelastic properties of selected materials and their relevance for applications to acoustic light modulators and scanners," *J. Appl. Phys.* 38, pp. 5149 (1967)

[11] A. A. Oliner, editor, *Topics in Applied Physics, Vol. 24: Acoustic Surface Waves*, Springer, Berlin (1978)

[12] J. M. Hammer, "Modulation and switching of light in dielectric waveguides," in *Integrated Optics*, edited by T. Tamir, Topics in Applied Physics, Vol. 7

[13] T. G. Giallorenzi and A. F. Milton, J. Appl. Phys. 45, pp. 1762 (1974)

[14] D. Mergerian and E. C. Malarkey, *Microwave Journal*, 23, pp. 37 (1980)

Practice Problems

1. Confirm the expression of Maxwell's equations as listed in Eq. 12.8 by deriving them from the general form of Maxwells Equations listed in Chap.2, using a single frequency plane wave with phase $\exp(-j(\omega t - kz))$.

2. Consider the waveguide phase modulator shown in Figure 12.23.

   (a) Assume that only the TE wave is to be modulated. What is the correct orientation for the GaAs crystal if the applied field is as shown? Make a sketch of the modulator, and identify the x, y, and z-axes.

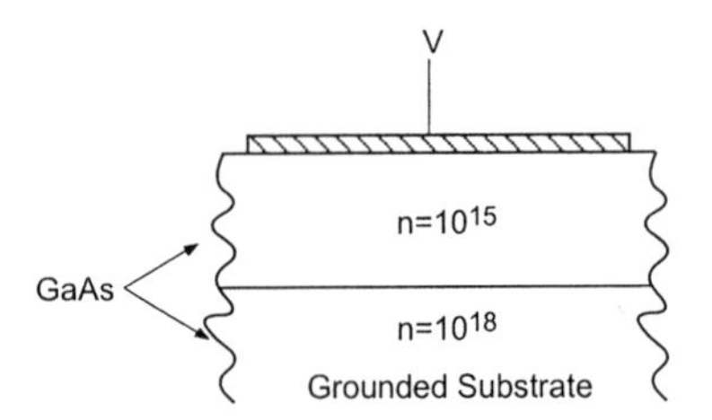

Figure 12.24. Intensity modulator for problem 3.

(b) Assume that the electric field strength in the waveguide is equal to $V/20\mu m$. What length should the electrodes be if a 10 V input is to produce a π phase shift?

(c) If you only had this device in your lab, and you wanted to make a polarization rotator using it, how would you do it?

3. An intensity modulator is built based on the concept of turning a single mode waveguide on and off via the electro-optic effect. Consider the semiconductor structure shown in Fig. 12.24. The top layer of GaAs is lightly doped, and is 5 μm thick. The substrate is heavily doped. A Schottky barrier is placed on the surface for a distance L. Due to the light doping of the top layer, a reverse biased field will develop most of the field in the thin layer.

   (a) What orientation should the crystal be if a positive voltage is to increase the index of the top layer?

   (b) How much voltage is required to increase the index sufficiently to cause the top layer to become a waveguide for the lowest order mode of an asymmetric waveguide?

   (c) How would you determine the length of the electrodes?

4. Find the effective index of refraction, n_{eff} for an extraordinary wave travelling at an angle of 50° with respect to the c-axis in a crystal of $LiNbO_3$. Use the data in Table 12.1 for refractive indices of the ordinary and extraordinary directions.

5. Consider the problem described in Example 12.2. How large an electric field, E_z is required to change the index of refraction by 0.0001? What voltage would be required to achieve this field across a waveguide 10 μm thick?

6. What is the index of refraction in a crystal of GaAs if an electric field is applied along the x direction of the crystal? Describe the index of refraction for GaAs when the applied field is directed along the xy axis.

7. The $LiNbO_3$ cystal has a crystal symmetry of 3m. The electroptic tensor for this crystal is given by

$$\begin{bmatrix} 0 & -r_{12} & r_{13} \\ 0 & r_{22} & r_{13} \\ 0 & 0 & r_{13} \\ 0 & r_{51} & 0 \\ r_{51} & 0 & 0 \\ -r_{22} & 0 & 0 \end{bmatrix}$$

Show that if an electric field is applied along the z axis, the index of refraction becomes

$$\begin{aligned} n_x &= n_0 - \frac{1}{2}n_0^3 r_{13} E \\ n_y &= n_0 - \frac{1}{2}n_0^3 r_{13} E \\ n_z &= n_e - \frac{1}{2}n_e^3 r_{33} E \end{aligned}$$

8. Develop an expression for the half-wave voltage similar to Eq. 12.23 for a $LiNbO_3$ modulator cut so that light propagates along the x-axis, and the modulating electric field is applied along the z-axis. Use the electrooptic tensor in Prob. 7. and data from Table 12.1 for this problem.

9. Design a TE to TM converted based on the electro-optic effect in GaAs. Assume the waveguide is 3 μm thick, has a core index $n_f = 3.6$, a substrate index $n_s = 3.59$, and a cover index of $n = 1$. Assume that electrodes are placed on both sides of the guiding dielectric, so that a planar electric field is established in the guide. Choose the correct orientation of the GaAs crstal so that an applied field will rotate the polarization of the field.

10. A Fabry-Perot modulator is made as shown in Fig. 12.25.

    (a) Plot the reflectivity of this device as a function of λ over two free spectral ranges (FSR) of the device.

    (b) How much does the index of the Fabry-Perot have to change to shift the transmission peak by 1/2 FSR?

    (c) What E-field is necessary to create the shift decribed in part b?

    (d) Will this modulator work as well with with unpolarized light as with polarized light? If not, which polarization is preferred?

    (e) What is the spectral bandwidth of this device?

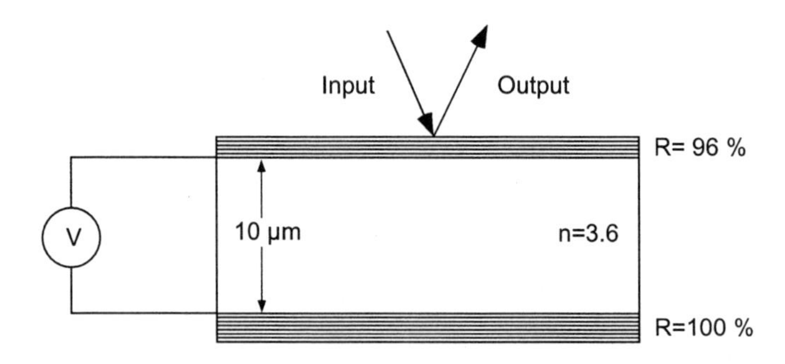

Figure 12.25. Fabry-Perot structure for problem 10.

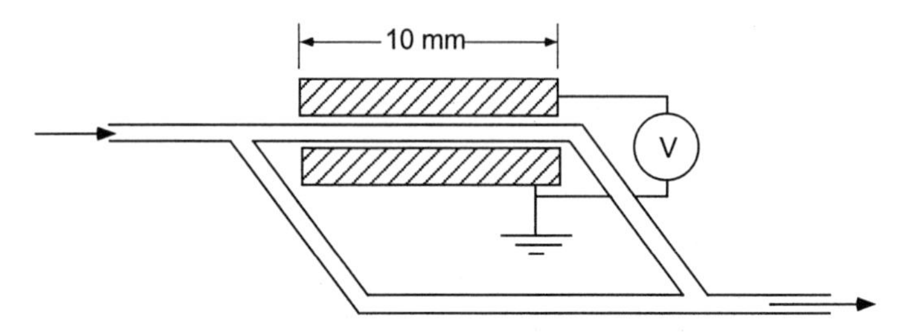

Figure 12.26. Mach-Zender structure for problem 11.

11. The Mach-Zender interferometer shown in Fig. 12.26 is made of GaAs. What E-field is required to make the device switch from "on" to "off"?

12. A Franz-Keldysh modulator is built using GaAs. If the contrast ratio between "on" and "off" is to be 10 dB, and the operating wavelength is 900 nm, how thick should the GaAs device be made? Use the data in Fig. 12.18. What is the minimum insertion loss for this device?

13. How much acoustic power is required to change the index of refraction in a GaAs layer 2 μm thick and 100 μm wide by 0.0001?

14. An acousto-optic modulator is built using $PbMO_4$. If an RF signal at 100 MHz drives the acoustic wave in this crystal, and an optical wave with $\lambda = 0.6328$ nm is Bragg reflected of the acoustic wave, what is the angle between the incident and reflected optical beam?

Chapter 13

PHOTONIC CRYSTALS

1. Introduction

In the previous chapters of this text we have explored how light can be guided by high index regions of a dielectric. Many types of waveguide were examined, including graded index slabs, circular step-index fibers, and rectangular ridge structures. In all of these cases we found that light can be guided by high index regions and as a result can go around bends, couple to other waveguides, and be manipulated to perform certain functions. The guiding mechanism in all cases arose from the "attraction" light has toward higher index regions. We described this attraction in a variety of terms, including total internal reflection, the eikonal equation, or spatial resonances forming from reflections on index changes, but in all cases the basic mechanism is the same: higher index regions of a dielectric act like a "potential well" for light, and so long as things change slowly (bends are gentle, or dimensions vary slowly), light tends to remain trapped in that potential well.

In this chapter we explore a dramatically different method for guiding light. Instead of using materials with high refractive indices to attract and trap light, we will consider materials called photonic crystals that actually repel light. Optical waveguides are made in photonic crystals by removing material, creating channels through which light can propagate. Without sounding too anthropomorphic, this is similar to the difference between doing something because you *want* to do it and doing something because its the only thing that you possibly can do. The difference may sound subtle, but we will see that by using photonic crystals it is possible to guide light through low index materials (even vacuum), and to make light turn extremely sharp corners. Because photonic crystals operate by a different mechanism than index-guided structures, they offer new opportunities for devices and systems.

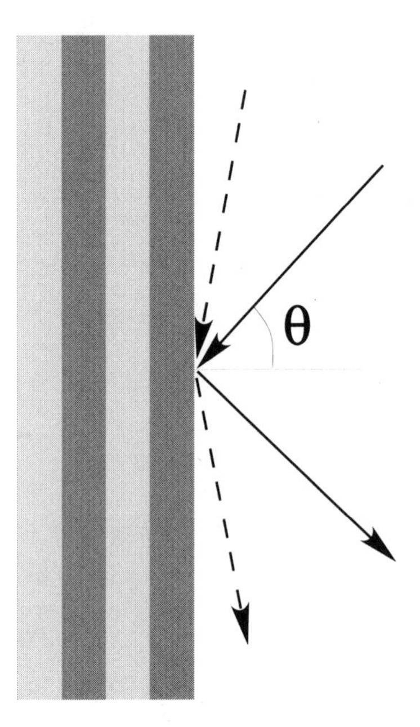

Figure 13.1. Cross-section of a one dimensional photonic crystal, with alternating high and low index dielectric layers.

2. Basic Physics of the Photonic Crystal

In the broadest sense photonic crystals are periodic dielectric structures composed of alternating high and low index of refraction dielectric materials, with periodicity on the order of the wavelength of light. In principle the periodicity can be in one-, two-, or three-dimensions, so they resemble large-scale versions of the crystalline structure of many solid state materials. This periodic structure is one reason they are called "crystals". We will discuss one- and two-dimensional photonic crystals in this chapter. Using conventional lithographic processes developed originally for the semiconductor electronics industry, it is possible to fabricate planar two-dimensional structures today, but extending this technology to the third dimension is still a challenge.

Because light experiences a small reflection each time it crosses a dielectric boundary (recall the discussion on Fresnel reflections in Chapter 2), light travelling through a periodic structure will experience multiple reflections that constructively and destructively interfere with one another. As we will see, when the period of the structure is comparable to $\lambda/2$ of the incident light, these reflections dominate the optical behaviour of the material. In order to develop the principle of operation for a photonic crystal we will first analyze the

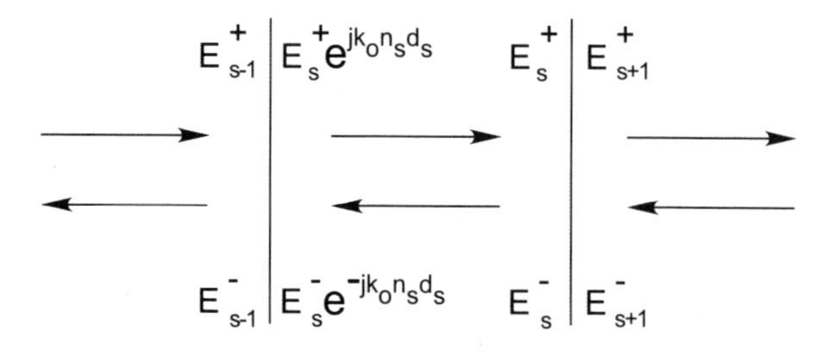

Figure 13.2. Field amplitudes at the interface between layer s and $s-1$

one-dimension (1-D) case as shown in Fig. 13.1. This is a stack of alternating layers of high and low index material. If the indices and layer thicknesses are chosen properly, the multilayer film will reflect light from all directions, not just along the single axis that forms the stack.

Each layer of the stack is characterized by refractive index n_s and thickness d_s, where s indicates a the sth layer. We will first consider light incident on the multilayer stack perpendicular to the layer's surface plane, along the z-axis. In each layer there will be two waves propagating in the forward and backward direction with complex amplitudes E_s^+ and E_s^-. As a wave propagates across a layer, it accumulates a phase equal to $k_o n_s d_s$. As shown in Fig. 13.2, if E_s^+ and E_s^- are the amplitudes on right side of layer s, the amplitudes at the left side are given by $E_s^+ \exp(jk_o n_s d_s)$ and $E_s^- \exp(-jk_o n_s d_s)$.

The electric field is continuous at the interface between layer s and layer $s-1$

$$E_{s-1}^+ + E_{s-1}^- = E_s^+ e^{jk_o n_s d_s} + E_s^- e^{-jk_o n_s d_s} \tag{13.1}$$

The magnetic field amplitudes in the forward and backward direction can be found using the impedance of the medium, yielding $H_s^+ = E_s^+/\eta_s$ and $H_s^- = -E_s^-/\eta_s$. The tangential magnetic field also must be continuous at the interface, so

$$H_{s-1}^+ + H_{s-1}^- = H_s^+ e^{jk_o n_s d_s} + H_s^- e^{-jk_o n_s d_s} \tag{13.2}$$

If we define the total field at each plane $H = H_s^+ + H_s^-$ and $E = E_s^+ + E_s^-$, Eqs. 13.1 and 13.2 can be combined using some trigonometric identities to yield

$$\begin{aligned} E_{s-1} &= E_s \cos(k_0 n_s d_0) + j\frac{1}{\eta_s} H_s \sin k_o n_s d_s \\ H_{s-1} &= j\eta_s E_s \sin(k_0 n_s d_0) + H_s \cos k_o n_s d_s \end{aligned} \tag{13.3}$$

To simplify the evaluation of many multilayers, we can put this in matrix form

$$\begin{bmatrix} E_{s-1} \\ H_{s-1} \end{bmatrix} = \begin{bmatrix} cos(k_0 n_s d_0) & j\frac{1}{\eta_s}\sin(k_o n_s d_s) \\ j\eta_s \sin(k_0 n_s d_0) & \cos(k_o n_s d_s) \end{bmatrix} \begin{bmatrix} E_s \\ H_s \end{bmatrix} = M_s \begin{bmatrix} E_s \\ H_s \end{bmatrix} \tag{13.4}$$

The incident field amplitudes are related to the field amplitudes at layer n by multiplication of a series of matrices where each matrix contains the thickness d_s and the index n_s of each layer in the film

$$\begin{bmatrix} E_0 \\ H_0 \end{bmatrix} = \prod_{s=1}^{n} M_s \begin{bmatrix} E_s \\ H_s \end{bmatrix} \tag{13.5}$$

The total reflection and transmission of a general multilayer film with a total of N layers can now be calculated. [2]

Example 13.1 Reflectivity of a 10 layer stack

Most dielectric mirrors are made by depositing a stack of alternating high and low index layers onto a substrate such as glass. Let's calculate the reflectivity and transmissivity of a film consisting of 5 alternating quarter wavelength thick layers of Si and SiO_2.

Since this stack repeats a sequence of low-index/high-index layers, let's first calculate the transmission and reflection of a pair of quarter wavelength thick layers of high and low index of refraction using Eq. 13.5. The matrix describing one period consisting of a pair of high and low index layers is given by:

$$\begin{aligned} M &= \begin{bmatrix} cos(k_0 n_H d_H) & j\frac{1}{\eta_H}\sin k_o n_H d_H \\ j\eta_H \sin(k_0 n_H d_H) & \cos k_o n_H d_H \end{bmatrix} \\ &\times \begin{bmatrix} cos(k_0 n_L d_L) & j\frac{1}{\eta_L}\sin k_o n_L d_L \\ j\eta_L \sin(k_0 n_L d_L) & \cos k_o n_L d_L \end{bmatrix} \end{aligned} \tag{13.6}$$

For quarter wavelength thick layer, $k_o n_H d_H = k_0 n_L d_L = \pi/2$, and the matrix is equal to

$$M = \begin{bmatrix} 0 & j\frac{1}{\eta_H} \\ j\eta_H & 0 \end{bmatrix} \begin{bmatrix} 0 & j\frac{1}{\eta_L} \\ j\eta_L & 0 \end{bmatrix} = \begin{bmatrix} \frac{-\eta_L}{\eta_H} & 0 \\ 0 & \frac{-\eta_H}{\eta_L} \end{bmatrix} \tag{13.7}$$

From Eq. 13.5, the reflectivity and transmissivity of a multilayer stack with q periods, surrounded by air ($n_o = n_N = 1$), are then given by

$$\begin{bmatrix} E_0 \\ H_0 \end{bmatrix} = \begin{bmatrix} \frac{-\eta_L}{\eta_H} & 0 \\ 0 & \frac{-\eta_H}{\eta_L} \end{bmatrix}^q \begin{bmatrix} E_s \\ H_s \end{bmatrix} \tag{13.8}$$

We can directly calculate the power transmitted and reflected by this stack using $E_0 = 1 - R$, $H_0 = 1 + R$, and $E_s = H_s = T$, where R and T are the

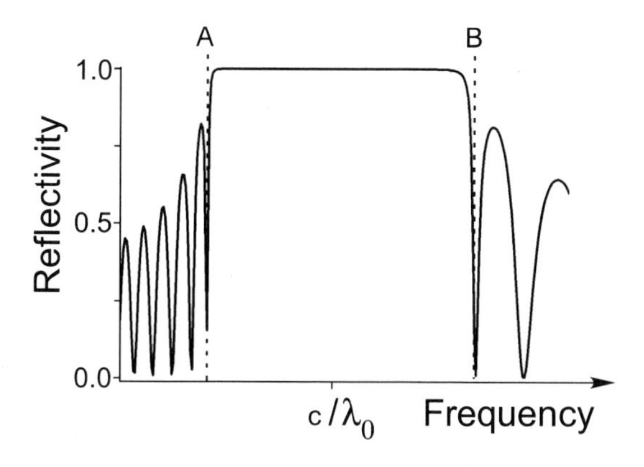

Figure 13.3. Reflectivity of the dielectric stack with quarter wave layers

transmission and reflection amplitudes. The reflectivity, $R^2 = 1 - T^2$ is given by

$$R^2 = \left[\frac{(-\frac{n_L}{n_H})^q - (-\frac{n_H}{n_L})^q}{(-\frac{n_L}{n_H})^q + (-\frac{n_H}{n_L})^q}\right]^2 \tag{13.9}$$

For $q = 5$ periods with $n_H = 3.5$ (Si layer) and $n_L = 1.5$ (SiO_2 layer), the reflectivity and transmissivity are equal to $R^2 = 0.999$ and $T^2 = 0.001$. Such multilayer films with high reflectivity are called Distributed Bragg Reflectors (DBR).

This is a very high reflectivity, far exceeding what can be achieved with a simple metal film. For example, a silvered mirror used commonly in residences has a reflectivity of approximately $R^2 = 0.92$, or effectively an 8% power loss for each reflection. So the multilayer mirror is desired when low absorption is needed. The comparative disadvantage of the multistack mirror is that it has a limited spectral bandwidth, because the layers are only quarter-wave thickness for one particular wavelength. However, the spectral width can be controlled by choosing the index difference, and by adjusting the layer thicknesses in the stack.

3. The Photonic Band Gap

Using Eq. 13.5 we can calculate numerically the reflection and transmission coefficient as a function of frequency. In Fig. 13.3 we have plotted the reflectivity of the multilayer film from Example 13.1 with optical thickness $n_H d_H = \lambda/4$ and $n_L d_L = \lambda/4$.

One can see that over a frequency range centered at $\nu = c/\lambda_0$ there is a spectral range in which light is totally reflected and cannot propagate through the crystal. This spectral range is called the Photonic Band Gap. Points A and B in Fig. 13.3 correspond to the upper and lower frequency limits of the reflectivity. The bandwidth of the reflectivity ($\Delta\nu = \nu_B - \nu_A$) depends on the index difference between the layers and on the relative thicknesses of the high and low index layers. A similar system, with total periodicity $n_H d_H + n_L d_L = \lambda/2$, but with thicknesses of the layers differing from $\lambda_0/4n_H$ and $\lambda_0/4n_L$, will form a photonic band gap, but its bandwidth will be reduced compared to the quarter wavelength layer system.

Intuitive picture of the photonic band gap

Since only one wavelength will match the period of the lattice, why does a band gap occur? Why isn't the reflection only at one wavelenth?

There is a simple reason why a broad range of wavelengths are reflected by the structure. From the previous section we saw that a photonic band gap occurs when the wavelength corresponds to twice the periodicity of the structure, $\lambda = 2a$, where $a = n_H d_H + n_L d_L$ is the period. In a periodic structure the field distribution will be periodic relative to the structure, i.e., at any point x the field must be the same as at the points $x + ma$, where m is an integer number and a being the periodicity of the structure. There are many ways to position a wave inside the grating, but we gain insight by looking at the two extreme cases. In one case (see Fig. 13.4a) the nodes of the field will be in the high index of refraction layers and the antinodes in the minima , and in the other case (see 13.4b) the nodes of the field will be in the low index of refraction layers and the antinodes in the high index of refraction layers (Fig 13.4b)

In Fig. 13.4a the wave's energy (proportional to E^2) is concentrated in the low index layers, while in Fig. 13.4b the energy is concentrated in the high index layers. Because the wavelength is shorter in the high index material, when the wave energy is concentrated in the high index layer it corresponds to a longer λ_0 (recall, λ_0 is the wavelength of the wave in vacuum) When the energy is concentrated in the lower index layer, the vacuum wavelength λ_0 will be shorter. Using the relation between frequency and wavelength, $\omega = (c/k_0)/\sqrt{\epsilon/\epsilon_0}$, the wave in case (a), corresponds to a field with lower frequency (point A in Fig. 13.3) and the wave in case (b) corresponds to a field with a higher frequency (point B in Fig. 13.3). For all frequencies in-between these values, the nodes will lie somewhere between the two extreme points and will still find themselves in-phase with the lattice. The difference between these two extreme frequencies, $\Delta\nu$, corresponds to the photonic band gap. This is why the spectral width of the quarter-wave stack depends on the index difference between the layers, and on the asymmetry of the layers.

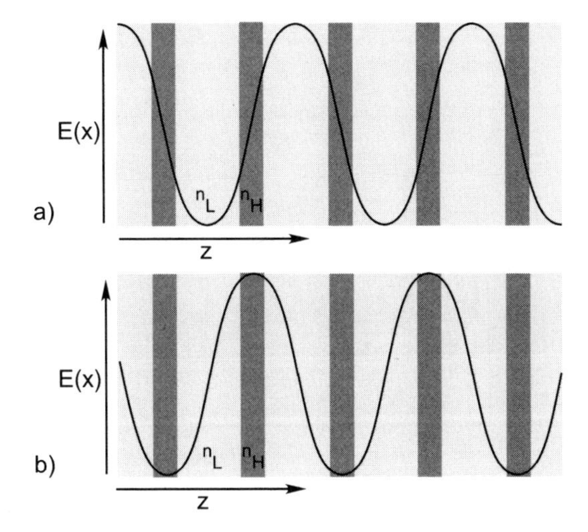

Figure 13.4. Field distribution inside a periodic structure. (a) the antinodes are in the low index region, (b) the antinodes are in the high index region

4. Photonic States of a 1D Photonic Crystal

So far all calculations have been done for normal incidence ($\theta = 0°$). Does the photonic gap exists for all angles incident on the multilayer film? Using Snell's law one can rigorously derive Eq. 13.5 for oblique angles. In practice, the frequencies for which light can propagate through the crystal are found using numerical calculations that solve the eigenvalues of the wave equations, in the same way that it is done for semiconductors. For greater angles the spectrum of the crystal shifts to higher frequencies. The blue shift of the spectrum is due to the fact that the effective k_x-vector in the direction perpendicular to the surface plane becomes smaller than k, given by

$$k = \frac{\pi}{(d_H n_H + d_L n_L)} \tag{13.10}$$

If $k_x < k$, for the same dimensions of the structure d_H and d_L, the condition of half-wave is satisfied at higher frequencies than ck_x. Fig.13.5 shows a calculated spectrum of a Distributed Bragg Reflector at several different angles.

To determine the angular reflectivity of the multilayer stack, let us now plot the dispersion of the multilayer stack, i.e., the lower and higher limiting frequencies of the photonic band gap, as a function of the k_y-vector in the direction parallel to the layers plane. k_y is given by

$$k_y = k \sin\theta = \frac{2\pi}{\lambda} \sin\theta \tag{13.11}$$

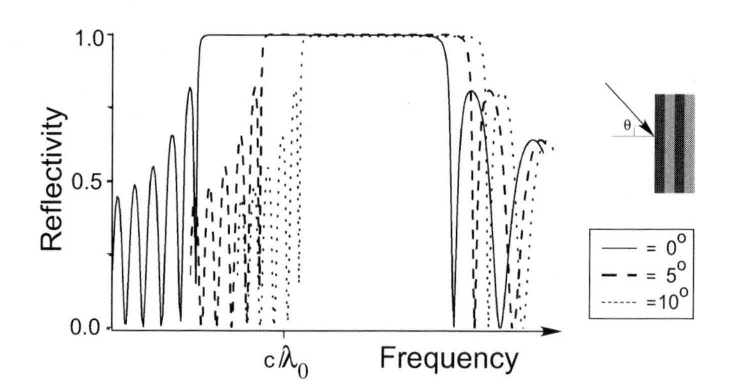

Figure 13.5. Reflectivity of a Distributed Bragg reflector calculated for different incident angles. The polarization difference is negligible for small angles.

The results of the calculation are shown in Fig. 13.6. The vertical line at $k_y = 0$ indicates an angle of $0°$. At this angle, the reflectivity of the structure is shown in Fig 13.5. Points A and B of Fig. 13.5 are shown on the dispersion diagram. One can see that due to the curved nature of the dispersion, for both polarizations, the band gap between ω_A and ω_B des not occur at all values of k_y (i.e. all angles). For example, for TM polarization, one can see that at a frequency ofω_1 light is reflected at angles ranging from 0 to $\theta = \sin^{-1}(k_y/k) = \sin^{-1} 0.8 = 53°$ At a frequency ω_2 light is reflected at a maximum angle of $\theta = \sin^{-1}(k_y/k) = \sin^{-1} 0.3 = 17.4°$.

Fig. 13.7 shows that for a 1-D photonic structure the photonic band gap does not necessarily exist for all angles. In the next sections we will see however that if the photonic states of the surrounding medium are taken into account, light can be made to totally reflect in all directions, despite the fact that the periodicity is only in one direction. This is called an omni-directional reflector. A truly omni-directional reflector would be useful for applications such as coating the inside of a tube to make a waveguide in which light propagates in a low-index core such as air or vacuum.

5. Photonic States of a Continuous Medium

Let us analyze the dispersion of light in a continuous medium. For light travelling at an angle θ with respect to the z axis as shown in Fig.13.7, the

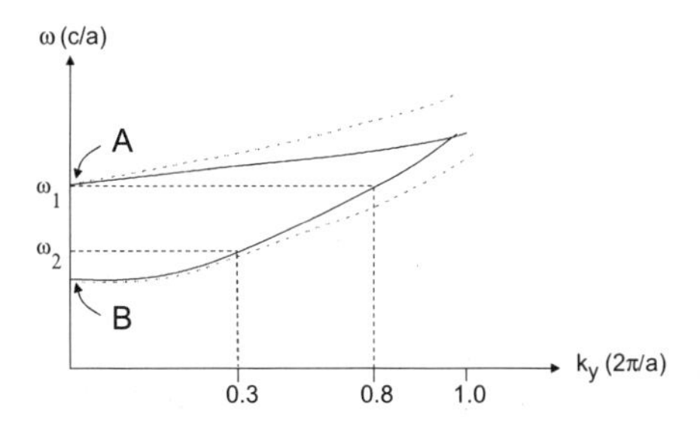

Figure 13.6. Dispersion diagram for a multilayer. Both the TM (solid) and TE (dashed) cases are plotted.

frequency of light in such a medium is given by:

$$\omega = ck = \frac{k_y}{\sin\theta_s} \tag{13.12}$$

The relationship in Eq.13.12 is called the light line. Consider light incident from a medium with a high index of refraction onto an interface with a surrounding medium of lower index. All states with $\theta_H < \theta_{crit}$ are states that propagate in the high index medium and are transmitted to the surrounding medium. All states with angle above the critical angle, $\theta_H > \theta_{crit}$, do not couple to the surrounding medium. From Eq. 13.12 all states with $\theta_H > \theta_{crit}$ obey

$$\omega < \frac{k_y}{\sin\theta_{crit}} \tag{13.13}$$

Therefore all states below the light line (Eq. 13.13) remain in the high index medium due to total internal reflection. Similarly, light incident from the surrounding medium with frequency obeying Eq. 13.13 does not couple to the high index medium and remains in the surrounding medium.

Fig. 13.8 plots the light line for light propagating in a high index media surrounded by air and by oxide. The photonic states of the surrounding medium that do not couple to the high index medium are shown by the shaded areas. For convenience we again use the normalized axis. Here the periodicity a is defined artificially for a continuous medium.

6. Onmnidirectional Photonic Band Gap of a Crystal in a Continuous Medium

In order to ensure that a photonic crystal reflects in all directions it is sufficient to ensure that only those states above the light line are indeed reflected. This

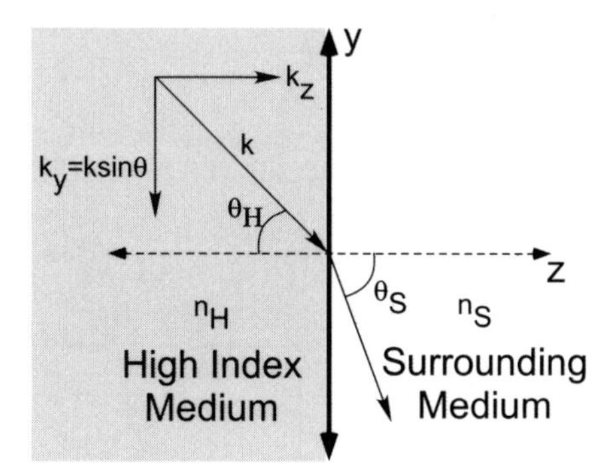

Figure 13.7. Light incident from a high index to a low index medium

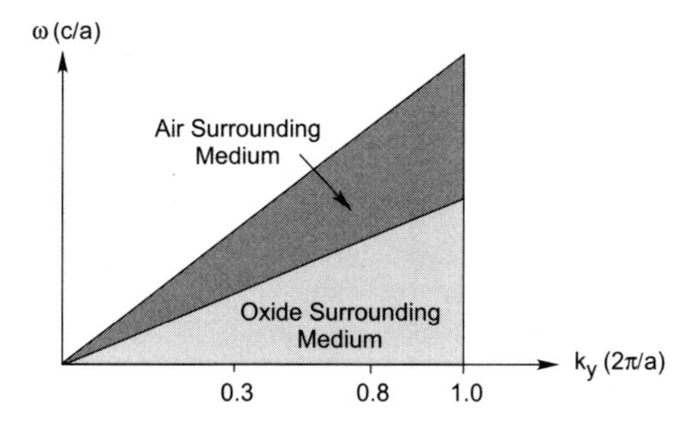

Figure 13.8. Dispersion diagram for a semiconductor based medium surrounded by a continuous medium of air and of oxide. The shaded areas show the photonic states of the surrounded medium

is because, as we saw in the last section, light travelling below the light line cannot couple to the surrounding medium and is totally reflected back into the high index medium due to total internal reflection.

Fig. 13.9 shows the dispersion diagram of multilayer film with indices $n_1 = 3.4$ and $n_2 = 1.46$ as a function of the incident k-vector parallel to the film plane. This dispersion diagram was calculated using the MIT Photonic-Bands software package [1]. The wave vector k and frequency are normalized in terms of a, where a is the thickness of one period in the DBR.

The white space between the two dark gray areas is the photonic band gap. All the modes in dark gray areas between the light lines $c\lambda/a < 2\pi k_y/a$ (note the left and the right sides of the plot are for TM and TE polarization, respectively) can propagate in the multilayer film.

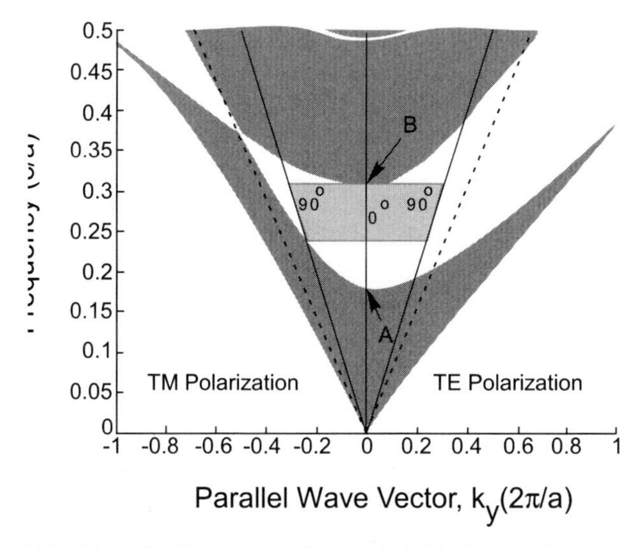

Figure 13.9. Normalized Frequency (a is the period of the Bragg reflector) versus the parallel wave vector for a distributed Bragg reflector. The light gray region is the omnidirectional band.

The light solid lines are for light incident from air corresponding to an angle of 90 , or, equivalently, light incident from the photonic crystal with a critical angle. The photonic band gap between the light lines shown in the light gray area is a range of frequencies where light *at any incident angle* is perfectly reflected. In this frequency range there are no allowed DBR modes in this region, regardless of angle or polarization. In this case the photonic band gap is called omnidirectional band, or complete band gap. Such thin films have very unusual applications. For example, imagine putting such a structure on the inside of a tube, as shown in Fig. 13.10. Light propagating down the tube would be totally confined to the center region due to the photonic crystal nature of the multilayer. Hollow guides such as this have been constructed, and display very unusual properties. The usual problems of material dispersion and optical nonlinearities are removed, creating the ability to make new types of optical device. Such fibers can be fabricated, for example, coating a glass rod followed by chemical removal of the supporting rod [3].

For a surrounding medium with a higher index of refraction than air, the light lines fall at a greater angle, as indicated by the dotted lines in Fig.13.9. Such structure is important for application such as waveguiding in photonic crystals, where the waveguide is composed by dielectric material with index higher than air. One can see in Fig. 13.9 that in this case, for TM polarization, there are

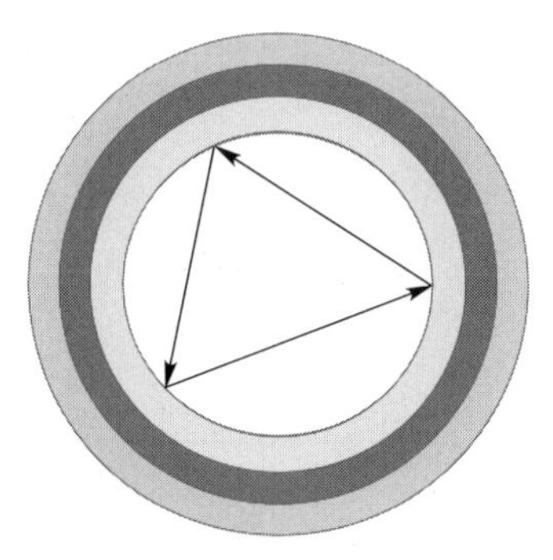

Figure 13.10. An omnidirectional reflector is placed on the inside of a hollow tube to create a waveguide that operates in air.

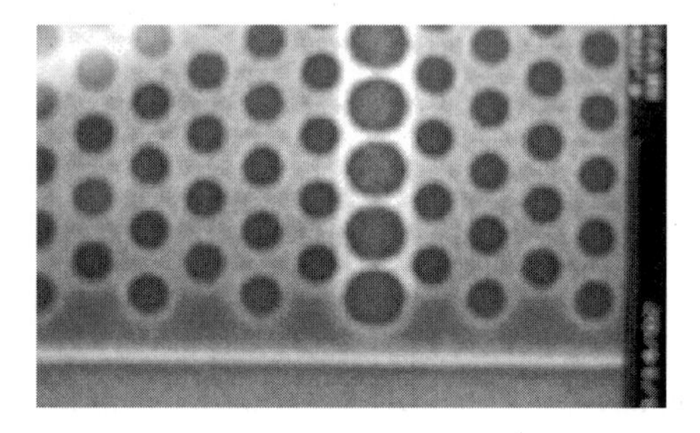

Figure 13.11. Photonic crystal designed as a periodic array of holes in Silicon on insulator

allowed modes at higher angles of approximately 80°, and therefore the gap is incomplete.

7. Two-Dimensional Photonic Band Gap Structures

To ensure that a complete omnidirectional photonic band gap can be created for light incident from high index of refraction materials, a photonic crystal with a periodicity in more than one direction is required. Planar slabs of 2-D photonic crystal waveguides have been fabricated using standard semiconductor lithographic techniques, however a true 2-D photonic crystal would have infinite extension along one axis. Planar two dimensional periodic structures typically are constructed as array of posts or holes in a semiconductor (see Fig. 13.11).

The applications for an omni-directional mirror are limited, but there is great interest in creating channels in such structures which would act as optical waveguides. Since any light travelling down such a channel would be completely reflected by the surrounding region of space, the light could be guided in man unconventional ways, including sharp bends as schematically shown in Fig.13.12.

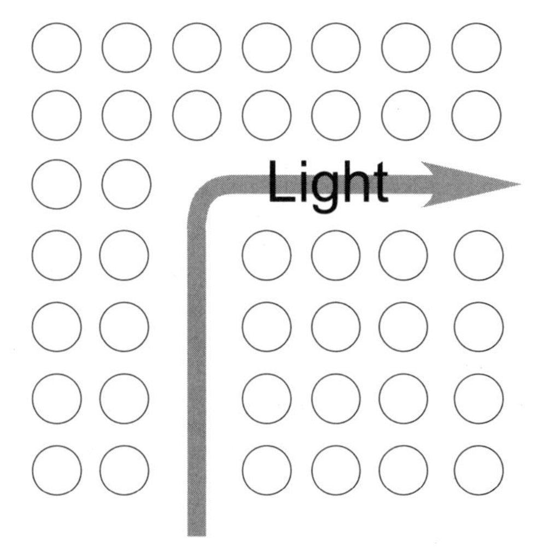

Figure 13.12. A schematic representation of a two-dimensional photonic crystal waveguide.

8. Summary

In this chapter we reviewed the basic physics of the photonic crystal. We emphasized the calculation of the reflectivity of the crystal as a function of angle, with the goal of identifying structures in which certain spectral wavelengths of light would not exist in any polarization or propagation direction. Such omnidirectional reflectors can be used to create impenetrable barriers to light. By using an omnireflector on the inside of a tube, an optical waveguide can be made where light propagates through vacuum or air, not through a high index material. By making two-dimensional slab photonic crystals, it should be possible to create waveguides on substrates which can be forced around sharp corners.

The key issues to be resolved concerning photonic crystals are primarily related to the fabrication of the crystals. It is difficult, if not impossible today, to lithographically create true 3-dimensional structures with the feature sizes needed to act as photonic crystals for visible and near infrared light. Also, because the light is constantly reflecting off of the crystal, scattering losses in photonic crystal devices has proven to be a significant issue. Nevertheless if there is progress on these issues, it is likely that photonic crystals will find unique applications in integrated photonics.

References

[1] S. G. Johnson and J. D. Joannopoulos, "Block-iterative frequency-domain methods for Maxwell's equations in a planewave basis," Optics Express 8, pp. 173-190 (2001)

[2] J. D. Joannopoulos, R. D. Meade, and J. N. Winn, *Photonic Crystals: Molding the Flow of Light*, Princeton University Press, Princeton, NJ (1995)

[3] Ibanescu et al., An All-Dielectric Coaxial Waveguide, Science 2000 289: 415-419

Practice Problems

1. Design a multilayer that has a reflectivity of more than 99% at a wavelength of 1.5 microns using SiO-2 (n=1.5) and SiON(n=1.6) . You will need to use Matlab, Matchcad or a similar numerical tool.

   (a) What are the layer thicknesses and the number of layers required for such high reflectivity?

   (b) Plot the reflectivity as a function of wavelength. Estimate the photonic bandgap range ($\Delta\nu$) in which the reflectivity is high.

   (c) Now vary the composition of the SiON, from $n = 1.6 \rightarrow 2$ in steps of $\delta n = 0.1$. Plot the photonic band gap as a function of index contrast. d. For a composition of n=1.5/n=1.6, vary the thickness of layers to decrease the photonic band gap range. Plot the reflectivity and compare it with the one in part (b).

2. Complete the calculation showing that Eq.13.5 follos from Eq. 13.4

Chapter 14

INTEGRATED RESONATORS AND FILTERS

1. Introduction

In optical telecommunication of today, wavelength division multiplexing and demultiplexing (WDM) is extensively used for increasing the accessible bandwidth in a single fiber. In WDM, a series of discrete wavelengths are transmitted through the same fiber (the bus), each one them encoded individually. Adjacent channels are separated from one another by 200 GHz, and typically are modulated to carry signals with 10 or 40 GB/s of information. Over one hundred separate wavelengths (or channels) can be carried on a fiber simultaneously, which means that terabytes of data can be carried on a single fiber. The key issue in WDM systems is finding ways to add and drop individual wavelengths from the fiber while letting the rest pass on to their ultimate destination. This add/drop process requires optical multiplexers and demultiplexers. Fig. 14.1 shows a schematic representation of an optical add/drop multiplexer. A single waveguide carrying a number of discrete wavelengths enters the multiplexor. It is desired that one channel be extracted ("dropped") while the rest pass through without loss. Conversely, it is necessary that information at one particular wavelength be able to be put onto the waveguide without interfering with the other channels. This is called an "add". Finding an effective and inexpensive way to create ad/drop filters for WDM applications is a major issue today. In this chapter we will look at some of the key technologies involved in this problem, including the fiber Bragg grating, Mach-Zender interferometers, and Hi-Q resonators made from integrated waveguide structures.

2. Fiber Bragg Gratings

The fiber Bragg grating has probably pushed the telecommunication industry toward wavelength division multiplexing more than any other development in

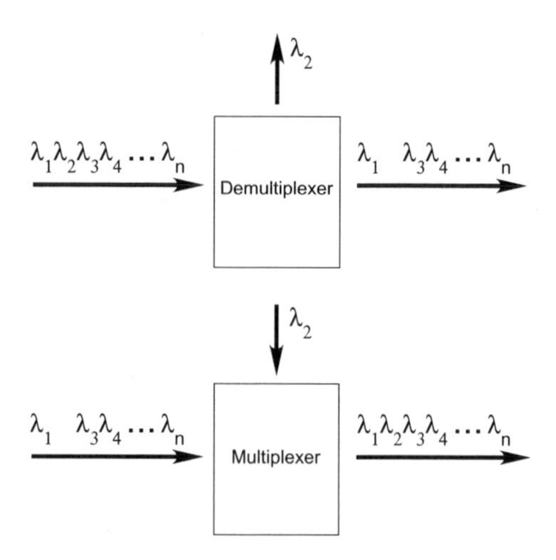

Figure 14.1. A multiplexer on a WDM system has to add or drop one specific wavelength channel on the waveguide, and let the rest of the channels proceed unattenuated

the last decade. Ten years ago it was an open question time division multiplexing or wavelength division multiplexing would be the best choice for the future. The biggest problem facing WDM systems was the need for stable wavelength sources and references that could be installed in many situations, and still operate within a certain bandwidth. The fiber grating, when properly packaged, can provide an absolute wavelength stability to better than a nanometer, which is sufficient for the WDM standards today.

In Chapter 10.5 we discussed how coupled mode theory can be applied to the analysis of the Bragg grating. While rigorous, the analysis is cumbersome. Today most Bragg grating design is performed using matrix methods as developed in Chapter 12 for the photonic crystal. The fiber Bragg grating can be formed in fibers either by UV exposure, by ion implantation, or by photolithography. Most commercial processes use UV exposure through a phase mask to create a spatially modulated index within the core. The UV light slightly increases the local index of the fiber's core. The index modulation must have a period equal to one half the wavelength they are intended to reflect.

Waveguide gratings are generally formed lithographically by etching into the cladding near the core. The period of the grating has to be at one-half the wavelength of interest, however this is not the vacuum wavelength, it is the wavelength inside the waveguide given by $\lambda = \lambda_0/n_{eff}$ where n_{eff} is the effective index of the mode in the waveguide. For glass substrates operating at telecommunication wavelengths ($\approx 1.5\mu$m), the wavelength in the waveguide will be $\approx 1\mu$m, and thus the period of the grating will be on the order of 0.5μm.

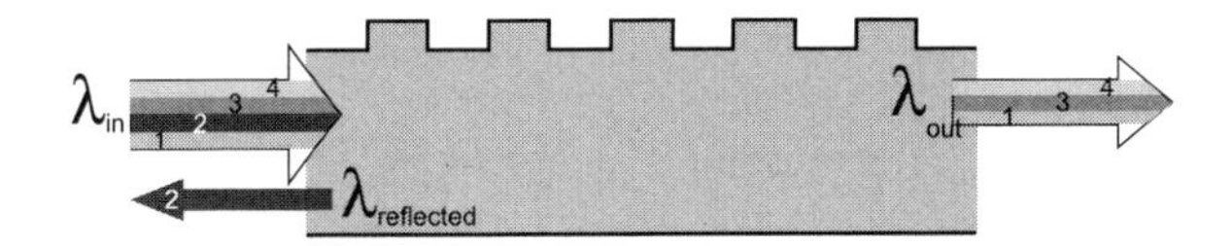

Figure 14.2. A waveguide Bragg grating with an etched core has an index modulation with a period of $\lambda_n/2$ to reflect a desired wavelength λ_n. Non-resonant waves travel through the grating without reflection

This dimension is right at the limit of what conventional optical lithography can provide, so most waveguide gratings are written with e-beam lithography. In high confinement systems such as Si/SiO_2 this is even more of an issue. A cross section of a planar waveguide with a Bragg grating is shown in Fig.14.2

There are several ways to analyze this grating, including coupled mode theory. We would like to apply the powerful matrix multilayer analysis to this structure, but in order to do that we would need to know the effective index of each region in the spatially modulated grating. Let us consider two slab waveguides with thickness d_A and d_B. We will consider here a single polarization and assume that both thicknesses are such that the slabs are single mode. For each slab waveguide, the modes are slightly different, with different effective indices $n_A = \beta_A/k_0$ and $n_B = \beta_B/k_0$. The fiber grating can be viewed as being equivalent to a multilayer dielectric stack as discussed previously in Section 13.2, with periodicity of half a wavelength and layers of indices n_A and n_B. We already saw in Section 13.2 that the reflectivity and bandwidth of a multilayer structure is a function of the index difference and of the number of periods. The reflectivity increases with both index contrast and the number of layers, while the spectral bandwidth narrows (which is to say, becomes more selective) as the index difference decreases. In fiber Bragg gratings, the effective index difference is usually on the order of 10^{-3}, so the bandwidth can be made to be narrow. The small index difference leads to small reflections at each stack, so many "stacks" are needed to get a substantial reflection. In fibers, a typical grating will be two millimeters long, which means there will be approximately 2000 quarter-wave stacks in the waveguide. The combination of a small Δn and a long interaction region provide the narrow, highly selective, spectral bandwidth that is necessary for WDM systems.

Waveguide gratings have many applications other than spectral filters. It is possible to write a grating that has a period that slowly increases with position. This is called a "chirped" grating, and provides a spatially modulated spectral response. Chirped gratings are often used in telecommunication links to equalize a dispersive medium. A pulse which has become spectrally dispersed due to waveguide and material dispersion effects can be almost restored back to its original shape by reflecting off of a chirped grating.

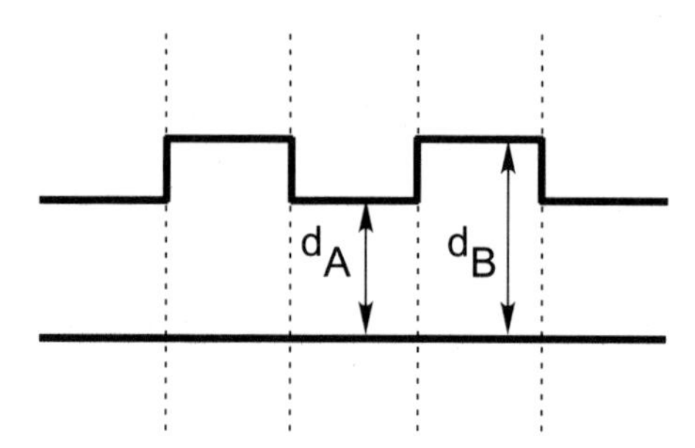

Figure 14.3. A cross-section of a spatially modulated waveguide.

3. Resonators

In this section we will describe a general resonator and its quality factor. The analysis follows the development of Haus [1]. The most general and familiar resonance phenomena is the one of circuit theory, namely the LC circuit. The basic characteristic of a resonator is that it will build up energy at a specific frequency (or wavelength. The dissipation of this power occurs due to losses, or through coupling to the outside world. In the case of integrated optics, the resonator is usually a reflective cavity where light is confined, and power dissipation occurs through coupling to nearby waveguides or layers, and losses due to scattering and absorption.

The field in the cavity oscillates in time with a frequency ω_0 and decays exponentially in time with a lifetime of τ. The basic equation for the time dependence of the amplitude of the field inside the cavity resonator is given by:

$$\frac{da}{dt} = (j\omega_0 - \frac{1}{\tau})a \tag{14.1}$$

The total energy in the cavity is proportional to $|a|^2$. The rate at which power dissipates from the cavity is given by the rate of change of this energy

$$\frac{d|a|^2}{dt} = 2|a|\frac{d|a|}{dt} = \frac{2}{\tau}|a|^2 \tag{14.2}$$

The last step in Eq. 14.2 is derived from Eq. 14.1

The quality factor Q of a resonator is given by the ratio of the total energy stored in the cavity divided by the total power dissipated in one cycle of the oscillation

$$Q = \frac{|a|^2\omega_0}{\frac{2}{\tau}|a|^2} = \frac{\omega_0\tau}{2} \tag{14.3}$$

We can also define the quality factor in terms of its spectral properties. The lifetime τ can be defined as the time that it takes for the amplitude to decay to

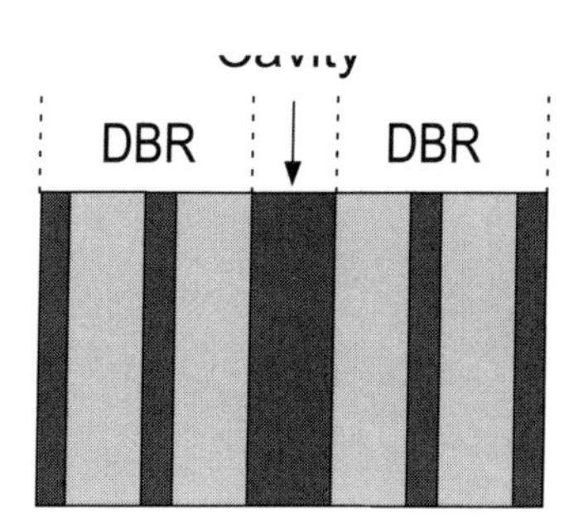

Figure 14.4. Schematic of a 1-D cavity, formed by surrounding a thin layer of material by two Distributed Bragg Gratings.

its half value. In the frequency domain, Eq. 14.1 becomes

$$a = \frac{a_{max}}{j(\omega - \omega_0) + 1/\tau} \tag{14.4}$$

From Eq. 14.4,$\Delta\omega_{1/2} = 2/\tau$, where $\Delta\omega_{1/2} = |\omega - \omega_0|$ is the spectral full width at half maximum. Therefore Eq.14.3 in the spectral domain is given by:

$$Q = \frac{\omega_o}{\Delta\omega_{1/2}} \tag{14.5}$$

In the next sections we will consider two examples of cavity resonators: i)a one-dimensional multilayer stack that forms a standing wave, and 2) a ring resonator, which is a two dimensional, travelling wave device. Both devices are used for Wavelength division multiplexing applications.

4. 1-D Cavity Resonator

An example of an integrated 1-D resonator is the interference filter formed by embedding a “defect" layer (the cavity) between two distributed Bragg reflectors discussed in Section 12.3. The cavity layer has a thickness equal to the periodicity of the multilayer.

The multilayer in Fig.14.4 can de described symbolically as $(HL)^2HH(LH)^2$, in which the two consecutive H layers make the half-wave layer. The transmission and reflection of such a multilayer can be derived using the matrix formalism discussed in Section 12.3. The calculated reflection of such a multilayer is shown in Fig. 14.5. One can see that the reflection is very similar to a plain DBR, but with a deep notch in the reflectivity at the center of the band at a frequency $\omega_0 = c/\lambda_0$ where λ_0 is equal to twice the thickness of the cavity. This frequency is called a "cavity mode" frequency. At the cavity mode the

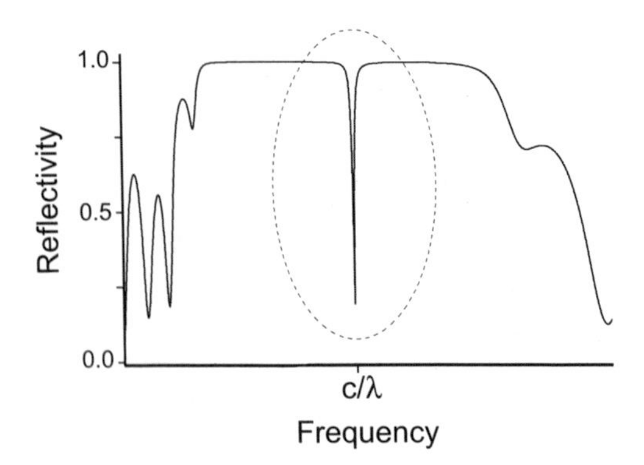

Figure 14.5. Numerically calculated reflection of a multilayer described in Fig.14.4

transmission through the entire structure is equal to one, and the field inside the cavity layer is built up. In order to estimate the lifetime of this resonator, let us consider only the spectral region around the cavity mode, shown in dashed line. In this region, the transmission can be derived analytically if we assume that the mirror reflectivity is $R_{DBR} \approx 1$ and if we recognize the structure is exactly like a Fabry-Perot interferometer with mirror reflectivity R_{DBR} separated by the cavity thickness d_{cav}. The transmission is then given by [2]:

$$T(\lambda) = \frac{1}{1 + \frac{4R^2_{DBR}}{(1-R^2_{DBR})} \sin^2 \left| \frac{\pi n_H d_{cav}}{\lambda} - \frac{\pi n_H d_{cav}}{\lambda_0} \right|} \tag{14.6}$$

The transmission decreases to half of its value when

$$\frac{4R^2_{DBR}}{(1 - R^2_{DBR})^2} sin^2 \left| \frac{\pi n_H d_{cav} \Delta\lambda_{1/2}}{\lambda\lambda_0} \right| = 1 \tag{14.7}$$

. Around the cavity mode, $\Delta\lambda = \lambda - \lambda_0 \approx 0$ so

$$Q = \frac{\omega_0}{\Delta\omega_{1/2}} = \frac{\lambda_0}{\Delta\lambda_{1/2}} = \frac{4\pi R}{1 - R^2} = \frac{\pi(n_H^{4q} - n_L^{4q})}{n_H^{2q} n_L^{2q}} \tag{14.8}$$

The last step in Eq. 14.8 is obtained from Eq. 13.9 for the reflectivity of a DBR. One can see that the when the reflectivity of the mirrors increases the quality factor of the cavity increases. This is because the cavity becomes more isolated from the outside world and the coupling to the continuum modes of the surrounding medium decreases.

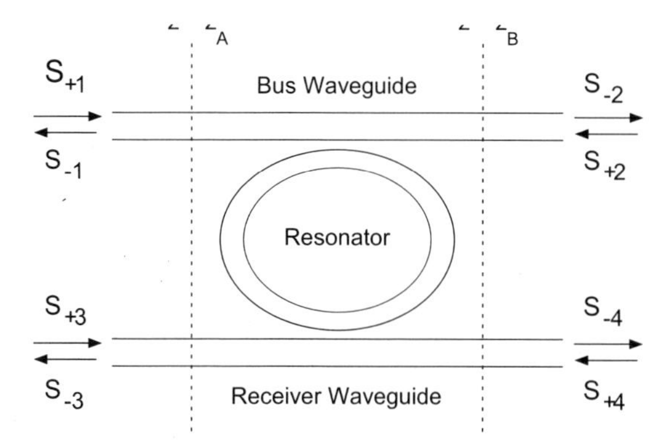

Figure 14.6. Schematics of a 2D resonator coupled to a bus waveguide and a receiver waveguide.

5. 2-D Cavity Resonators

In this session we will expand the concepts discussed previously for 1D cavity to 2D cavities using the coupled mode theory. Examples of 2D resonators are rings, racetracks , and disk-shaped waveguiding structures. Here we will consider a general resonator with any geometry and analyze its transmission properties as a function of frequency. The exact geometry of the resonator is accounted for only by the coupling coefficients between the resonator and the adjacent waveguides.

Consider the configuration shown in Fig. 14.6 of a resonator coupled to two waveguides. Light carrying information is coupled to the top waveguide, called the bus. The information is then either transmitted or coupled to the bottom waveguide through the resonator. The bottom waveguide is called the receiver.

Consider light incident from the left side of the waveguide with amplitude s_{+1}. The incoming and outgoing fields from the bus waveguide are denoted as s_{+1} and s_{-1} at $z = z_A$ and s_{+2} and s_{-2} at $z = z_B$. Similarly the fields at the receiver waveguide are denoted as s_{+3} and s_{-3} at $z = z_A$ and s_{+4} and s_{-4} at $z = z_B$. Here we will assume that the waveguides are single mode.

Using the coupled mode theory formalism, the outgoing amplitudes s_{-1},s_{-2},s_{-3} and s_{-4} are given by

$$\begin{aligned} s_{-1} &= e^{-j\beta_B l}(s_{+2} - k_2^* a) \\ s_{-2} &= e^{-j\beta_B l}(s_{+1} - k_1^* a) \\ s_{-3} &= e^{-j\beta_R l}(s_{+4} - k_4^* a) \\ s_{-4} &= e^{-j\beta_R l}(s_{+3} - k_3^* a) \end{aligned} \tag{14.9}$$

Where k_1, (k_2),k_3 and (k_4) are the coupling coefficients of the waveguide to the resonator in the forward (and backward) direction.

Let us assume that the only incoming field is the s_{+1}, i.e., (s_{+2}= s_{+3}= s_{+4}=0). One can see from Eq.14.9 that in the absence of the resonator, the waveguide mode in the bus propagates undisturbed with a wavector β_B and the phase delay between the two points 1 and 2 separated by $l = (z_A - z_B)$ is simply $\beta_B l$. Similarly to the analysis in Section 10.3, the coupling coefficients are given by the overlap of the modes of the waveguides with the modes of the resonator

$$
\begin{aligned}
\mathcal{K}_1 &= -\frac{j\omega\epsilon_0}{4}\int_{Z_1}^{Z_2} dz \int\int dxdy(n^2 - n_r^2)\mathcal{E}_r\mathcal{E}_+^B e^{-j\beta_B(z_1-z_2)} \\
\mathcal{K}_2 &= -\frac{j\omega\epsilon_0}{4}\int_{Z_1}^{Z_2} dz \int\int dxdy(n^2 - n_r^2)\mathcal{E}_r\mathcal{E}_-^B e^{-j\beta_B(z_1-z_2)} \\
\mathcal{K}_3 &= -\frac{j\omega\epsilon_0}{4}\int_{Z_1}^{Z_2} dz \int\int dxdy(n^2 - n_r^2)\mathcal{E}_r\mathcal{E}_+^R e^{-j\beta_R(z_1-z_2)} \\
\mathcal{K}_4 &= -\frac{j\omega\epsilon_0}{4}\int_{Z_1}^{Z_2} dz \int\int dxdy(n^2 - n_r^2)\mathcal{E}_r\mathcal{E}_-^R e^{-j\beta_R(z_1-z_2)}
\end{aligned}
\quad (14.10)
$$

where $\mathcal{E}_r$, $\mathcal{E}^B$, and $\mathcal{E}^R$ are the mode profiles of the resonator and the bus and receiver waveguides. Here for simplicity we are dropping the indices for the different resonator modes. The evolution of the resonator in time is given by a modified Eq. 14.1, including the coupling of the resonator to the incoming waveguide modes s_i

$$\frac{da}{dt} = (j\omega_0 - \frac{1}{\tau_o} - \frac{1}{\tau_e} - \frac{1}{\tau_f})a + k_1 s_{+1} + k_2 s_{+2} + k_3 s_{+3} + k_4 s_{+4} \quad (14.11)$$

where $1/\tau_0$ is the decay rate due to loss, $1/\tau_e$, $1/\tau_f$ are the decay rates due to the coupling to the bus and the receiver. The change in energy in the resonator mode is equal to the difference between the incoming and outgoing powers.

$$\frac{d|a|^2}{dt} = \sum_{i=1}^{4} |s_{+i}|^2 - \sum_{i=1}^{4} |s_{-i}|^2 - L \quad (14.12)$$

Where L is the power lost due to loss. Substituting Eq.14.9 into Eq.14.12, we have

$$\frac{d|a|^2}{dt} = -\sum k_i^2 |a|^2 + k_1 s_{+1} a^* + k_1 s_{+1} a - L \quad (14.13)$$

The change in energy can also be derived from Eq.14.11

$$\frac{d|a|^2}{dt} = 2|a|\frac{d|a|}{dt} = -\frac{2|a|^2}{\tau_e} - \frac{2|a|^2}{\tau_f} + k_1 s_{+1} a^* + k_1 s_{+1} a - \frac{2|a|^2}{\tau_0} \quad (14.14)$$

Comparing Eq.14.13 and Eq.14.14 we have

$$k_1^2 + k_2^2 = \frac{2}{\tau_e} \quad \text{and} \quad k_3^2 + k_4^2 = \frac{2}{\tau_f} \tag{14.15}$$

The quality factor of the resonator is given by Eq.14.3

$$Q = \frac{\omega\tau}{2} \tag{14.16}$$

where $1/\tau = (1/\tau_e) + (1/\tau_f) + (1/\tau_0)$ From Eq. 14.17 and Eq. 14.15 one can see that the higher the overlap between the resonator and the waveguide modes, i.e. the smaller the gap between the resonator and waveguides, the smaller the quality factor of the cavity.

Now we need to determine the amplitudes a , s_{-1} and s_{-2}. From Eq.14.4 we have

$$a = \frac{k_1 s_{1+}}{j(\omega - \omega_o) + (1/\tau_e) + (1/\tau_f) + (1/\tau_0)} \tag{14.17}$$

Therefore Eq. 14.9 becomes:

$$\begin{aligned}
s_{-1} &= -e^{-j\beta l} \frac{k_1 k_2^*}{j(\omega - \omega_0) + (1/\tau_e) + (1/\tau_f) + (1/\tau_0)} s_{+1} \\
s_{-2} &= e^{-j\beta l} \left(1 - \frac{k_1 k_1^*}{j(\omega - \omega_0) + (1/\tau_e) + (1/\tau_f) + (1/\tau_0)}\right) s_{+1} \\
s_{-3} &= -e^{-j\beta l} \frac{k_1 k_4^*}{j(\omega - \omega_0) + (1/\tau_e) + (1/\tau_f) + (1/\tau_0)} s_{+1} \\
s_{-4} &= e^{-j\beta l} \frac{k_1 k_3^*}{j(\omega - \omega_0) + (1/\tau_e) + (1/\tau_f) + (1/\tau_0)} s_{+1}
\end{aligned} \tag{14.18}$$

6. 2D Resonator Coupled to a Single Waveguide

Let us consider the configuration in Fig 14.6, with a negligible coupling of the bus to the resonator $1/\tau_f = k_4 = k_3 = 0$. For a travelling wave, coupling to the backward direction is zero, therefore $k_2 = 0$. From Eq. 14.18, the transmission response at resonance $\omega_0 = \omega$ is given by

$$T = \frac{|s_{-2}|^2}{|s_{+1}|^2} = \left| \frac{|k_1|^2}{j(\omega - \omega_0) + 1/\tau_o + 1/\tau_e} \right|^2 \tag{14.19}$$

Off resonance, $|\omega - \omega_o| >> 1/\tau_0$ the transmission is equal to 1. At the resonance wavelength the transmission from Eq.14.15 is equal to:

$$T = \frac{|s_{-2}|^2}{|s_{+1}|^2} = \left| \frac{|k_1|^2}{1/\tau_o + 1/\tau_e} \right|^2 = \left| \frac{1 - \tau_e/\tau_o}{1 + \tau_e/\tau_0} \right|^2 \tag{14.20}$$

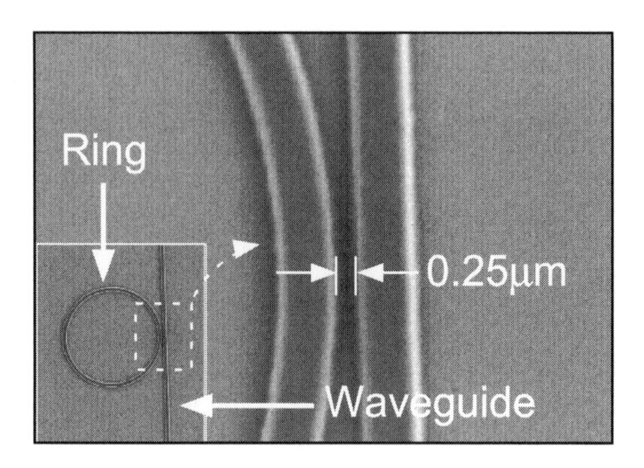

Figure 14.7. SEM photo showing the topview of a ring resonator coupled to a single waveguide. Inset shows the whole ring structure.

Therefore the effect of the loss on the transmission response is strongly dependent on the cavity geometry. Let us consider high losses. For a cavity with high Q, or $\tau_o \approx \tau_e$, the loss affects strongly the transmission response: $T \approx 0$. For a cavity with low Q, or$\tau_0 > \tau_e$, then the effect of the loss is small, i.e.,$T \approx 1$. This explains why cavities with high Q's are so difficult to fabricate. If Q increases, the effect of the losses becomes stronger. This is precisely why achieving high Q resonators in practice is challenging: once the Q is increased, any small loss becomes very significant and decreases the transmission response.

An example of a 2-D cavity is a ring resonator. In such a ring resonator, the resonance wavelengths satisfy the condition:

$$\frac{\Re[n_{eff}(\lambda_0)] \cdot L}{\lambda_0} = m \tag{14.21}$$

where n_{eff} is the complex effective index of the eigenmode inside the resonator, λ_0 is the free-space wavelength, $L = 2\pi r$ is the ring resonator perimeter, and m is an integer. Figure 14.7 shows an example of a fabricated ring resonator with Si core and SiO2 cladding and a diameter of 10μ m. Fig 14.8. shows the spectral response of the ring resonator. One can see that, as predicted from Eq.14.20, the transmission is close to 1 except for the wavelengths that correspond to the resonances. At these resonances, light is scattered off the sidewall roughness of the waveguides and the transmission decreases.

7. Ring Resonator as an Add/Drop Filter

Ring resonators coupled to two waveguides can be used an add/drop filter for WDM. In such filter, light is coupled in to the bus waveguide with several

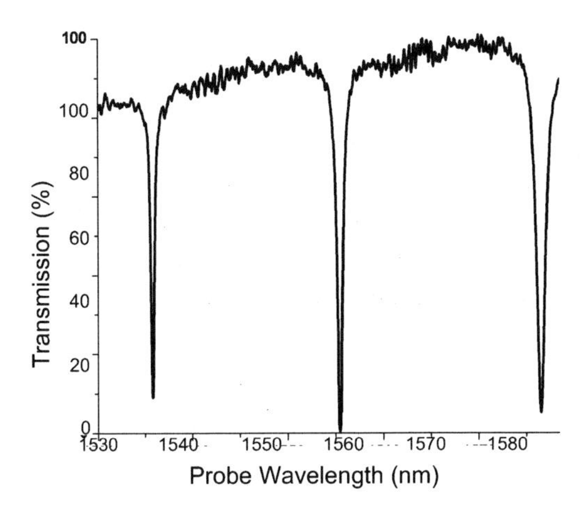

Figure 14.8. TM spectral response of a ring resonator coupled to a single mode waveguide.

wavelength. A small spectral range of those wavelengths coupled to the structure are coupled to the resonator and "dropped" into the receiver waveguide. Similarly, light coupled into the Bus can be added to the Bus waveguide (see Fig. 14.9). By cascading several of these resonators, one can demultiplex or multiplex signals for WDM applications. In order to understand the principle of operation of the add/drop filter, we will first analyze the ring resonator as a demultiplexer with an incoming traveling wave as an input as described in Fig. 14.9(a). In this case, the waves propagating backward in the bus and the waves propagating forward in the receiver can be neglected. Therefore, from Eq.14.9 $k_2 = k_3 = 0$. From Eq.14.14,$s_{-4} = s_{+1} = 0$. The coupling rate of the resonator mode to the bus and receiver waveguides is then

$$k_1^2 = \frac{2}{\tau_e} \quad \text{and} \quad k_4^2 = \frac{2}{\tau_f} \tag{14.22}$$

The light intensity I_T transmitted through the bus waveguide and the light intensity I_D exiting the receiver waveguide in the “drop" channel, are given by:

$$I_T = \left|\frac{s_{-2}}{s_{+1}}\right|^2 = \left|1 - \frac{2/\tau_e}{j(\omega - \omega_0) + 1/\tau_0 + 1/\tau_e + 1/\tau_f}\right|^2 \tag{14.23}$$

$$I_D = \left|\frac{s_{-3}}{s_{+1}}\right|^2 = \left|\frac{\sqrt{\frac{4}{\tau_e \tau_f}}}{j(\omega - \omega_0) + 1/\tau_0 + 1/\tau_e + 1/\tau_f}\right|^2 \quad (14.24)$$

Off resonance, $I_T = 1$ and $I_D = 0$. If a maximum power-transfer condition hold,

$$\frac{1}{\tau_0} = \frac{1}{\tau_e} - \frac{1}{\tau_f} \quad (14.25)$$

then at resonance Eqs.14.23 and 14.24 become:

$$\begin{aligned} I_T &= 0 \\ I_D &= 1 - \frac{\tau_e}{\tau_0} \end{aligned} \quad (14.26)$$

Therefore, if loss can be neglected, the channel coupled into the ring resonator with a wavelength corresponding to the resonance frequency of the ring is coupled to the bottom channel. All the other channels with wavelength that do not correspond to the resonance frequency of the ring are transmitted through the bus waveguide. The net result is a drop of a channel from the bus waveguide. From symmetry consideration, the response of the multiplexer shown in Fig 14.8 can be understood from eq. (23) as well. Light traveling in the bus with wavelength that do not correspond to the resonance frequency of the ring are transmitted through the bus waveguide. Light coupled to the add channel in the bottom waveguide, with a frequency corresponding to the one of the rings resonances, will couple to top waveguide and will be transmitted through the bus in addition to the off-resonance signals propagating through the same bus. The net result is an addition of a channel to the bus waveguide.

8. Sharp Bends Using Resonators

Waveguide bends are basic structures for optical interconnects, and are therefore very important photonic integrated circuits. For highly dense photonic circuits, these bends are required to be extremely sharp in order to minimize real estate and maximize integration. However, as we saw in Section 8.6, any abrupt directional change in the dielectric waveguide cause mode conversion into radiation losses. Here we will show how resonators can be used for sharp waveguide bends with low loss.

The transmission of the resonator given by Eq. 14.19 is solely governed by the resonator properties and the coupling of the waveguides to the resonator. The relative configuration of the bus and receiver waveguides is not included in the expression. For example, one could envision two waveguides in a sharp angle, connected by a resonator (see Fig. 14.10). If the resonator is lossless,

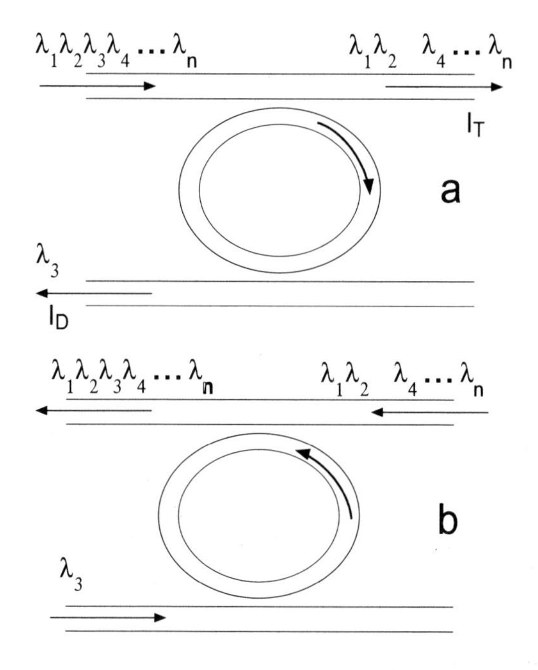

Figure 14.9. A ring resonator can serve as drop ((a), or as an add (b).

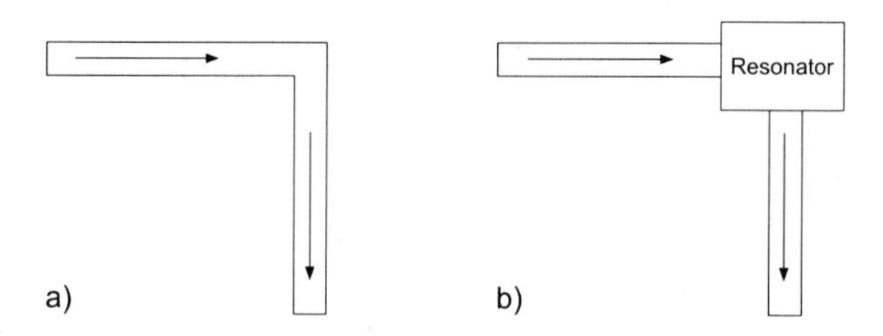

Figure 14.10. Two possible configurations for a waveguide bend. A) a harsh 90° bend that will suffer tremendous radiation losses, and b) a bend with an impedance matching resonator which couples one waveguide to the other.

at the resonance wavelength the transmission is equal to one). Therefore a resonator can "bridge" between the incoming field and the outgoing field in a configuration of the waveguides that would otherwise induce high losses. Figure 14.10 shows an example of a 90° bend (a) and a 90° bend modified into a cavity (b). In case (a) most of the light at the bend will be coupled into radiation modes. In case (a) light will be totally coupled to the neighboring waveguide through the resonator involving only loss of the resonator.

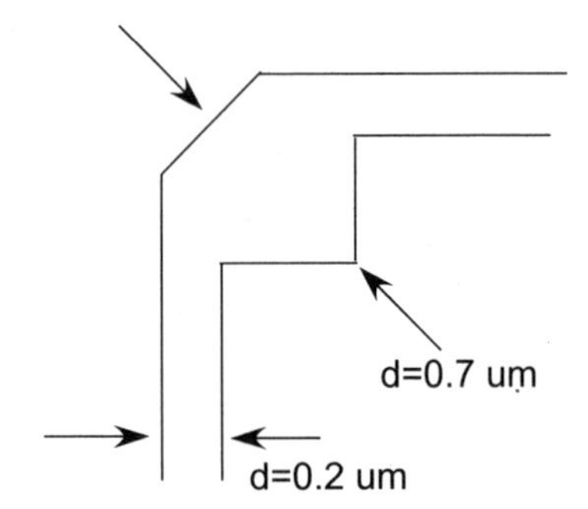

Figure 14.11. A further modification on the resonator idea.

In practice, the resonator has losses, both due to radiation and due to fabrication losses. In this case, from Eq.14.26 one can see that in order to minimize the effect of the losses, the coupling of the waveguide to the resonator $1/\tau_e$ must be increased relative to radiation losses $1/\tau_0$, so that the transmission of the bend is equal to $T_b = 1 - \tau_e/\tau_o \approx 1$. A better coupling between the incoming waveguide to resonator can be obtained by pushing the mode of the resonator inward. The structure shown in Fig. 14.11 is a square resonator, with a "cut" in the corner of the bend for better resonator-waveguide matching. The structure is designed for Si (n=3.2) surrounded by air. The waveguide widths are 0.2 μm and the dimension of the resonator d shown in the figure is equal to 0.7 μm. Using such structures, the authors in [1] have shown a 90° bed with submicron dimensions and with losses less than 1% per bend.

9. Summary

The use of high index waveguides provides the ability to make very small optical structures, but the issues of bending the light around corners without suffering excess loss has always been a limitation. We tried to show here that by taking advantage of resonant structures one could dramatically enhance the usefulness of integrated optical structures. The ability to couple between waveguides, and to make resonant structures which show high wavelength selectivity is a critical step in eventually making complex optical circuits completely based on couplers and filters.

References

[1] Manolatou and Haus " Passive components for dense optical integration ", Kluwar, 2001

[2] S. G. Lipson and H. Lipson, *Optical Physics*, Cambridge University Press, New York (1981).

Appendix A
The Goos-Hänchen Shift

When describing total internal reflection of a plane wave, we developed expressions for the phase shift that occurs between the incident and reflected waves as a function of angle of incidence. We explained the phase shift as being due to the fact that some energy is stored in the evanescent field of the interface before returning to the reflected wave. This relative storage delay introduced a phase shift.

A second way to look at this problem is using rays, and to describe the phase shift as being due to the ray actually travelling a small distance into the lower index medium before being reflected [?]. Figure **??** shows a ray incident on an interface at an angle greater than the critical angle. An incident ray behaves as if it were laterally displaced upon reflection.

If we examined this problem with a plane wave, we would see no shift simply because there is no lateral variance or structure in a plane wave. We will study the reflection using a packet of plane waves that form a beam of light. The incident beam is labelled A. A simple way to describe a spatial beam is to superimpose two plane waves with slightly different angles. If the z-component of the corresponding wave vectors are $\beta \pm \Delta\beta$, then the incident wave can be described (at $x = 0$) as

$$\begin{aligned} A(z) &= [e^{j\Delta\beta z} + e^{-j\Delta\beta z}]e^{-j\beta z} \\ &= 2\cos(\Delta\beta z)e^{-j\beta z} \end{aligned} \tag{A.1}$$

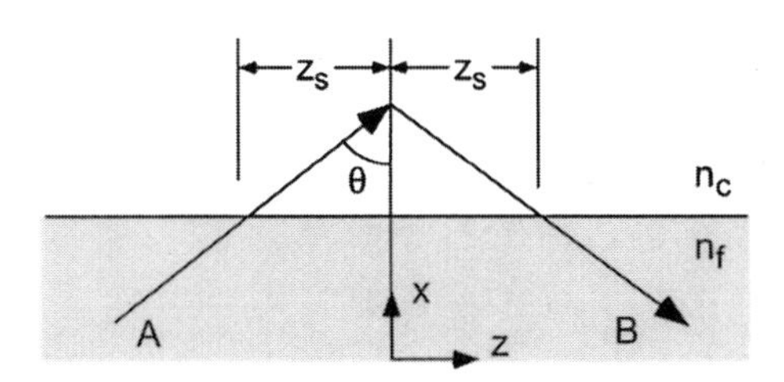

Figure A.1. A ray that undergoes total internal reflection is laterally shifted a distance $2z_s$.

The phase shift that occurs upon reflection is a function of θ and β. For small $\Delta\beta$ we can use the expansion

$$\Phi(\beta + \Delta\beta) = \Phi(\beta) + \Delta\Phi = \Phi(\beta) + \frac{d\Phi}{d\beta}\Delta\beta \tag{A.2}$$

Applying the appropriate phase shift to each component of the superposition, the reflected wave at $x = 0$ is

$$\begin{aligned} B(z) &= [e^{j(\Delta\beta z - 2\Delta\Phi)} + e^{-j(\Delta\beta z - 2\Delta\Phi)}]e^{-j(\beta z - 2\Phi)} \\ &= 2\cos\Delta\beta(z - 2z_s)\, e^{-j\beta z} \end{aligned} \tag{A.3}$$

where

$$z_s = d\Phi/d\beta \tag{A.4}$$

Thus, the phase shifts accumulated by the various components of the wave packet have the net effect of displacing the beam along the z axis a distance of $2z_s$. The spatial shift is largest where the derivative of the phase shift with respect to incident angle is largest. Inspection of Fig. 2.18 in Chapter 2 shows that this shift is largest near the critical angle. In terms of wave parameters, we can evaluate the derivative in Eq. ?? using Eq. 2.108 to get

$$k_0 z_s = \frac{\tan\theta}{\sqrt{\beta^2 - n_c^2}} \tag{A.5}$$

for the TE modes. The TM modes are described by

$$k_0 z_s = \frac{\tan\theta}{\beta^2 - n_c^2}\frac{1}{(\beta^2/n_f^2 + \beta^2/n_c^2 - 1)} \tag{A.6}$$

This lateral shift is called the Goos-Hänchen shift. Notice that the ray picture can be used to describe how far the field penetrates into the low index medium. From Fig. ??, this distance is

$$x = z_s/\tan\theta = \frac{1}{\beta^2 - n_c^2} = \frac{1}{\gamma} \tag{A.7}$$

The depth described by the Goos-Hänchen shift is exactly the same as the characteristic length described by the decaying evanescent field using the wave picture.

All of this may seem like a needlessly complicated way to look at phase shifts. But there is a practical application. The phase shifts that occur on reflection really do lead to an effective displacement of the beam. This is critical in optical waveguides which use reflecting bends. Consider the waveguide structure shown in Fig. ??. A rectangular waveguide is bent by using TIR to redirect the guided light around a corner. Such a structure saves a great deal of area on an integrated optical circuit.

We know that coupling efficiency depends on having the input mode profile match the output mode profile. To take account of the Goos-Hänchen shift, the reflecting facet must be moved toward the inside corner of the bend. Calculations and measurements have shown losses on the order of 1 dB due to this shift [?]. These losses are critical in photonic integrated circuits, and in laser designs. In practice the position of the reflecting facet is adjusted based on Goos-Hänchen calculations to maximize the overlap of the reflected and guided mode.

The "impedance matched" bends described in Chapter 13 alleviate some of the issues presented here.

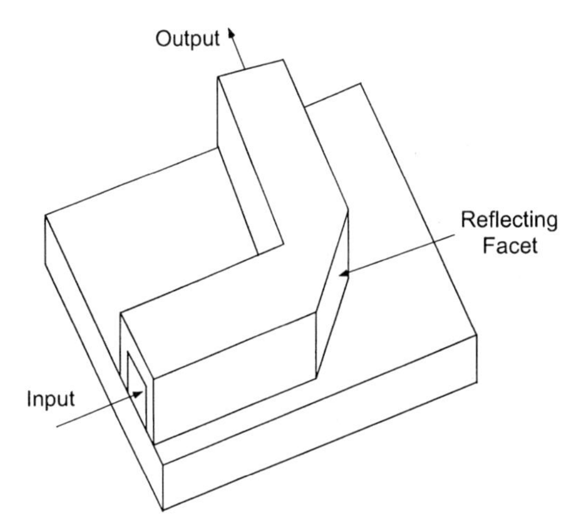

Figure A.2. A surface waveguide bend that operates using total internal reflection on the reflecting facet.

References

[1] H. Kogelnik, "Ch.2: Theory of dielectric waveguides," in *Integrated Optics, Vol. 7, Topics in Applied Physics*, T. Tamir, editor, Springer-Verlag, Germany (1979)

[2] A. Himeno, H. Terui, and M. Kobayashi, "Loss measurement and analysis of high-silica reflecting bending optical waveguides,", Jour. Lightwave Tech. 6, 41 (1988)

Appendix B
Bessel Functions

1. Bessel Functions of the First Kind

Bessel Functions are the solution to the differential equation (in real variables)

$$x^2 \frac{d^2 y}{dx^2} + x \frac{dy}{dx} + (x^2 - \nu^2) y = 0 \tag{B.1}$$

where ν is called the *order*. When ν is not an integer, there are two independent solutions to the equation, $J_\nu(x)$ and $J_{-\nu}(x)$,

$$J_\nu(x) = \sum_{k=0}^{\infty} \frac{(-1)^k}{k!\Gamma(\nu + k + 1)} \left(\frac{x}{2}\right)^{\nu+2k} \tag{B.2}$$

For the cases examined in this book, ν is always an integer, so the Γ function, $\Gamma(\nu + k + 1)$, can be replaced by a simple factorial, $(\nu + k)!$. In such a case, the solutions have the form

$$J_\nu(x) = \frac{x^\nu}{2^\nu \nu!} \left[1 - \frac{x^2}{2^2 \cdot 1!(\nu + 1)} + \frac{x^4}{2^4 \cdot 2!(\nu + 1)(\nu + 2)} - \cdots\right] \tag{B.3}$$

The integer ν can be positive or negative, and the relevant solutions are related through

$$J_{-\nu}(x) = (-1)^\nu J_\nu(x) \tag{B.4}$$

Using the recurrence relations, it is possible to relate various solutions. Some of the relations we found useful in deriving certain formula in this text are

$$\begin{aligned}
J_{\nu-1}(x) + J_{\nu+1}(x) &= \frac{2\nu}{x} J_\nu(x) \\
J_{\nu-1}(x) - J_{\nu+1}(x) &= 2J'_\nu(x) \\
\nu J_\nu(x) + xJ'_\nu(x) &= xJ_{\nu-1}(x) \\
\nu J_\nu(x) - xJ'_\nu(x) &= xJ_{\nu+1}(x) \\
J'_0(x) &= -J_1(x)
\end{aligned} \tag{B.5}$$

Explicit differential forms of the Bessel functions are derivable from the above equations:

$$\begin{aligned} J'_\nu(x) &= J_{\nu-1} - \frac{\nu}{x} J_\nu(x) \\ J'_\nu(x) &= -J_{\nu+1} + \frac{\nu}{x} J_\nu(x) \end{aligned} \tag{B.6}$$

2. Modified Bessel Functions

The modified Bessel functions of the second kind that are used to describe evanescent waves can be derived from the integral representation

$$K_\nu(x) = \frac{\sqrt{\pi}(z/2)^\nu}{\Gamma(\nu+1/2)} \int_0^\infty e^{-x\cosh t} \sinh^{2\nu} t \, dt \tag{B.7}$$

For integer values of ν, the following recurrence identities can be used to convert from one order to another:

$$\begin{aligned} (-1)^{\nu-1}K_{\nu-1}(x) - (-1)^{\nu+1}K_{\nu+1}(x) &= \frac{2\nu}{x}(-1)^\nu K_\nu(x) \\ (-1)^\nu K'_\nu(x) &= (-1)^{\nu-1}K_{\nu-1}(x) - \frac{\nu}{x}(-1)^\nu K_\nu(x) \\ (-1)^{\nu-1}K_{\nu-1}(x) + (-1)^{\nu+1}K_{\nu+1}(x) &= 2(-1)^\nu K'_\nu(x) \qquad \text{(B.8)} \\ (-1)^\nu K'_\nu(x) &= (-1)^{\nu+1}K_{\nu+1}(x) + \frac{\nu}{x}(-1)^\nu K_\nu(x) \end{aligned} \tag{B.9}$$

3. Asymptotic Expansions

For fixed order ν ($\nu \neq -1, -2, -3, \ldots$), and $x \to 0$,

$$J_\nu(x) \approx \frac{(x/2)^\nu}{\Gamma(\nu+1)} \tag{B.10}$$

For fixed order ν and $|x| \to \infty$,

$$J_\nu(x) \approx \left(\frac{2}{\pi x}\right)^{1/2} \cos\left(x - \frac{\nu\pi}{2} - \frac{\pi}{4}\right) \tag{B.11}$$

For fixed ν and large $|x|$,

$$K_\nu(x) \approx \left(\frac{\pi}{2x}\right)^{1/2} e^{-x} \left[1 - \frac{4\nu^2 - 1}{8x} + \frac{(4\nu^2-1)(4\nu^2-9)}{2!(8x)^2} + \ldots\right] \tag{B.12}$$

Appendix C
Optical Power Limit of a Waveguide due to Stimulated Raman Scattering

Consider an optical fiber with area A and length L. Let a pump wave with frequency ω_p be injected at $z = 0$ with power P_p. The pump intensity in the waveguide is therefore $I_p = P_p/A$. In the absence of any nonlinear interaction, the pump propagates as

$$P_p(z) = P_p(0)e^{-\alpha_p z} \tag{C.1}$$

The Stokes wave is described by the following differential equation

$$[d/dz + \alpha_s]P_s(z) = G_r I_p(z) P_s(z) \tag{C.2}$$

where G_r is the Raman gain coefficient for the medium and wavelength. If we make the simplifying assumption that the pump power is not depleted by nonlinear processes (this will give us an upper limit), the Stokes wave then follows

$$[d/dz + \alpha_s]P_s(z) = G_r P_s(z) I_p(0)e^{-\alpha_p z} \tag{C.3}$$

The solution to this is

$$P_s(z) = P_s(0)\exp(-\alpha_s z + \frac{G_r I_p(0)}{\alpha_p}[1 - e^{-\alpha_p z}]) \tag{C.4}$$

If we assume that the fiber is very long, so that $\alpha_p L \gg 1$, then Eq. ?? becomes

$$P_s(L) = P_s(0)\exp[-\alpha_s L + \frac{G_r I_p(0)}{\alpha_p}] \tag{C.5}$$

The gain term in Eq. ?? is equivalent to saying the gain is produced by the incident pump power over an effective length, $L_{eff} = 1/\alpha_p$.

If no Stokes wave is injected at $z = 0$, then all output appearing at $z = L$ will be due to amplified spontaneous Raman scattering. The summation over the length of the fiber for all the spontaneous emission weighted by its net gain is equivalent to assuming an input flux of 1 photon per mode of the fiber. Ref. [?] shows that this is equivalent to defining the input power as

$$P_s(0)_{eff} = (h\nu_s)(B_{eff})(\text{number of transverse modes}) \tag{C.6}$$

where

$$B_{eff} = \frac{\sqrt{\pi}}{2} \frac{\Delta\nu_{fwhm}}{[I_p(0)G_r/\alpha_p]^{1/2}} \tag{C.7}$$

Now, to ensure that nonlinear conversion of pump power into Stokes power is not a problem, we demand that the *Stokes* power at $z = L$ is less than the *signal* power at $z = L$

$$P_s(0)_{eff} \exp[-\alpha_s L + \frac{I_p(0)G_r}{\alpha_s}] < P_p(0) \exp[-\alpha_p L] \tag{C.8}$$

The absolute upper limit to the pump power will be Eq.**??** is satisfied with an equality. This power is defined as P_{crit}. For a single mode fiber, and assuming $\alpha_s = \alpha_p$, the relation becomes

$$\frac{\sqrt{\pi}}{2}(h\nu_s)\left(\frac{G-r}{A\alpha_p}\right)\Delta\nu_{fwhm} = \left(\frac{G_r P_{crit}}{A\alpha_p}\right)^{3/2} \exp\left(-\frac{G_r P_{crit}}{A\alpha_p}\right) \tag{C.9}$$

This is a fairly complex equation. It turns out that the critical power is only weakly dependent on the choice of $\Delta\nu_{fwhm}$, but is critically dependent on α, $G - r$, and A. For the range of parameters used for fused silica ($\Delta\nu_{fwhm} \approx$ 6THz, $\alpha \approx 10^{-5}$ (which corresponds to about 4 dB/km), and G_r in the range of 10^{-13} W/m, the critical power can be described to a very good approximation as [**?**]

$$P_{crit} = 16(A\alpha_p/G_r) \tag{C.10}$$

References

[1] RGSmith R. G. Smith, "Optical power handling capacity of low loss optical fibers as determined by stimulated Raman and Brillouin scattering," Applied Optics 11, 2489-2494 (1972)

Appendix D
Useful Data

Table D.1. Physical Constants

Name	Symbol	Value
Velocity of light in vacuum	c	2.99792×10^8 m/sec
Permittivity of vacuum	ϵ_0	8.8542×10^{-12} Farad/m
Permeability of vacuum	μ_0	$4\pi \times 10^{-7}$ Henry/m
Electron charge	e	1.60219×10^{-19} C
Electron mass	m_e	9.1095×10^{-31} kg
Planck's constant	h	6.6262×10^{-34} J-sec
Proton mass	m_p	1.67265×10^{-27} kg
Bohr radius	a_0	$0.528\mathring{A}$
Avogadro's number	N_A	6.023×10^{23}
Boltzmann's constant	k	1.380×10^{-23} J/K

Table D.2. Energy Conversion factors

1 eV	$= 1.602 \times 10^{-19}$ J
1 eV	$= 2.42 \times 10^{14}$ Hz
1 eV	$= 8.07 \times 10^3 cm^{-1}$
300 K	$= 2.59 \times 10^{-3}$ eV $\approx \frac{1}{40}$ eV

Table D.3. The Electromagnetic Spectrum

	Typical Wavelength	Frequency	Photon Energy
(cm)	(Hz)	(eV)	
sphline AM radio	3×10^4	10^6	4×10^{-9}
FM radio	3×10^2	10^8	4×10^{-7}
Radar	3	10^{10}	4×10^{-5}
Infrared	3×10^{-4}	10^{14}	0.4
Visible	6×10^{-5}	5×10^{14}	2
Ultraviolet	1×10^{-5}	3×10^{15}	12
X-rays	3×10^{-8}	10^{18}	4000

Index